Alarmist:

A person who needlessly alarms or attempts to alarm others, as by inventing or spreading false or exaggerated rumors of impending danger or catastrophe.

—**American Heritage Dictionary**

We, the public, are told the science on global warming is "settled," "certain," and beyond debate or discussion. The planet is warming, humans are the main reason, and the results will be catastrophic.

In fact, as the climate alarmists know full well but do not wish the public to know, the planet isn't warming right now and hasn't since the late 1990s. How can humans be primarily responsible for warming that isn't occurring? And there is no empirical evidence at all to indicate that if the planet does begin to warm again, the results will be disastrous—the outcome may even be a greener planet since carbon dioxide acts as a fertilizer. The only evidence for impending catastrophe exists in computer models. In fact, as this book shows, global warming alarm is based more on politics, environmentalist beliefs and psychological groupthink than science.

False Alarm is for the concerned citizen who wants to learn all sides of the climate change issue, not just that of the climate alarmists who currently dominate the media.

WHAT OTHERS SAY ABOUT SCIENCE, CLIMATE SCIENCE, AND THE HUMAN-CAUSED WARMING HYPOTHESIS

We need to get some broad based support, to capture the public's imagination. That, of course, entails getting loads of media coverage. So we have to offer up scary scenarios, make simplified, dramatic statements, and make little mention of any doubts we might have.

—**Climatologist Stephen Schneider**

A scientist is never certain.

—**Physicist Richard P. Feynman**

Review by experts and governments is an essential part of the IPCC process.

—**IPCC Mission Statement**

By the standards of the geological past, we live in an Ice Age. The world has rarely been as cold as it is today.

—**Science writers John & Mary Gribbin**

It would be impossible for any investigative reporter ... to objectively delve into global warming and conclude that the science was settled.
—Journalist Lawrence Solomon

The majority of the earth's history has been marked by conditions warmer than those of today.
—Climatologists Edward Aguado & James E. Burt

In climate research and modeling, we should recognize that we are dealing with a coupled non-linear chaotic system, and therefore that long-term prediction of future climate states is not possible.
—IPCC 2001 report

It's almost naive, scientifically speaking, to think we can give relatively accurate predictions for future climate. There are so many unknowns that it's wrong to do it.
—James Lovelock

The idea that human beings have changed and are changing the basic climate system of the Earth through their industrial activities and burning of fossil fuels—the essence of the Greens' theory of global warming—has about as much basis in science as Marxism and Freudianism. Global warming, like Marxism, is a political theory of actions, demanding compliance with its rules.
—Historian Paul Johnson

Even a succession of professional scientists—including famous astronomers who had made other discoveries that are confirmed and now justly celebrated—can make serious, even profound errors in pattern recognition.
—Astronomer Carl Sagan

People are not the enemy of the environment. Nor is affluence the enemy. Affluence does not inevitably foster environmental degradation. Rather, affluence fosters environmentalism. As people become more affluent, most become increasingly sensitive to the health and beauty of their environment. And gaining affluence helps provide the economic means to protect and enhance the environment.
—Geographer Jack M. Hollander

Everyone supports reducing our production of greenhouse gases—until they are told how much it will cost the economy. Maybe that's why they're never told.
—Climatologist Roy W. Spencer

When people come to know what the truth is (about global warming alarmism), they will feel deceived by science and scientists.
—Physicist Syun-Ichi Akasofu

The threat of global warming will eventually recede. The need for an apocalyptic vision, however, will not. The next threat will contain many of the characteristics of the global warming threat. It will predict the end of the world. It will be based on "scientific facts." It will require massive counseling for the psychological distress it will cause. It will require the creation of a massive bureaucracy. And it will require the transfer of massive amounts of money to the hypothesized victims of the future crisis.
—Historian Paul Johnson

FALSE ALARM

Global Warming –
Facts Versus Fears

PAUL MACRAE

SPRING BAY PRESS

SPRING BAY PRESS
PO Box 55043
3825 Cadboro Bay Road
Victoria BC Canada V8N 4G1

For rights information and bulk orders, please contact:
Email: springbaypress@gmail.com
Website: www.paulmacrae.com

Trade distribution in Canada by
Sandhill Book Marketing Ltd.

False Alarm
ISBN 978-0-9864862-0-3 (trade paperback)
Cataloguing information available from
Library and Archives Canada

Printed in Canada

10 9 8 7 6 5 4 3 2 1

DEDICATIONS

To Sheila,
now and forever, my toughest critic and best editor,
my best friend, and my beloved wife.

And to

Dr. Andrew Weaver,
who motivated me to learn more about climate science.

[Warming alarmism is] the worst scientific scandal in history.

—IPCC contributor Kiminori Itoh

TABLE OF CONTENTS

Global warming has some science at its core. But it has been overlaid with a vast engine of continuous alarmism, propaganda, relentless campaigning, facile projections, and not a little bullying righteousness by some of its celebrity proponents. It is, for all that is shouted to the contrary, more a cause than a science.

—Rex Murphy[1]

Fear is the most powerful enemy of reason.

—Al Gore, Assault on Reason[2]

"Well," he continued, "there are a lot of things that want saying which no one dares to say, a lot of shams which want attacking, and yet no one attacks them. It seems to me that I can say things which not another man in England except myself will venture to say, and yet which are crying to be said."

I said: "But who will listen?"

—Samuel Butler, *The Way of All Flesh*[3]

INTRODUCTION

In August 2007, *Newsweek* magazine ran an article entitled "The Truth About Denial" that attacked those who believe global warming is a "hoax."[4] The magazine also ran a poll on its website with the question: "Do you believe that global warming is a major threat to life on earth?"[5] The question got 48,583 responses, which is much larger than the sample for most professionally conducted opinion polls. Of those who replied, 57 per cent said "yes," global warming is a major threat to life on earth, 37 per cent said "no," and six per cent weren't sure. In the discussion thread that followed the poll, one *Newsweek* reader argued that global warming fears were overblown. He got this response from another reader: "Would you kindly provide me with the source of your secret hidden knowledge that we are not all going to die?"

Similarly, on May 11, 2007, a letter to the Toronto *Globe and Mail* opined: "Global warming threatens our entire species and a great many others."[6] A personal email to me, responding to a newspaper column I'd written that was critical of global warming fears, was subject-lined: "Stop giving soapboxes to the planet burners." This writer obviously believed the planet was going to "burn up" if it got much warmer. Finally, a *Globe and Mail* article quoted a lobbyist for nuclear energy, John Ritch, warning that greenhouse gases will lead to consequences "that are—quite literally—apocalyptic: increasingly radical temperature changes, a worldwide upsurge in violent weather events, widespread drought, flooding, wildfires, famine, species extinction, rising sea levels, mass migration, and epidemic disease that will leave no country untouched."[7] Even the head of the United Nations, Ban Ki-Moon, has said that without curbs on carbon dioxide, we face "oblivion."[8]

Clearly, many people are terrified that global warming is going to kill us all and take the planet with it. And these fears aren't surprising: scary books and articles about global warming dominate the media these days. Why do scary books dominate? Because scary sells. Scary stories about global warming sell magazines and newspapers and TV documentaries; scary also sells books. In a bookstore near you are titles on global warming like:

- *Apocalypse 2012*
- *The Heat is On*
- *Earth in the Balance*
- *Heat: How to Stop the Planet Burning*
- *When the Rivers Run Dry*
- *High Tide: The Truth about Our Climate Crisis*
- *Saving the Century*
- *2030: Thermageddon in Our Lifetime*, and
- *Six Degrees: Our Future on a Hotter Planet.*

Scary also sells tickets to films, as former U.S. vice-president and green crusader Al Gore proved with his blockbuster 2006 movie *An Inconvenient Truth*.

So I hope you'll read on when I tell you that this book is not trying to scare you about global warming. Quite the opposite: this book argues that most if not all of the "consensus" global-warming hypothesis is misleading, exaggerated, or just plain wrong.

The consensus global-warming hypothesis has three parts: first, global warming is "real"; second, global warming is *principally* due to increased levels of carbon dioxide caused by human activity; and third, this human-caused warming will have drastic, even apocalyptic, results "whose ultimate consequences could be second only to global nuclear war."[9] How true is this three-part hypothesis?

For a start, global warming *has been* "real," but it is *not* real as of the time of writing (2010): most climatologists now accept that the planet has not warmed on average since the late 1990s, and a leading climate alarmist, Phil Jones, former head of the Climatic Research Unit of East Anglia University, even acknowledged in February 2010 that there had been "no statistically significant" warming since 1995 and that there may even have been a slight *cooling* since 2002.[10] This is worth thinking about: for more than a decade, during which we've been repeatedly and stridently warned about the apocalyptic perils of global warming, *there has been no warming.* (In the Appendix on page 369 is the URL for a U.S. National Oceanic and Atmospheric Administration—NOAA—gadget that allows readers to discover for themselves whether the planet has been cooling or warming for any two dates between 1895 and 2009.) A decade's lack of warming is one reason why the climate doomsayers have switched from declaring that "global warming is real" to "climate change is real"[11]—it's absurd to claim "global warming is real" when the planet isn't warming. Indeed, given that climate change is always going on, the now-familiar phrases "stop climate change" or "climate change is real" are equally absurd; you might as well try to "stop weather" or state that "weather is real."

Second, while more than six billion human beings must inevitably be affecting the climate, there is ample evidence that our carbon emissions are not the "principal" source of warming—natural factors like the sun and oceans still predominate, while other human activities, including cities and agriculture, are probably more powerful warming agents than carbon dioxide emissions. Indeed, carbon dioxide, whether human-caused or otherwise, is not the principal cause of planetary warming in general; that honor goes to water vapor. We will look at the evidence for this statement in Chapter 3.

Third, what is the source of my "secret hidden knowledge" that we're not all going to die in a climate apocalypse? That's easy—the earth's geological history. Global warming is a natural phenomenon in the life of our planet, now and in the past: *for 90 per cent of the past 570 million years, the earth has been warmer than it is now* and life not only evolved but thrived.[12] Furthermore, as we'll see in Chapter 2, for most of this 570 million years, carbon dioxide levels have been higher, and often five to 10 times higher, than today's. In fact, there is not a shred of *empirical* evidence that increased carbon emissions will cause a climate catastrophe; *all* of the evidence for an apocalypse comes from climate computer models that are, as we will see in Chapter 4, highly flawed. Therefore, this book argues that warming, at least at present levels and for the foreseeable future, is not a major threat to us, to other species, and certainly not to the planet. This book also argues that, if climate must change, as it does and always has and always will, then it's better if that change is toward a warmer rather than a cooler climate, because a cooler climate—leading inevitably to the return of the glaciers—truly would be catastrophic for humanity and civilization.

Finally, this book argues that to cope with global warming, we don't have to risk wrecking our technology-based economy by drastically reducing carbon emissions. Yes, we should be more efficient in our use of energy, but that's just common sense—didn't our parents tell us not to waste electricity? However, if we damage our economy with poorly thought-out responses to potential global-warming threats—threats that still aren't well-researched or understood and may even be non-existent—then we will be less economically capable of dealing with the problems of climate change that do inevitably occur.

So, yes, of course, there will be problems ahead in dealing with a warmer planet, if warmer is how the climate is tending in the long term. But even warming isn't a sure thing given that, in the first decade of the 21st century and perhaps even earlier, the planet hasn't warmed. Some of the problems of a warming planet, such as rising oceans, may be very difficult to adapt to for some nations, but the oceans aren't rising nearly as quickly as the global warming alarmists would like the public to believe. And, based on past geological history, sea levels will rise at least another four to six metres (13 to 20 feet) no matter what we do or don't do. Human beings have handled climate change in the past, we're quite technologically capable of dealing with it now, and we will be even more capable in the future.

Humanity's problems at present have to do more with overpopulation and poverty than climate. Well-off countries are much better positioned to deal with global warming than poor countries, so we won't solve the warming problem by making ourselves poorer—the alarmist "solution"—but, paradoxically, by making poor countries richer. In the short term, that may mean *more* carbon emissions; in the long term, all of humanity will be better off even if the planet is a bit warmer, as we will see in Chapter 9.

However, this book does want to warn you and, yes, maybe even scare you, about two issues.

First, you should be concerned that the Canadian and other national governments may be wasting a colossal, even scary, amount of money, resources and time trying to fix a problem—global warming—that probably can't be fixed because it is part of the natural climate cycle of our planet. The billions of dollars wasted on curbing carbon would be far better spent on, for example, curbing population, reducing poverty, eliminating *toxic* pollutants (carbon dioxide is not a pollutant), and finding energy alternatives to diminishing fossil fuels.

Second, you should be concerned about serious problems in some of the "science" around climate change. "Science" is in quotation marks because much of what you read about climate change from "consensus" climate scientists as well as environmental groups is not based on empirical evidence; as noted above, their views are based *entirely* on computer climate models coupled with firmly held self-interest, belief, opinion, political agendas, ideology, and even religion. As a result of errors in these models and arrogance by consensus climate scientists, much of what we, the public, have been told about global warming is misleading, exaggerated, or just plain wrong.

Indeed, in November 2009 there was startling confirmation of this corruption of climate science when more than a thousand emails from the Climatic Research Unit (CRU) at East Anglia University in England, written by some of the top figures in climate research, were either hacked or, more likely, released by an internal whistleblower. These emails, dubbed "Climategate," revealed:

- a concerted campaign to manipulate climate research data to conceal the lack of 21st century warming ("hide the decline") or to exaggerate warming,
- alarmists' concern that the consensus computer models had failed to predict the 21st-century halt to warming, and
- attempts and threats to harm the careers of and prevent academic publication by scientists ("deniers") who dared to challenge the alarmist hypothesis.[13]

The emails showed clearly that the so-called scientific "consensus" on global warming—the belief that warming is happening, it's all humanity's fault, and it's going to be a disaster—functions more like the Spanish Inquisition than constructive scientific debate. As a result, those who don't conform to the consensus are labeled as heretics, although the term used is "deniers" in an offensive comparison to Holocaust deniers. These "heretics" are not only publicly derided but also, far too often if they are scientists themselves, cut off from publication and grants for research that might contradict the official consensus.

But what if the empirical evidence shows that human carbon emissions *aren't* the main cause of climate change and that we *aren't* heading for disaster? Or at least, what if the empirical evidence shows that the alarmist claims of human blame and impending doom are far from the certainty that consensus climate science claims? Wouldn't knowing these facts significantly change our approach to dealing with global warming or, for that matter, cooling? However, at the moment, only one point of view is permitted by the official climate consensus—global warming due almost entirely to human carbon dioxide emissions that will end in catastrophe. Other interpretations are actively discouraged and, if possible, suppressed. As I hope to show, this one-sided approach is bad science leading to possibly very bad, very expensive, and ultimately useless public policy.

So, I hope I've scared you, or at least worried you, enough to read on. I also hope that, if you do read on, you'll be less concerned about the alarmist climate bogeyman and more concerned about real issues, including ideologically based science and the folly of wasting precious resources trying to "stop" climate change, which is no more possible than stopping the tides.

Paul MacRae
June 2010

A NOTE ON SOURCES

My sources are in the notes at the end of the book and whenever necessary, I have included an Internet address. In almost all cases, however, articles cited from newspapers, journals and magazines, or postings on websites can be located using the article's title and author through any Internet search engine. Often, online academic journal articles can't be read in full without subscribing to the journal or having university access, although you can usually read an abstract of the article online without a subscription. If you use the computers in the library of your local college or university, these academic articles may be available to you in full.

Many of the books cited in *False Alarm* are online, in whole or in part. In particular, readers can often check the accuracy of my quotations by going to Google Books, finding the book named, and then searching for a word or phrases used in the quotation. Google Books does not reproduce the whole book for copyrighted publications, but often a "Search" will find the relevant passage in pages that are previewed or offered as "snippets." Books out of copyright are often complete on Google Books and relevant passages can easily be found using Search. Another good source of out-of-copyright books is the Gutenberg Project.

The works cited list for *False Alarm* is extensive and to save paper I've put my bibliography online at paulmacrae.com. On the website you will also find supplementary material that couldn't fit into a 370-page book.

All silencing of discussion is an assumption of infallibility.
John Stuart Mill[1]

Dogma, regarded as unquestionably true, … is incompatible with the scientific spirit, which refuses to accept matters of fact without evidence, and also holds that complete certainty is hardly ever attainable.
—Bertrand Russell[2]

What distinguishes knowledge is not certainty but evidence.
—Walter Kaufmann[3]

A scientist is never certain.
—Richard P. Feynman[4]

We can be absolutely certain only about things we do not understand.
—Eric Hoffer[5]

Chapter 1

CONSENSUS CLIMATE SCIENCE'S 'DOCTRINE OF CERTAINTY'

On May 22, 2007, the Victoria *Times Colonist* newspaper published a column I'd written entitled "The $8-billion global warming swindle." The headline was inspired by a British television documentary called *The Great Global Warming Swindle* and the column gave several examples of exaggeration, misinformation and out-and-out lies ("swindles") about global warming coming from such diverse sources as a Canadian Conservative member of Parliament and former U.S. vice-president turned Green Crusader Al Gore.

The MP was Mark Warawa, member of Parliament for Langley, B.C., and a comment he made on CBC radio sparked the column. Warawa told the radio audience: "Some have even called it [climate change] the greatest environmental challenge the world has *ever* faced."[6] The greatest environmental challenge the world has *ever* faced? Greater than the ice age that has dominated the world's climate for the past two million years? Greater than the asteroid that killed the dinosaurs 65 million years ago? Greater than the Permian-Triassic Extinction of 250 million years ago that killed more than

90 per cent of all species on the planet? Just to be sure I'd heard him right, I contacted Warawa's office to get the transcript of the broadcast. There the statement was, in print, with "ever" underlined and in bold type: "The greatest environmental challenge the world has **ever** faced."

Perhaps Warawa meant the greatest environmental challenge *human beings* have ever faced, but that's patently not true, either. As University of California archaeologist Brian Fagan writes:

> About 15,000 years ago, the ice sheets began to melt in earnest. The dramatic environmental changes associated with this warm-up make the climatic changes predicted by doomsday-sayers for the next millennium look puny in comparison.[7]

Fagan adds: "It is safe to say that people living on earth after about 12,000 years ago confronted global warming on a scale which we have not even remotely begun to experience in our own times."[8]

Warawa went on to say: "Science has shown that Canada could be affected by climate change more than other nations." This may, of course, be true, but the statement is misleading since it implies that the consequences for Canada must be negative. In fact, the overall effect of warming on this country will almost certainly be positive. After all, Canada is a *northern* nation; warming would open up northern lands to cultivation and habitation and allow extraction of natural resources that have previously been inaccessible. Global warming problems may be severe for some countries, and that is a major concern, but probably not for Canada. That a member of the governing party could make such obviously wrong and/or misleading statements, not in private but on the CBC for everyone to hear, is astounding, and shows how poorly informed Canadians, including MPs, are about the so-called global warming "threat."

MORE PUBLIC EDUCATION AND DEBATE NEEDED

To counter this misinformation, my May 22 column called for more public education, discussion and debate on the causes and possible effects of global warming before the Canadian government spent a projected $8-billion a year to stop something—then called global warming, now rechristened "climate change" because the planet has not been warming—that almost certainly can't be stopped. And even if warming could be stopped, it may not be the catastrophe it's made out to be. But in calling for public education, debate and discussion, I meant education, debate and discussion from more than one side of the global warming issue.

At the moment, global warming alarmists have a stranglehold on public opinion, to the point where an acquaintance asked me, after the May 22 column, whether I "really believed" what I'd written, as if no other point of view were possible. Indeed, the only major Canadian media outlet I am aware of that consistently reports on the skeptical side of the global warming issue is the *National Post* newspaper.[9]

And, as I'd discovered from previous columns in the *Times Colonist*, any remark critical of global warming alarmism, or critical about anything to do with the environmental movement, period, touches off a firestorm of angry letters to the editor. The May 22 "Global warming swindle" column was no exception. Some of the letters printed on May 24 were favorable but most were not. My personal emails, on the other hand, were supportive by a ratio of about two to one, in part because people who agree with a column don't tend to write to the newspaper's letters page, they contact the writer directly.

As mentioned in the Introduction, one hostile personal email was subject-lined: "Stop giving soapboxes to the planet burners." This writer clearly didn't know that the planet has been much warmer in the past—indeed, it's been warmer for 90 per cent of the 600 million years that multicellular life has existed on Earth—and didn't "burn up." His email also appeared as a letter in the *Times Colonist* and said: "Shame on the *Times Colonist* editors for approving the publication of Paul MacRae's climate change denier claptrap." Why bother discussing all sides of an issue? His mind was made up so let's hear no more debate.

A columnist quickly becomes thick-skinned, but one letter had to be taken seriously. It came from the University of Victoria's Dr. Andrew Weaver, one of Canada's leading computer climate modelers. He called the column an example of "unsubstantiated, poorly researched diatribes masquerading as scientific critiques" and as displaying "a woeful lack of scientific literacy."[10] Weaver's comment puzzled me since I based my *Times Colonist* climate columns, and more recently this book and my blog articles,[11] on scientific research—the data isn't pulled out of thin air. Also, I've met Dr. Weaver and know him to be a scientist of great integrity, so I wondered how he could make a comment like this. After all, my sources would hardly be considered scientifically illiterate—in error, perhaps, as any scientist can be, but not uninformed about either science or climate. And there are literally thousands of accredited scientists who oppose the global warming "consensus" and have said so publicly. For example, the Oregon Petition against the Kyoto Accord has more than 31,000 signatures; at least 9,000 of the signers are identified as scientists with PhDs, some in climatology, some not, but very few, presumably, suffering from a "lack of scientific literacy."[12] And yet they are, like me, skeptical about the exclusively human causes of global warming and its supposed catastrophic consequences. This book will cite many other scientists who don't buy into the overwhelming "consensus" that Weaver and others believe exists, or would like the public to believe exists.

No sensible person doubts that climate change is "real," as the global warming lobby likes to put it (although this statement is meaningless since climate change is always going on). My May 22 column said, "True, Canada will be affected by climate change," and no column or blog of mine has ever said that climate change wasn't happening. But surely an issue as complicated as the *causes* of climate change and how to *respond* to global warming should still be open to debate, as most scientific issues are, especially when billions of dollars in public money are involved for both climate research and Kyoto Accord-type interventions. Clearly, however, at least as far as Weaver and other consensus supporters are concerned, this view is wrong and even lacking in "scientific literacy." The climate-change issue, for the alarmists, is settled. No need for further discussion. Case closed.

THE DOCTRINE OF CERTAINTY

In their book *Taken by Storm*, which challenges the consensus climate-change orthodoxy and particularly its reliance on computer models, mathematician Christopher Essex and economist Ross McKitrick call this one-sided attitude the Doctrine of Certainty. They note: "The Doctrine of Certainty … is a collection of now familiar assertions about climate, all of which are to be accepted without question, because, as the Doctrine's supporters say, 'The time for questioning is over'."[13] Yet, as Essex and McKitrick add, echoing philosopher Bertrand Russell in this chapter's epigraph, "Certainty is anathema to modern science,"[14] which is always open to revision as more knowledge accumulates, and as even consensus climate scientists will admit, there is much about climate that we still don't understand and may never understand. In short, the Doctrine of Certainty is the exact opposite of the true scientific attitude.

Weaver presents himself as a practitioner of the Doctrine of Certainty in his 2008 book *Keeping Our Cool: Canada in a Warming World* when he writes:

> There is no ... debate [over the human causes of global warming] in the atmospheric or climate scientific community, and ... making the public believe that such a debate exists is precisely the goal of the denial industry. ... Scientific debate over global warming *would therefore imply uncertainty*.[15] [italics added]

For Weaver there is only one possible scientific position: global warming is happening, it's humanity's fault, it's going to be a disaster, and let's not hear any conflicting opinions. Weaver's position against public debate on climate is stated explicitly in a 2003 interview with *The Ring*, the University of Victoria newspaper:

> Why is everybody confused about global warming? I think the media plays a central role. One newspaper headline last year said "Study Deflates Global Warming." A week later the same paper ran another story, headlined "Global Warming Severity Grows." So to the average person, the science is swinging like a pendulum between "It's real and it's not."[16]

It would be far better, in Weaver's view, not to let the public hear any opinions by people critical of the so-called climate consensus to avoid possible confusion. There is, after all, only one true position and it happens to be that of the alarmist "consensus."

David Suzuki, Canada's most prominent environmental alarmist, also believes the public should not be troubled by critical debate about the reality and causes of global warming. For example, he stormed out of a Toronto radio interview because the interviewer, John Oakley, dared to suggest that global warming might not be a "totally settled issue."[17] In his 1992 book *Earth in the Balance,* Al Gore also strongly endorses the Doctrine of Certainty:

> If, when the remaining unknowns about the environmental challenge enter the public debate, they are presented as signs that the crisis may not be real after all, it undermines the effort to build a solid base of public support for the difficult actions we must soon take.[18]

Yet, as we will see, the "remaining unknowns" on the climate issues are legion—climate research is, *par excellence*, the study of unknowns—and Gore, Weaver, Suzuki, et al., beg the question when they assume, *a priori,* that there is a "crisis" that is "real" when the reality of the crisis is precisely what the alarmists need to prove and have not proved. That is, so far they have not presented any *empirical* proof that humans are the principal source of warming or that warming will be catastrophic. Their only "proof" comes from computer models. How do we know they don't have empirical proof and are relying completely on computer models? There are three reasons. First, *alarmist climatologists say so themselves*. For example, a prominent climate-science alarmist, Stanford University's Stephen Schneider, has written: "Computer modeling is our *only available tool* to perform what-if experiments such as the human impact on the future."[19] [italics added] Second, as we will see in Chapter 8, the climate alarmists admit that they are relying on "intuition" or "a sense" that we are facing disaster rather than hard evidence. Indeed, science writer Lydia Dotto, in her 1999 alarmist book *Storm Warning: Gambling with the Climate of Our Planet*, states explicitly: "There is *no proof* that global warming will cause adverse, even catastrophic damages around the world."[20] [italics added] Third, we know that climate alarmists lack proof because they so often fall back on the Precautionary Principle, as Dotto does in her book: if something *might* be disastrous, we must treat it as if it *is* disastrous, even if we don't have proof.[21]

And, so, lacking empirical proof, the climate "consensus" wants us, the public, to accept its alarmist position *before* we see any proof beyond fallible computer studies (we'll discuss the problems with climate computer models in Chapter 4). As MIT astrophysicist Richard Lindzen puts it, global warming "is the only subject in atmospheric science where a consensus view has been declared before the research has hardly begun."[22] And even if the consensus climate alarmists establish scientifically that there is a "crisis"—so far, again, their only "proof" comes from computer models—they surely then must prove that their anti-carbon strategies for stopping the global warming "crisis" will work. This they have not proven and can never prove because, as they like to say, we can't put the earth in a test tube.

Where does the burden of proof in the climate change debate lie? For Gore, "Those who argue that we do nothing until we have completed a lot more research are trying to shift the burden of proof even as the crisis deepens."[23] But surely Gore has it backwards. Changing our technological society along the lines that Gore, Suzuki, Weaver and others want—that is, radically—would cost tens of billions of dollars and damage the livelihoods of millions of people, as we will see in more detail in Chapter 13. Surely it's the global warming alarmists with extreme views, not the skeptics urging caution and empirical evidence, who should justify beyond a reasonable doubt the massive economic and social transformations—the "difficult actions"—being demanded. So far, the alarmists have not provided this proof, as we will see, although, as practitioners of the Doctrine of Certainty, they believe they have.

THE 'CONSENSUS' HYPOTHESIS

What, exactly, is alarmist climate science so certain about that no other scientific views are to be heard by the public? Gore, in his usual convoluted syntax, puts the global warming hypothesis this way in the book version of his movie *An Inconvenient Truth*:

> There is a misconception that the scientific community is in a state of disagreement about whether *global warming is real*, whether *human beings are the principal* cause, and whether its consequences are *so dangerous as to warrant immediate action*. In fact, there is *virtually no serious disagreement* remaining on any of these central points that make up the *consensus* view of the world scientific community.[24] [italics added]

So, putting this into more straightforward English: for Gore, global warming is "real," its *principal* cause is humanity—that is, the planet wouldn't be warming if it wasn't for our carbon emissions—and the results will be catastrophic. Not only is there, for Gore, "virtually no serious disagreement," but he (falsely) claims "100 per cent" consensus for this view among peer-reviewed scientific articles on climate.[25] We'll look at Gore's inflated and almost always inaccurate claims in more detail in Chapter 6.

Climate scientists R.A. Warrick, E.M. Barrow and T.M.L. Wigley cast the hypothesis into more scientific language as follows:

> The potential rates and magnitudes of the GHG [greenhouse gas]-induced change ... give rise to legitimate concerns about the future. These concerns include the following:
>
> - first, that *humankind may now be a potent factor in causing unidirectional global changes* which could dominate over natural changes on the decade-to-century time scale;
> - secondly, that, in terms of recent human experience, changes in climate and sea level *could accelerate to unprecedented rates*;
> - thirdly, that human tinkering with the global climate system could have *unforeseen catastrophic consequences* (e.g., "runaway" warming or sea level rise from strong positive feedbacks); and
> - finally, that the quickened rates of change could exceed the capacity of natural and human systems to adapt without undue disruption or cost.[26] [italics added]

It's worth noting that the above *is* an hypothesis, which is a tentative scientific explanation that has not yet successfully passed enough experimental tests to qualify as a theory. The only "tests" the anthropogenic (human-caused) global warming hypothesis has passed are in computer models; the AGW hypothesis has not passed any significant tests in the real world. That is, there is no *empirical* proof that humans are now the driving force behind climate; there is, as we will see, far more empirical proof that we are *not* the driving force, including the fact that the planet was not on average, as of 2010, warming, and might even be cooling slightly. And yet, the AGW hypothesis is presented to the public not just as an hypothesis and not even as a theory but as a "certain" scientific fact—hence, "the science is settled."

The AGW hypothesis may, of course, be true. But the hypothesis may also be false. Surely at least some empirical evidence is needed before the alarmist view should be uncritically accepted, as the "consensus" wishes.

ALARMISM AND SKEPTICISM

A word about two of the terms used in this book: "alarmist" and "skeptic."

In the first drafts of this book I used the word "pessimists" for anthropogenic global warming believers. However, although they are pessimists, I believe a more accurate term is "alarmists," for two reasons. First, this is how consensus climate scientists describe themselves. For example, Warrick, Barrow and Wigley refer in their book to "the *alarming* rate of atmospheric concentrations of carbon dioxide."[27] [italics added] Weaver, Suzuki, Gore and many others in the global warming consensus admit they are trying to create an air of alarm about climate change, at least in their writings for the public—their writings for scientists are usually much more restrained and nuanced. Secondly, the dictionary defines an alarmist as "a person who needlessly alarms or attempts to alarm others, as by inventing or spreading false or exaggerated rumors of impending danger or catastrophe." Since this book argues that rumors of impending catastrophe from global warming lack any empirical justification—i.e., are false and/or exaggerated—"alarmist" seems to be fair comment, although "pessimists" is also accurate in most contexts.

On the other hand, for those like myself who doubt that human beings are the principal cause of global warming and that global warming will be disastrous, Gore coined the term "deniers." Gore thereby, unethically and illogically, lumped those who deny an historical fact, the Holocaust of the Second World War, with those who question highly speculative computerized guesses about the planet's future climate. Bill McGuire, an earth sciences professor at University College, London, made the connection between Holocaust and climate "deniers" explicit during an television interview:

> Every time you address the Holocaust, you don't bring somebody in that says it didn't happen. And we're at that stage now. We have Holocaust deniers; we have climate-change deniers. And to be honest, I don't think there's a great deal of difference.[28]

Little difference between those who deny the deaths of six million human beings and those who challenge some very, as we will see, dubious science? This is an astonishing statement, especially from a scientist.

Let's be as clear as possible about what being a "denier" means. Nobody denies that global warming has occurred—the planet has been warming for at least 15,000 years since the end of the last glaciation; there's a full consensus about that, at least. But note: the warming that began 15,000 years ago, and that melted millions of tonnes of glacial ice, began without any human

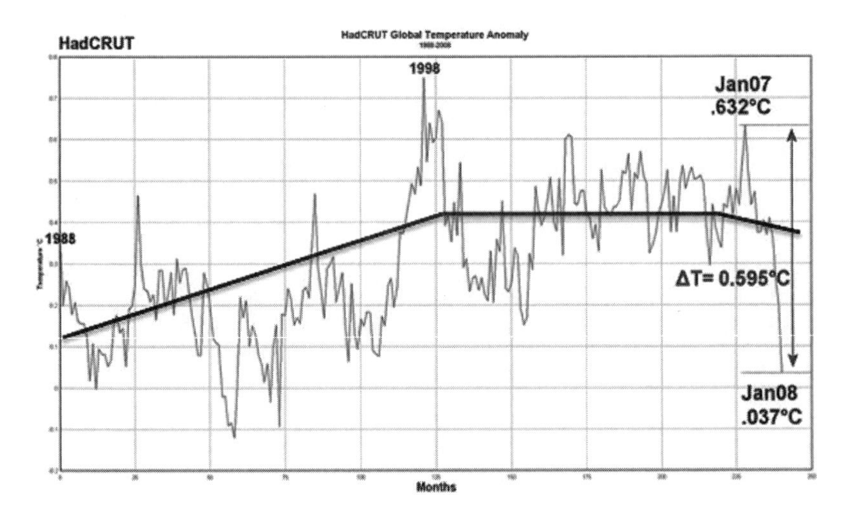

Figure 1.1: Average temperature from 1988 to 2008. Black line shows average temperature. Data source: Hadley Climatic Research Unit. Graph: Anthony Watts

causes—human beings hadn't even developed agriculture yet. Moreover, as of the time of writing (2010), land, ocean and satellite temperature readings show that global warming has stalled, and some climate scientists even believe the planet may be on a cooling trend that could last a decade and perhaps longer.[29] The planet hasn't warmed on average since the late 1990s and perhaps even before that, depending on how the numbers are crunched, and the global temperature in 2008-2009 plunged (see Figure 1.1[30] I've added a line to show the average temperature).

In Chapter 2 we'll look at several other times in the past 15,000 years when the climate has swung toward colder rather than warmer and back again, and changed sometimes several degrees Celsius within decades—the current so-called "accelerated" climate change is not, thus, in any way "unprecedented." And, as we will also see in Chapter 2, the long-term trend in earth's climate isn't toward "runaway greenhouse warming," as, say, Weaver fears when he writes: "Under [the AGW climate science] scenario, global warming *continues unbounded for centuries* with consequences that would not bode well for humanity."[31] [italics added] On the contrary, as climatologist William F. Ruddiman warns: "Earth's long-term forecast over tectonic time scales calls for colder temperatures and more ice."[32]

Nor does a "denier" deny that human beings are contributing to climate change—how could it be otherwise? There are six billion-plus of us and counting, after all, so we're bound to affect the planet to some extent. However, it's likely that global warming—even slightly accelerated global warming—is still *mainly* a result of natural factors that we can't control, given that only two to five per cent of yearly carbon dioxide emissions come from human causes.[33] Warming is also significantly influenced by other human activities that can't be eliminated, such as cities (the Urban Heat Island effect) and agriculture (which is a source of methane, a greenhouse gas that is more powerful than carbon dioxide). If this is so, curbing fossil-fuel emissions will have little or no effect on warming.

What's denied in this book is the following:

- It is denied that carbon dioxide is the *principal* cause of global warming, as the consensus claims. Yes, CO_2 causes some warming, but, as we will see in Chapter 2, the geological record over millions of years shows little correlation between carbon dioxide levels and temperature. Also, the much feared "runaway greenhouse effect" supposedly caused by excess carbon dioxide has so far never occurred except under very special conditions that do not apply today (e.g., massive volcanic eruptions). Finally, the main greenhouse gas is not carbon dioxide, it's water vapor.
- Even if carbon dioxide had the warming effect that the climate alarmists claim, skeptics do not believe *human* carbon emissions are the *principal* cause of warming. Why? Because human carbon emissions are less than five per cent of yearly natural carbon emissions. Yes, humanity has an effect on climate, but natural factors—slightly amplified by human carbon dioxide emissions but also by cities and agriculture—are far more powerful warming (and cooling) agents, as we will see in Chapter 5.
- It is denied that, even if humans were the primary cause of warming, we can do anything meaningful about global warming without wrecking our industrial, technological civilization, which *would* be a catastrophe. In other words, "deniers" believe the cure would

be worse than the disease: a strong economy can respond better to climate change than a weak economy given that climate change, whether cooler or warmer, is inevitable. We will discuss the potentially dire economic effects of severe carbon curbs in detail in Chapter 13.

- It is denied that, however caused, global warming must be the catastrophe that doomsayers like Gore, Suzuki and Weaver would have us believe. First, there is no empirical evidence for this extreme position; the apocalyptic visions are all the product of fallible computer models, overactive imaginations and political agendas. And second, as we will see in Chapter 2, in the past 15,000 years warming has almost always been beneficial for humanity; there is no reason to believe future warming will be any different. Further, we have all the technological tools we need to cope with warming, if that is what the planet does. If we're going to spend money on climate change, we should spend it on *adapting* to warming (or cooling) rather than on a futile effort at trying to *mitigate* (stop) "climate change."

This doesn't mean we can all relax and say, "Great, there's no problem at all." "Deniers" do not deny that human beings will have to make adjustments, and sometimes major adjustments, if global warming continues or, worse, if global warming ends, which would plunge us back into ice age conditions. However, humanity as a species has always had to adjust to climate change so there's nothing unusual about the current situation. Nor, indeed, is there anything to fear in climate change beyond the anxiety humans always feel in the face of natural forces such as earthquakes, floods, volcanoes, droughts, etc., all of which have had a far more powerful impact on us and the planet than the gradual warming of the past 150 years.

If warming continues, some parts of the earth may become less habitable and people will have to migrate, as they have many times in history. In other regions, particularly the northern parts of Canada and Russia, vast areas may open up to cultivation and settlement. Even Greenland has arable land, as the Vikings discovered when they settled there during the Medieval Warm Period a thousand years ago.[34] And, yes, this population movement could lead to wars because, unlike major climate changes in the past, the earth has less empty land to move to. Fresh water may also become a problem as glacial meltwater decreases,[35] but this problem would arise anyway given that the glaciers have been melting for the past 15,000 years—that's why what we're in is called an "interglacial"—long before humans could be blamed. Similarly, sea-level rise may also create difficulties but, again, sea levels have risen 120 metres (400 feet) in the last 15,000 years and still haven't reached the levels of previous interglacials. We are not at fault for sea-level rise, either. At worst, our activities may be accelerating sea-level rise and the global temperature by a percentage point or two.

In other words, there is nothing new or unusual about climate change or the human need to respond to it. The most pressing problems associated with climate change are primarily political and demographic, not climatic, and these problems require political and demographic solutions. If the world is overpopulated to the point where global warming (or cooling) causes severe hardship, such as a flooded Bangladesh, then overpopulation, not climate, is the problem that needs to be addressed. If the inability of poor countries to cope with global warming (or cooling) is the issue, then the problem is poverty, not climate. The billions of dollars spent trying to "stabilize the climate" would be far better used to curb population and reduce world poverty, problems we can actually do something about.[36] We'll look at these issues in Chapter 13.

NOT 'I DENY' BUT 'PROVE IT'

A better term than global warming "denier" is skeptic. The skeptic's attitude isn't, "I blindly oppose or deny" but "prove it to me"—a *scientific* approach. Based on this scientific approach, skepticism is a more reasonable position than climate change alarmism.

For a start, there may be nothing to be alarmed about. As alarmist climatologist Phil Jones has acknowledged, there has been no "statistically significant" warming since at least 1995, and since 2002 the planet has cooled slightly.[37] That's at least a decade of non-warming, yet during this whole time climate alarmists have been whipping up fears about "runaway" warming and the Intergovernmental Panel on Climate Change claims that "warming of the climate system is *unequivocal* as is now evident from *increases* in global average air and ocean temperatures"[38] [italics added]—increases it is now evident, in 2010, that did not occur. As this book tries to show, many of the arguments put forward by global warming alarmists are riddled with errors, exaggerations, poorly supported assumptions and outright fabrications.

A skeptic considers the psychological, social and even genetic factors that might lead us, as a species, to be alarmist rather than optimistic about the future and to fall back on "religious" rituals, like the Kyoto Accord, when faced with natural conditions that we can't control. Chapters 11 and 12 discuss some of the psychological and sociological elements of global warming belief.

A skeptic considers the other ideological agendas that are working within the environmental movement. In particular, fears about industry-caused global warming have been a boon for the political far-left, which is trying to keep the anti-capitalist faith alive amidst the ruins of Marxism. As former *National Post* columnist Elizabeth Nickson wrote, correctly: "[The] Kyoto [climate accord] is a stalking horse for socialism."[39] Those who support the now-defunct Kyoto Accord, or its more recent incarnation, the Copenhagen Protocol, as purely *environmental* measures should be aware of the *political* and *economic* implications as well. Chapter 13 has more detail on this topic.

A skeptic believes there are better ways to spend billions of dollars a year than on a doomed and wasteful attempt to stop the climate from changing. As any high-school geography student should know, climate change is always going on, always has, always will. The call to "stabilize the climate,"[40] as the Sierra Club puts it, is absurd: there is no such thing as a climate that remains perpetually in stasis. And yet many Canadians, and especially those in the environmental movement, think climate stability is a state we are moving away from and something we can, and should, strive for. Because they are often aware of the geological and climatological history of the planet, skeptics know that the 20th- and 21st-century climate is far from "unprecedented" and that there are many possible drivers for climate change apart from carbon dioxide. Chapters 2 and 5 have details on these points.

However, unlike the climate alarmists, skeptics do not believe they are inarguably right; they do not, or at least should not, subscribe to the Doctrine of Certainty. And skeptics understand that humanity will need to cope with climate change, hopefully in useful and creative ways; skeptics don't have their heads in the sand on this issue.

Skeptics also firmly believe that criticisms and questions about the alleged global warming "crisis" should not be blithely dismissed, as Weaver, Suzuki, Gore, et al., wish to do. The public needs *more* information on global warming, not less, and from many points of view, not just the alarmist side. Why? Because the case for a global warming "crisis" or "catastrophe" is far from proved; indeed, as we will see in Chapter 8, the warming alarmists say, outright, that they expect the public to accept their views based on their "intuition" of catastrophe, *without proof*. The danger, therefore, is not that we will under-react, as Gore asserts, but that we will over-react, spending billions of dollars and damaging our economy in a foolish, King Canute-like attempt to turn back the tide of climate change rather accept our human limitations when faced with natural forces and adapt to them.

BRINGING THE PUBLIC INTO THE DEBATE

One final point: are non-scientists—lay people, including journalists— fit to make informed judgments about climate science? Or should we leave these questions to the scientific experts, as the alarmist consensus would prefer? For example, Richard Somerville of the Scripps Institute of Oceanography has written: "People who are not experts, who are not trained and experienced in this field, who do not do research and publish it following standard scientific practice, are not doing science."[41] In other words, the public has no right to comment on climate matters. And it's true, as journalist Dan Gardner points out in his book *Risk: Why We Fear the Things We Shouldn't—and Put Ourselves in Greater Danger*, that it's very difficult for non-scientists to understand the complexities of any scientific discipline. To get up to speed on a scientific question requires, if nothing else, an enormous amount of time, time that most non-scientists don't have. Gardner writes:

> For [non-scientists], the only way to tap the vast pools of scientific knowledge is to rely on the advice of experts—people who are capable of synthesizing information from at least one field and making it comprehensible to a lay audience. This is preferable to getting your information from people who know as little as you do, naturally, but it too has limitations. For one thing, experts often disagree. Even

> when there's widespread agreement, there will still be dissenters who make their case with impressive statistics and bewildering scientific jargon.[42]

Gardner could have had climate science in mind when he wrote this paragraph, because within climate science there is sometimes passionate disagreement among experts, even if most of the experts subscribe to the overall alarmist consensus. How are lay people to know which approach to believe? Philosopher of science Morris Cohen offers the following guideline:

> To be sure, the vast majority of people who are untrained can accept the results of science only on authority. But there is obviously an important difference between an establishment that is open and invites every one to come, study its methods, and suggest improvement, and one that regards the questioning of its credentials as due to wickedness of heart. ... Rational science treats its credit notes as always redeemable on demand, while non-rational authoritarianism regards the demand for the redemption of its paper as a disloyal lack of faith.[43]

In a similar vein, the late astronomer and writer Carl Sagan wrote:

> Part of the duty of citizenship is not to be coerced into conformity. ... If we can't think for ourselves, if we're unwilling to question authority, then we're just putty in the hands of those in power. But if the citizens are educated and form their own opinions, then those in power work for us.[44]

So what's required, according to Cohen and Sagan, is a scientific establishment that is open to criticism and dialogue, and a public that is willing to listen respectfully to the experts, but that refuses to passively, uncritically accept the experts' authority without seeing proof. But for lay people to get the information they need, open, honest, public discussion of the many points of view around climate is essential. So far, in Canada at least, this discussion has not occurred on any meaningful level, including in the nation's Parliament and provincial legislatures. Nor, incredibly, has there been, to my knowledge at the time of writing (April 2010), a nationally televised debate on climate change as usually occurs with other major public policy issues. Instead, the reports of the Intergovernmental Panel on Climate Change have been accepted as gospel, without criticism or due diligence by politicians or the public. Worse: the consensus climate-science position has been accepted uncritically by almost all of the Canadian media. This lack of critical public discussion at even at the highest levels of decision-making is perhaps why a member of Parliament felt comfortable making factually incorrect and even absurd statements on CBC radio.

Of course, alarmist climate scientists who discourage discussion are doing this for, by their lights, the best of all possible reasons: they believe they are right and that we are facing catastrophe. They are bolstered by the belief that since science is objective and disinterested, "one need have no qualms about excluding other people from decision-making since they would, in any event, have arrived at the same conclusions as oneself."[45] This exclusivist view might be justified if consensus climate science tried to prove its hypotheses through objective, empirical data; there would then be fewer skeptics. In practice, however, "global warming is a phenomenon generated by complex computerized climate models."[46] Further, these models are neither objective nor disinterested—in fact, as we will see in Chapter 4, they are highly susceptible to programmer bias. In short, alarmist climate science has become suspiciously like, in Cohen's terms, a "non-rational authoritarianism." Alarmist climate science is non-rational because it does not reach its conclusions empirically but through "intuition" backed by computer models that are also informed by intuition, and authoritarian since it seeks to ignore or suppress skeptical criticism. Why this suppression? Because consensus climate science does not trust the public to come to the "right" conclusion after hearing *both sides* of the climate issue, alarmist and sceptical.

In his famous essay *On Liberty*, Victorian philosopher John Stuart Mill (1806-1873) offered the definitive answer to this closed-minded position. He wrote:

> The peculiar evil of silencing the expression of an opinion is that it is robbing the human race, posterity as well as the existing generation, those who dissent from the opinion, still more than those who hold it. If the opinion is right, they are deprived of the opportunity of exchanging error for truth: if wrong, they lose what is almost as great a benefit, the clearer perception and livelier impression of truth, produced by its collision with error.[47]

The Victorian historian, politician and essayist Thomas B. Macaulay (1800-

1859) put the case even more simply: "Men are never so likely to settle a question rightly as when they discuss it freely."[48] However, if the consensus climate establishment is not open to critical public debate, then lay people must take it upon themselves to *become* informed on climate issues, to get as much information as they can from all sides of this complex topic, and then to think for themselves. There is a wealth of material on the alarmist side of the climate issue; information from the skeptical perspective is harder to find. This is the information gap that, I hope, this book will help to fill, but there are many other skeptical resources that I've cited in this book.

But does a journalist, or in my case, ex-journalist, have the credentials to criticize the climate science establishment? I am not, of course, a climate scientist—but then, neither are Al Gore or Rajendra Pachauri, the head of the IPCC at time of writing,[49] and both won a Nobel Prize for their "climatology." I'm a former editor and writer with *The Globe and Mail* and Victoria *Times Colonist*, among others, now a teacher of professional writing at the University of Victoria. But is a journalist (or former journalist) automatically disqualified from critically examining climate science? The answer is clearly no: many journalists have written on climate science from the *alarmist* side, including Ross Gelbspan, Eugene Linden, Lydia Dotto, Geoffrey Simpson and Andrew Revkin, to name just a few, without drawing complaint from climate scientists. Presumably, an journalist can tackle this subject from the skeptical side as well without drawing criticism just for not having a PhD in climatology.

Moreover, unlike most lay people and journalists, an academic schedule gave me the time and resources to investigate in depth the claims of both consensus climate science and the skeptics. This book is the result of more than two years of research, and it comes down on the side of the climate skeptics because there is simply more evidence for the skeptical position—real, empirical evidence, rooted in the planet's geological history, not just evidence from computer models—than for the consensus climate position.

Also, while I may not be a climate scientist, as a journalist for 35 years I believe I have developed an ability to sense when the public is being told partial truths or falsehoods by public and private interests, including scientists with a non-scientific agenda. My journalistic "intuition"—a word climate alarmists also like to use, as we will see—as well as my research tells me we are not being told the truth or anything close to the truth about climate change. In fact, we are being actively misled by alarmist climate science allied with political, environmental and religious interests.

Another journalist who has examined, critically and in depth, a scientific discipline is Toronto's Elaine Dewar. Her excellent book *Bones: Discovering the First Americans* probes the controversy in paleo-anthropology over how North America's first peoples came here: was it via a land-bridge across Alaska 12,000 years ago or by sea thousands of years earlier? Anthropologists who believe the latter are, Dewar reports, treated much the same as climate skeptics: they are "heretics" who are shunned and suppressed. On her credentials, Dewar writes:

> So why, you might ask, should a journalist investigate a question that has bedeviled scientists for 150 years? Journalists are willing to go anywhere, to be passionate fools, to ask innocent questions, to ignore barriers, to look for patterns that connect disciplines and solitudes, and to have no vested interest in anybody's intellectual capital. Science must be public and transparent or it loses all meaning, and journalists bring evidence out of the lab and spread it before the general public. We are also a little like bees. We dip into everybody's business and carry the news along. We cross-pollinate. We fertilize. Sometimes we sting.[50]

So, I hope this book cross-pollinates, fertilizes, and stings. I hope it can be part of a *real* debate in Canada about the origins of global warming, whether human or natural; about whether global warming is a potential catastrophe for this country and the world or, perhaps, a boon; and about how Canadians can and should respond to climate change—an unavoidable fact of life on this planet—before the nation's energies and resources are fully committed one way or the other.

The majority of the earth's history has been marked by conditions warmer than those of today.
—Edward Aguado & James E. Burt[1]

By the standards of the geological past, we live in an Ice Age. The world has rarely been as cold as it is today.
John & Mary Gribbin[2]

Chapter 2

GLOBAL WARMING: THE GEOLOGICAL BACKGROUND

Before we can understand global warming today we need to put it into the context of the past. Has global warming ever occurred previously in our planet's history, before human beings could have caused it? If so, what caused this warming? And, based on past experience, is today's warming as extreme and "unprecedented" as the global warming alarmists claim?

Ice Age, a small book by British science writer John Gribbin and his wife Mary, begins: "By the standards of the geological past, we live in an Ice Age. The world has rarely been as cold as it is today." We're in an Ice Age? Why do so many alarmist books and articles on global warming ignore or gloss over this important fact? The Gribbin book goes on: "We think it is normal to have ice at both poles of our planet ... but, in the long history of the Earth, polar ice caps are rare, and having two polar ice caps at the same time may be unique."[3] This fact is also rarely if ever mentioned in the alarmist writings. Knowledge about the long-term geology of climate change reveals even more surprising facts that cast doubt on whether humans are the main cause of global warming, on our ability to "stabilize the climate," and on whether we are facing the climate catastrophe that Al Gore and alarmist climate science claim.

The major surprise, as noted above, is that warming is *unusual* at this point in the planet's geological history. The earth's default position right now is Ice Age and this has been so for the past two million years, with occasional and relatively short-lived periods of warming called interglacials. The interglacial we're in now, called the Holocene, started about 15,000 years ago. The initial Holocene warming had no input from humans at all—we hadn't even invented agriculture—so we can be absolutely certain right from the start that global warming (like cooling) is primarily a *natural*, not human-made, phenomenon.

A second surprise: the planet has been through about ten 100,000-year ice age cycles in the past million years as part of an era geologists call the Pleistocene, with 41,000-year cycles that were slightly less extreme in the million years before that. That is, we're in a single ice age made up of two million years of glacial and interglacial cycles. In the past million years, the cold, glacial parts of the cycle have lasted an average of 80,000-90,000 years. The warm interglacials are, on average, a mere 10,000-20,000 years. This is worth pausing to think about: in the past million years, glacials have lasted, on average, *four to five times* as long as the warm periods. It's also worth recalling that 700 million years ago, before complex life began on this planet, the earth may have been almost entirely covered with ice—the so-called Snowball or Slushball Earth.[4] Could this happen again? Possibly: we are within a mere five degrees Celsius of "runaway" global cooling, i.e., glaciers once again covering large portions of the northern continents.

Earth began its slide into ice-age conditions about 50 million years ago, for reasons we'll discuss below, and the slide continues (Figure 2.1[5]). In fact, far from being worried about global warming, we're very lucky, first, to be in an interglacial at all and, second, that the interglacial has lasted as long as it has: 10,000 years is the best-before date for most interglacials over the past million years and we're past that point.

A third surprise is that, contrary to what the global-warming alarmists want the public to believe, the peak temperatures in several previous interglacial periods were 1-3 degrees Celsius *higher* than today's temperatures, at a time when human beings were not a factor (see Figure 2.2. The horizontal line is the temperature below which the earth goes into a glacial[6]). For example, in the most recent interglacial 125,000 years ago, called the Eemian, hippopotamus and other tropical wildlife lived in what is now Britain. The IPCC notes that the climate then was an estimated 1-3° Celsius warmer than our interglacial so far and sea levels were four to six metres (13 to 20 feet) higher than today's levels.[7] Indeed, "the last interglacial ... might

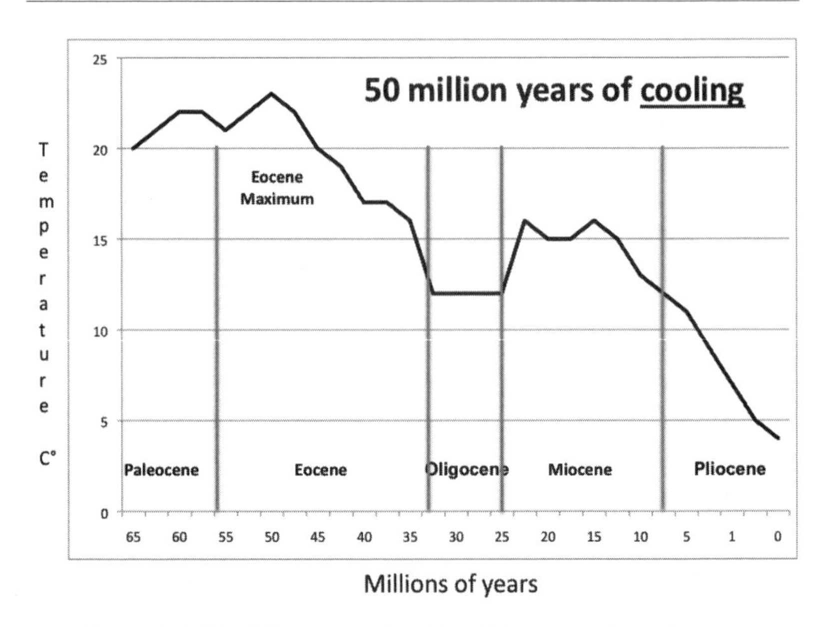

Figure 2.1: 50 million years of cooling. We are now in an ice age.
Source for temperature data: Zachos, et al. (2001)

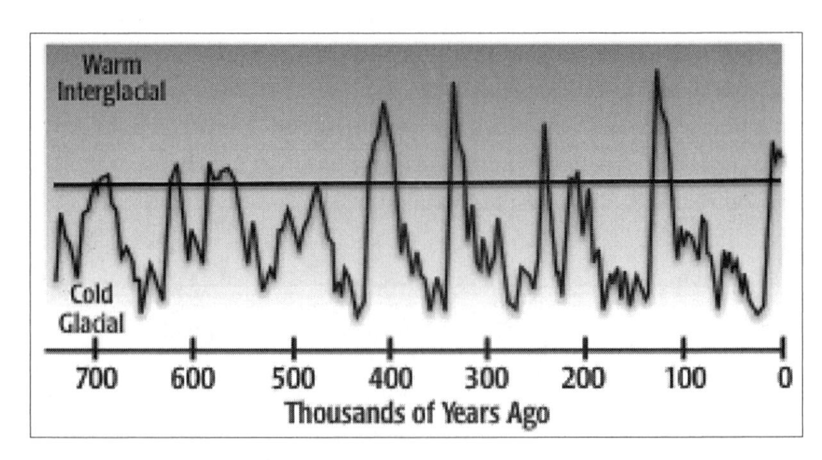

Figure 2.2: Glacials and interglacials over the past 750,000 years. The horizontal line is the threshold below which the earth goes into glaciation.

have set the record for Pleistocene warmth,"[8] that is, warmth over the past two million years. The interglacial of 350,000 years ago reached a higher maximum temperature than today's temperatures,[9] and the interglacial that began 430,000 years ago—and which was the longest at 28,000 years— was also "warmer than today and global sea levels were up to 15 metres (49 feet) higher."[10] In other words, the (slowly) rising temperatures and sea levels of our interglacial are not in any way unusual in geological terms, and they would have occurred whether humans emitted carbon or not. It's worth again being reminded that, in previous interglacials, *no humans were involved* in producing warming greater than today's by several degrees Celsius, with sea levels up to 15 metres higher than we've reached so far today. At worst, any human influence on climate would amplify these natural changes by a few percentage points.

The global temperature within our Holocene interglacial was also, about 7,000 years ago, 1-3° Celsius *higher* than today's temperature (see Figure 2.3).[11] Using this Holocene Optimum (also called the Holocene Maximum) as a starting point, the planet has on average been *cooling* in the past 7,000 years, not warming as the alarmists claim. And the average global temperature was at least a half-degree Celsius *warmer* than today a thousand years ago during the Medieval Warm Period (roughly 800-1300 AD), when the

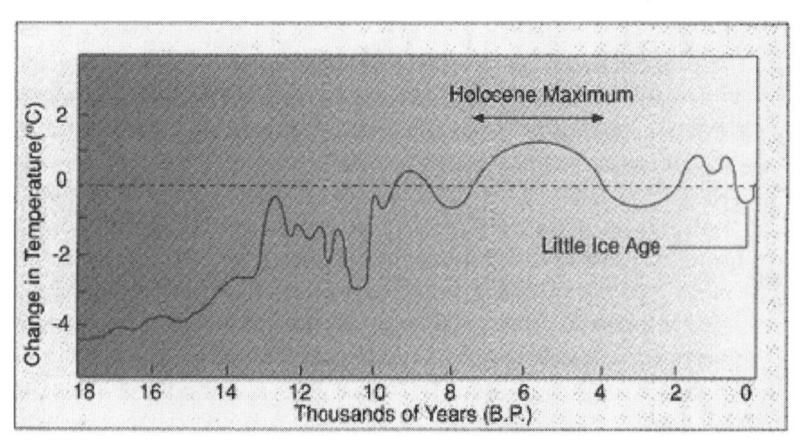

Figure 2.3: Temperatures over the past 18,000 years (the Holocene interglacial). Source: IPCC 1995 report.

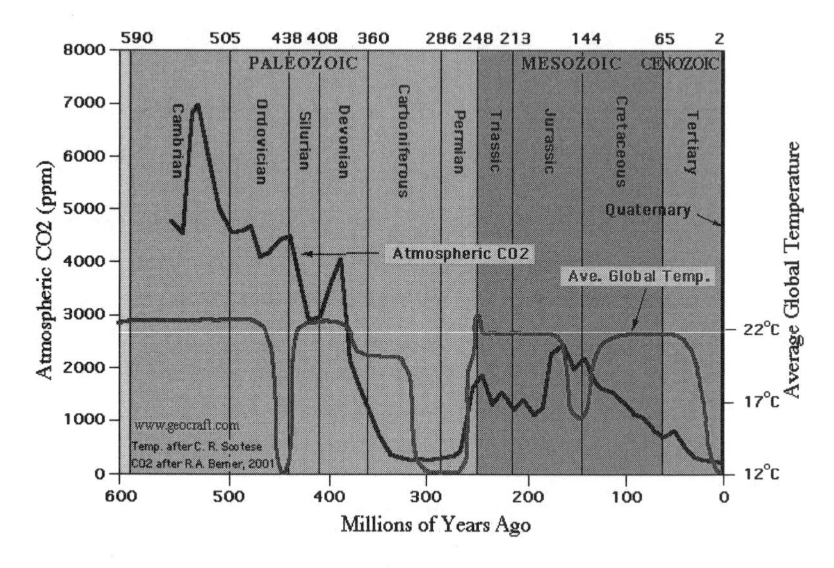

Figure 2.4: Carbon dioxide and temperature over 600 million years.

Vikings colonized Greenland.[12] In other words, the geological record shows that today's global temperatures are *not* extreme, "unprecedented," or beyond "natural variation," but are only nearing the top "normal" range for an interglacial, with or without human input.

Here's another surprising fact: *for 90 per cent of the past 570 million years*—that is, for most of the time that there has been complex life on earth—the climate has been *warmer*, and usually much warmer, than today (see Figure 2.4[13]).[14] In other words, the default condition for the earth, over deep geological time, is *warmer* than today, with much higher levels of carbon dioxide than today's 388 parts per million (ppm). As Edward Aguado and James. E. Burt note in their university-level textbook on climate:

> We tend to think of our own time as "normal," but this is really not the case at all. Looking at the broad span of Earth's history, we would have to describe the present climate as highly unusual, because *most of the time our planet has been considerably warmer than it is today.* Unlike today, when the Arctic Sea is mostly frozen all year and huge ice sheets cover the bulk of Antarctica and Greenland, *for most of its life Earth has been largely free of permanent (year-round) ice.*

> A more accurate depiction is one of a warm planet, punctuated by perhaps seven relatively brief ice ages.[15] [italics added]

Even though the planet's average temperature has in the past been up to 15° Celsius *higher* than the current global average temperature that the alarmists are so worried about,[16] and even though levels of carbon dioxide in the atmosphere have often been five and 10 times *higher* than today's levels—levels that alarmist climate science considers "dangerous"—life somehow managed to evolve, survive, and thrive, as we shall see in detail in Chapter 3. The last age of the dinosaurs, the Cretaceous, was about 10° C warmer than today, had no ice caps at all, and has even been dubbed "sauna earth." That is undoubtedly hotter than most of us would like, but if the dinosaurs could thrive at high temperatures and CO_2 levels, probably humans could, too. In other words, we're not going to "burn up" if the planet gets a few degrees warmer. Another warm time in the planet's recent geological history was the Eocene, the age of mammals that began about 55 million years ago. The Eocene, which lasted 20 million years, was much warmer than today with much higher levels of CO_2, and, again, the Eocene warming can't be blamed on humans.

'PARADISE LOST'

Tim Flannery is author of the best-selling book *The Weather Makers,* which predicts global catastrophe from warming. But he had this to say of the Eocene in a 2002 book, *The Eternal Frontier*, which describes the geological and biological history of North America:

> When Earth is warm (in greenhouse mode)—as it was around 50 million years ago—*North America is a verdant and productive land.* Almost all of its 24 million square kilometers, from Ellesmere Island in the north to Panama in the south, is covered in luxuriant vegetation.[17] [italics added]

The heading for this section of Flannery's book is: "In Which America Becomes a Tropical Paradise." Similarly, paleoclimatologist Donald R. Prothero notes that the Eocene temperature at what is now London, England, was 25°C, about 10°C warmer than today, with no ice caps. CO_2 levels were eight times today's levels when the Eocene began (that is, 3,000 ppm), and still four times higher than today's levels (that is, 1,600 ppm) 23 million years ago at the end of the Oligocene, the geological era that followed the Eocene.[18] Prothero describes the early and middle Eocene as a "lush, tropical world."[19] In the title of his book, Prothero calls the Eocene-Oligocene boundary, about 33 million years ago when the planet began to get colder in earnest, "paradise lost." If the Eocene, which was at least 10° C warmer than today with CO_2 levels four to eight times today's levels, wasn't a catastrophe—if it was a "paradise" that we "lost"—why should warming now be a catastrophe rather than, at worst, a challenge to human ingenuity?

Indeed, as Flannery warns in *The Eternal Frontier*, the real problem, at least for North America, isn't a climate that's too warm but one that's too cold:

> As the Earth cools, North America's capacity to amplify change rapidly drives it to a break point, beyond which it falls into the frigid grip of the poles. It can then be said to be in icehouse mode, *a mode that characterizes the present*."[20] [italics added]

Note, again, that Flannery is writing about North America, which includes Canada. And if he is right, based on past geological history, global warming will be *beneficial* for Canada and probably most of North America, not to mention northern Europe and Asia. Whether warming will be beneficial for more southerly countries we will consider in Chapter 13. One thing we can be sure of: global warming is infinitely preferable to global cooling. During the most recent glacial, which ended only 15,000 years ago, all of Canada and much of the northern United States, northern Europe and northern Asia were under up to three kilometers of ice (see Figure 2.5[21]).

So, looking at the geological record, is our time unusually, unprecedentedly warm? No, it is not, even for an interglacial. "Over the past *half-million years*, and within the context of the most recent *five full interglacials*, it is clear that the average near-surface air temperature of the earth during the 1990s was not unusually *warm*, but unusually *cool*."[22] [italics the author's] And, again, the warmth of previous interglacials had no human influence whatsoever. Yet, the global-warming alarmists would have us believe that our time is, somehow, going off the charts in terms of warmth.

Thanks to their in-depth knowledge of earth's history, geologists tend to be more skeptical of this computer-generated alarmism than the climatologists. For example, in March 2008, a poll of Alberta's 51,000 geologists

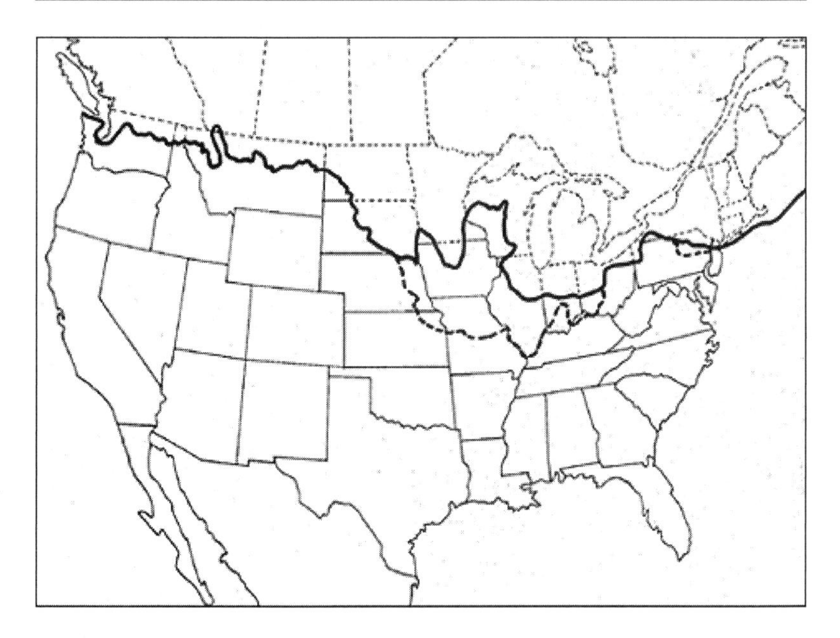

Figure 2.5: Extent of most recent glacial in North America

found that only 26 per cent of the 1,077 geologists who responded believe humans are the main cause of global warming. Forty-five per cent believe both humans and nature are causing climate change, and 68 per cent don't think the debate is "over," as alarmist climate science would like the public to believe.[23] Thus, Peter Sciaky, a retired geologist, notes:

> A geologist has a much longer perspective [than the greenhouse theorists]. There are several salient points about our earth that the greenhouse theorists overlook (or are not aware of). The first of these is that the planet has never been this cool. … [Also,] this is hardly the first warming period in the earth's history. The present global warming is hardly unique. It is arriving pretty much "on schedule."[24]

Should we fear a warming planet?

Perhaps there will be a time when panic and extreme measures to deal with global warming are called for, but so far there is no—repeat, no—evidence outside of computer-generated predictions that this time has arrived. The time to panic will be when the planet starts to get *colder*….

MILANKOVITCH CYCLES

Al Gore, in his movie *An Inconvenient Truth,* strongly implies, without actually stating it directly, that the massive ups and downs of climate over the past 650,000 years were caused by increases and decreases in carbon dioxide, thereby begging the question: if changes in carbon-dioxide levels caused changes in temperature levels, *what caused the changes in carbon-dioxide levels*? As we will see in more detail in Chapter 6, Gore has reversed cause and effect. In reality, first the temperature changes, then, *several hundred years later*, carbon-dioxide levels change. But this still begs the same question: what caused the temperature levels to fluctuate? What caused massive global warming and cooling, not just 15,000 years ago when the latest warming started but dozens of times over the past two million years, long before human activity had any effect on the planet, indeed, before modern humans had even evolved?[25]

There are several theories, but most scientists now think glacial-interglacial climate change is triggered by the Milankovitch cycles, which produce small changes in the earth's relationship to the sun. These cycles are based three factors. One is the earth's tilt, which varies between 22.1 degrees and 24.5 degrees on a 41,000-year cycle and causes the seasons. Secondly, the tilt wobbles like a top on a 23,000-year cycle, which is called the precession

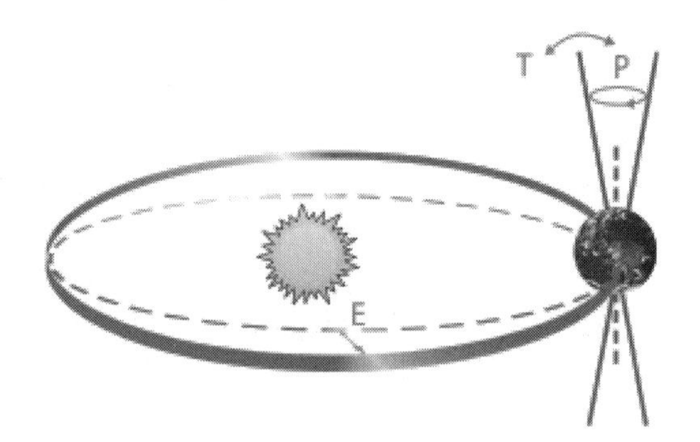

Figure 2.6: The Milankovich Cycles. E is the earth's elliptical orbit, on a 100,000-year time scale; T is the planet's tilt, which varies over 41,000 years; P is the earth's wobble, or precession, with a 23,000-year cycle. Source: IPCC 2007 report.

of the equinoxes. Finally, there is earth's slightly elliptical orbit around the sun, which has a 100,000-year cycle (see Figure 2.6).

While many climatologists don't think the Milankovitch cycles alone are enough to create glacials and interglacials, these cycles do seem to act as the trigger for other, amplifying factors (all, incidentally, natural, and including, in a minor role, carbon dioxide) that have led to the series of Pleistocene glaciations.[26] However, as skeptical Australian geologist Ian Plimer notes, how these cycles work is still poorly understood, and he adds: "If we cannot understand the biggest climate changes of all times, then we have to be very circumspect about claiming that we understand modern climate."[27]

That said, these three factors—elliptical orbit, tilt, and wobble—seem to underlie a 100,000-year cycle that matches fairly well what we know about the series of glacials and interglacials in the past million years, that is, one cycle of cooling and warming every 100,000 years or so, and a 41,000-year cycle of glaciations and interglacials in the million years before that.[28] Glacials begin when the three factors act together to produce cool summers in which, each year, less of the winter ice melts. This can occur, for example, when the planet's elliptical orbit takes it farther from the sun in summer at a time when the earth is also highly tilted. Over thousands of years the ice builds up to a tipping point and a full-blown glaciation begins. On the other hand, unusually warm summers produced by a combination of the three factors begin to push back the ice, leading eventually to an interglacial.

But what touched off the Milankovich cycles? After all, this ice age only began in earnest about two million years ago, while the three cycles have been going on for many millions of years. Why did the planet, after hundreds of millions of years of warmth, start to turn cold 50 million years ago despite carbon dioxide levels at least eight times higher than today's? Why didn't a 3,000 ppm "blanket" of carbon dioxide keep the planet warm, as we would expect from the anthropogenic warming hypothesis? Several factors seem to be involved.

CONTINENTAL DRIFT

A major amplifying factor for the Milankovitch cycles is continental drift. First, about 50 million years ago the Indian continental plate hit Asia and, over millions of years, this massive, slow-motion collision began to push up the Himalayan mountains and Tibetan Plateau. At the same time, the Rocky Mountains were also forming. Over time, the creation of these two major mountain chains disrupted the planet's warm air currents, leading to colder conditions overall. The uplift also created thousands of square kilometres of permanent snow cover at the mountain peaks that radiated solar heat back into space rather than allowing it to heat the planet.

A second tectonic factor was the slow drift of the Australian and South American continents away from Antarctica that began about 40 million years ago. Previously, an ocean current cycle took cold water from the south polar region into the tropics and brought warm water back, keeping temperatures in Antarctica mild enough for tropical conditions 40 million years ago and plant growth as recently as 14 million years ago. However, the northward drift of the two continents allowed the creation of a current that circulated around Antarctica without going into the tropics. This cold current gradually cooled the south polar region until permanent ice began to accumulate about 12 million years ago (see Figure 2.1, above).[29] Also important here: a land mass has been centred on the south pole for more than 100 million years. The land mass allowed ice to accumulate without melting into the ocean during the summer months as happens at the Arctic pole, which is mostly water.

A third important factor was the joining of the North and South American continents at the Isthmus of Panama about three million years ago, which blocked the mingling of the Pacific and Atlantic currents. This blockage helped create conditions for the creation of ice sheets in the Arctic.[30]

These three tectonic factors—mountain formation, the creation of the Antarctic circumpolar current and the closing of the Panama channel—gradually led to colder oceans, and therefore a colder planet, and therefore greater and greater accumulations of ice at the two poles, which further enhanced the cold factor. These conditions led to greater and greater swings of temperature and created, a million years ago, a full-blown ice age with 100,000-year cycles of long glacials and relatively short interglacials, all triggered by the Milankovitch cycles.[31] The key climatic element here was not the falling level of carbon dioxide but the re-distribution of wind and ocean currents due to continental drift.

Of the effect of the 100,000-year Milankovitch cycles in the Holocene, John and Mary Gribbin write:

> The escape from Ice Age conditions, beginning 18,000 years ago, required the combined influence of all three astronomical cycles to drag the Earth into a peak of warmth [the Holocene Optimum] about 6,000 years ago. Orbital eccentricity changes combined with a shift in the wobble of the Earth ... made June the month of closest approach to the Sun, boosting the heat of Northern Hemisphere summers, just at a time when the tilt of our planet reached a maximum, putting that summer sun particularly high in the sky.

They add, however:

> Since 6,000 years ago, all these factors have turned around, and conditions for Northern Hemisphere summer warmth are becoming less favorable. The prospect is for a return of Northern Hemisphere glaciation, on a timescale of thousands of years.[32]

Our Holocene interglacial may last 30,000 years, as the IPCC predicts,[33] or the shift to a cycle of glaciation may have begun several thousand years ago, as the Gribbins and climatologist William F. Ruddiman believe. However long our Holocene interglacial lasts, the long-range climate prognosis is bleak. To again quote Ruddiman: "Earth's long-term forecast over tectonic time scales calls for colder temperatures and more ice."[34]

Several more causes of global warming and cooling will be discussed in Chapter 5, but what the Milankovitch cycles, or the sunspot cycles, or ocean currents, or the galactic cosmic cycles, or dozens of other potential global warming culprits mean for human beings is that when the conditions on and around our planet are ready to get warmer, the planet gets warmer. And when the earth is ready to cool, it gets cooler. In either case, human action is irrelevant or minimal. Furthermore, the most likely "principal" culprit for both warming and cooling is the change in the amount of solar radiation the earth receives. "Considering that the amount of energy emitted by the Sun is not truly constant," write Aguado and Burt, "this mechanism of climate change has considerable theoretical appeal."[35] Ocean currents are the second "principal" factor in creating climate: "Ocean currents have an enormously strong effect on global climate."[36] Carbon dioxide is not even close to the top of the list in terms of influencing climate, and human-caused carbon dioxide is way down the list.

In other words, climate is almost entirely determined by cosmological and geological factors and it is, above all, chaotic—the IPCC calls climate "a coupled non-linear chaotic system." Just as there is no way of predicting with certainty what the *weather* will do, thanks to this chaotic structure there is no way of predicting with certainty what the *climate* will do, although this view is hotly denied by alarmist climate scientists. Yet the United Nations' Intergovernmental Panel on Climate Change itself actually makes the point that its computer climate models are unreliable predictors, writing in its 2001 report: "Long-term prediction of future climate states is not possible."[37] But despite these disclaimers, what appears in the media and the writings of alarmist climatologists is predictive certainty.

WHAT WE KNOW

Knowledge of the earth's geological history reveals clearly that, while humanity undoubtedly plays a small role in climate change, humans are not and cannot be, as alarmist climate science claims, the "principal" cause of global warming: natural forces such as the sun and oceans are far more powerful than humanity when it comes to creating climate, in the past and now. If our role in climate is minor, then efforts to stop or reduce global warming through curbs on carbon emissions cannot succeed, although they could cripple the world economy, leaving us poorer. If we are poorer, we will be less able to deal with the effects of climate change, which is always making the planet either warmer or colder. Therefore, the climate skeptic believes that the global warming alarmists' faith that human beings can somehow shift the climate cycle as we wish simply by curbing carbon emissions is optimistic, and probably even deluded. It would be easier to stop an ocean liner by putting a child's sailboat in its path.

So, just to review, here are some of the geological, atmospheric and astronomical observations that don't support alarmist climate science's anthropogenic hypothesis:

- Our planet is in an ice age, the coldest it's been in 250 million years, and the long-term forecast is for more glacial cold, not less, regardless of how much CO_2 we put into the air. That is, the planet's "certain" climatic future is another glacial, not a permanent greenhouse.

- The warm interglacials last a mere 10,000-20,000 years; the cold spells last an average of 80,000-90,000 years.
- The planet is not unusually warm. Within this Holocene interglacial, the planet was warmer 1,000 years ago, centuries before industrial carbon emissions were an issue, and has been warmer several times before that in the past 15,000 years, including the Holocene Optimum 7,000 years ago.
- The planet is not unusually warm compared to previous interglacials. Temperatures and sea levels were higher in previous interglacials before *homo sapiens* had even evolved; temperatures and sea levels will likely go higher in the next few centuries or millennia, as they did in past interglacials—naturally. Rising temperatures and sea levels are not humanity's fault, as the alarmists would like us to believe. They are a natural part of an interglacial cycle.
- Similarly, it is *natural* for glaciers to melt in an interglacial, which is why they're called interglacials, and there is no evidence that the current glacial melting is worse than in previous interglacials. Also, the current glacial melting began in 1750, before industrial emissions were an issue, as the earth pulled out of a centuries-long cold spell known as the Little Ice Age (roughly 1350-1850 AD). Melting glaciers are not humanity's fault, as the alarmists would like us to believe.
- The planet is not heading toward uncontrolled greenhouse warming. In fact, a runaway greenhouse has never occurred in the time that complex life has been on earth, even when, in the Cambrian era of 550 million years ago, CO_2 levels were 7,000 ppm, or almost 20 times higher than today's levels (see Figure 2.4). The closest the planet has come to runway warming was at the Permian-Triassic boundary, about 250 million years ago, but that warming was caused by a series of massive volcanic eruptions. There is nothing even remotely like this expected today.
- Despite higher carbon dioxide levels, some of it due to humans, the planet is *not currently warming*, and hasn't since the late 1990s

and perhaps earlier. In fact, the planet is now (as of 2010) cooling and could continue cooling for the next decade or possibly longer.[38] In other words, planetary warming and cooling are part of *natural* cycles; humans and their carbon emissions are not key players.

- Polar ice caps are unusual in earth's history; the planet (and humanity) can survive without them. However, a disappearance of polar ice is unlikely: the Antarctic ice cap, for example, is growing, not melting away.[39]

All of the above are climate *facts*, not theories based on computer models. They are inconvenient facts for AGW alarmists because they show that carbon dioxide is not the major climate-forcing factor that alarmist climate science claims, and that the human role in the carbon cycle is not a major factor in global warming (or cooling).

We'll discuss the natural factors causing climate change in Chapter 5, and analyze Gore's arguments for human-caused global warming in Chapter 6. But first, in Chapter 3, let's look in greater detail at the role carbon dioxide plays (or doesn't play) in global warming, and then, in Chapter 4, at some of the myths and misconceptions surrounding the United Nations' Intergovernmental Panel on Climate Change (IPCC), the chief source of global warming alarmism.

To reduce modern climate change to one variable, CO_2, or a small proportion of one variable—human-induced CO_2—is not science. To try to predict the future based on just one variable (CO_2) in extraordinarily complex natural systems is folly. Yet when astronomers have the temerity to show that climate is driven by solar activities rather than CO_2 emissions, they are dismissed as dinosaurs undertaking the methods of old-fashioned science.

—**Ian Plimer**[1]

Weak causes are likely to have weak effects.

—**Aaron Wildavsky**[2]

In climate research and modeling, we should recognize that we are dealing with a coupled non-linear chaotic system, and therefore that long-term prediction of future climate states is not possible.

—**IPCC 2001**[3]

Chapter 3

IS CARBON DIOXIDE THE GLOBAL WARMING VILLAIN?

Climate alarmists blame human-caused carbon dioxide emissions for today's, they claim, accelerated warming. Stop carbon dioxide emissions, they say, and the planet will eventually stop warming. However, as we saw in Chapter 2, a warmer planet and higher carbon dioxide levels are not, in themselves, a danger or even a problem for humans, animals or plants—quite the opposite actually, since higher CO_2 levels increase plant growth which, in turn, is good for animal life, including humans. As we saw in Chapter 2, CO_2 levels during the Eocene (55-33 million years ago) were four to eight times higher than today's levels without making life on earth impossible or even difficult—this was the era in which mammals evolved and flourished and it has even been called a "paradise".[4] And, CO_2 levels have been as much as 18-20 times higher than today's at times in the distant past,[5] yet somehow life soldiered through. Higher CO_2 levels and warmer temperatures do not equal "oblivion" or a "burning" planet, as the alarmists claim.

Furthermore, at several times in earth's history when CO_2 levels were much higher than today's the planet *cooled* and at other times CO_2 levels

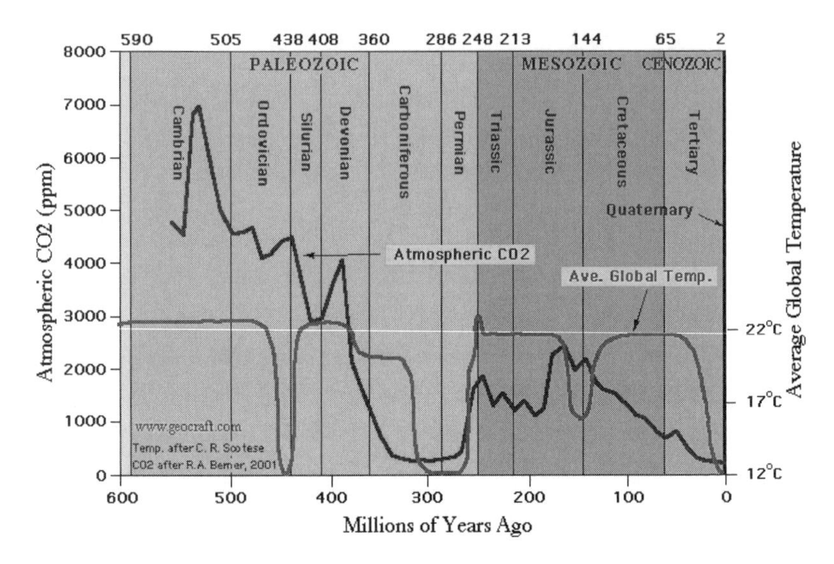

Figure 3.1: Carbon dioxide and temperature over 600 million years

were low while the temperature was high (see Figure 3.1). Clearly, high levels of carbon dioxide weren't keeping in the warmth in the geological past, as the anthropogenic global-warming hypothesis argues. Carbon dioxide levels began to decline 175 million years ago, but the temperature didn't begin to fall into the current ice-house conditions until about 50 million years ago. Thus, factors other than CO_2, including the position of the continents, which affected ocean and air currents, kept the planet warm as carbon-dioxide levels fell for that additional 125 million years. In other words there is not, as the AGW hypothesis claims, a strong causal connection between carbon dioxide levels and temperature and what connection there is shows carbon dioxide levels *following* temperature changes.

A QUIZ ON CARBON DIOXIDE

Here are some other carbon dioxide facts that citizens influenced by the anti-global-warming bandwagon might not know.

Many of us learned in high-school geography that earth's atmosphere is about 21 per cent oxygen, but after that the gaseous cast of characters and numbers may be blurry. There's nitrogen. Argon and some other trace gases.

Carbon dioxide is clearly important—we're "carbon-based" life forms, after all—but what exact percentage of the air we breath is carbon dioxide? Here's a short quiz:

The percentage of carbon dioxide in Earth's atmosphere is:

a. over 10 per cent
b. five to 10 per cent
c. one to five per cent
d. less than one per cent.

Given that carbon dioxide is supposed to be the key driver of global warming, most of us would probably guess "b"—five to 10 per cent—or at least "c". The correct answer is "d." Not only is carbon dioxide less than one per cent of the earth's atmosphere, it's way, way less than that—under .04 per cent at the moment (388 parts per million, to be precise). That's not 4 per cent, which would be four molecules per 100 molecules of dry air, which is the way atmospheric gases are measured. It's not 4 parts per 1,000, which would be .4 per cent. Carbon dioxide is 4 parts per 10,000, or 40 parts per 100,000, or, as the climate scientists like to measure it, 400 parts per million (ppm). In a sports stadium with 100,000 people representing the atmosphere, carbon dioxide would be a mere 40 people. In other words, far from being one of the "big" parts of the atmosphere like oxygen and nitrogen (which is 78 per cent, by the way), carbon dioxide is a trace gas—one of several trace gases that make up the remaining one per cent of the atmosphere that isn't oxygen and nitrogen. And yet, this is the gas, according to the climate alarmists, that is warming our world to catastrophic heights.

Then there's the question of exactly how much carbon dioxide we humans, as opposed to natural forces, contribute to the atmosphere every year. Humans must be putting out a *lot* of CO_2 to have the potentially catastrophic effect the alarmists claim. So here's another short quiz:

The *human* contribution to atmospheric carbon dioxide through fossil-fuel burning and deforestation every year compared to natural sources is:

a) over 20 per cent
b) 10-20 per cent
c) 5-10 per cent
d) 5 per cent or less.

Considering all the fuss over human carbon dioxide emissions, you'd think the number would be at least in the "b" range—10 to 20 per cent. Or, if not that, then at least in the "c" range—5-10 per cent. In fact, the answer is "d", 5 per cent or less. Environment Canada's website puts the human component of CO_2 emissions at a mere two per cent.[6] Other sources say from 3 to 5 per cent.

Carbon emissions are measured in gigatons: 1 gigaton = 1 billion metric tons. Natural CO_2 emissions from factors like vegetative decay and erosion equal about 62 gigatons per year; breathing by organisms (including humans) adds about 50 gigatons; human fossil-fuel use contributes 5 gigatons and forestry 2 gigatons. Summing up: natural sources = 112 gigatons a year; human sources = 7 gigatons, for a total of 119 gigatons per year.[7] Therefore, the human-caused portion of the yearly additional CO_2 is about 5 per cent. Of this extra seven gigatons, however, the IPCC estimates that just about half is reabsorbed by the oceans and by enhanced plant growth, since plants are stimulated by extra CO_2.[8] In that case, the mainly human-caused addition to the atmosphere each year is less than three per cent. In other words, natural forces such as vegetative decay, evaporation from the oceans, respiration (animal and human, so we're not the only creatures "polluting" the planet), volcanoes, etc., produce 95 per cent or more of the world's carbon dioxide emissions each year.

So let's do the math. Carbon dioxide is .04 per cent of the atmosphere. The human sources of carbon dioxide emissions are, let's say, about 4 per cent of that. Four per cent of .04 per cent equals .0016 per cent. This means the human contribution to carbon dioxide in the atmosphere each year is a mere 16 molecules per million (16 ppm), or 1.6 molecules per 100,000. In a sports stadium holding 100,000 cheering fans, that would be one person with a baby. It would be extremely difficult, if not impossible, to pick out the "signal"—that's the climatologists' term for the anthropogenic CO_2 contribution to warming—of one voice in such a throng. Similarly, it makes no sense to believe that an additional 16 parts per million of human-caused carbon dioxide per year could be doing the damage the climate alarmists claim. Yet, that's what we're told is gospel truth, utterly "certain," by the anthropogenic global-warming believers.

Ah, say the alarmists, but this anthropogenic carbon dioxide is *accumulating*. The IPCC tells us there is 30 per cent more CO_2 in the air today than the 280 ppm when the Industrial Revolution began in 1750, and much of this increase is human-caused. This may be true, but let's look at the 30 per cent (a scary number) from another perspective: 30 per cent of 400 ppm (slightly above the level of CO_2 in 2010) is 120 ppm. That is, humans have possibly added 120 ppm, which is just over .01 per cent of the atmosphere, or the equivalent of 12 people in a stadium of 100,000. This increase in carbon dioxide is not nothing, but again, it's difficult to believe that 12 extra people are giving off enough noise ("signal") to dominate all the other natural noises (solar variation, ocean currents, etc.) that combine to produce climate, and thereby throw the planet into uncontrolled greenhouse warming.[9] This makes little sense.

OTHER FACTORS CAUSING CLIMATE CHANGE

So, human-caused carbon dioxide is a small factor in the total amount of carbon dioxide added to the atmosphere each year. But carbon dioxide is also only *one* climate factor among hundreds of factors, all of which contribute to making climate. These factors, which we'll consider in more detail in Chapter 5, include sunspot activity, solar intensity, the position of the earth in relation to the sun, cosmic radiation, ocean currents, water vapor, the amount of ice cover, volcanic activity, desertification, urban heat islands, the positions of the roaming continents, cloud cover, the rise and fall of land masses, and even the movement of the solar system through the galaxy. Almost all of these factors are *variables*, not constants. Some can be predicted, like earth's position in relation to the sun, some cannot, like ocean currents, which are hugely important in determining climate and can suddenly change for no apparent reason. In other words, climate as a system is *extremely* unpredictable because there are so many fluctuating variables.

For example, the main "greenhouse gas" is not carbon dioxide but water vapor, which humans can't directly control. Most sources agree that water vapor and the clouds formed from water vapor are responsible for anywhere from 60 to 90 per cent of greenhouse warming on the planet, depending on the weather: more water vapor equals more clouds which usually equals more cooling; less water vapor equals fewer clouds equals more warming. Another way of putting this: greenhouse gases are responsible for making the planet 33° C warmer than it would otherwise be; water vapor accounts for 30-31 degrees of this warming.[10] Carbon dioxide and the other greenhouse gases (mainly methane, nitrous oxide and chlorofluorocarbons, or CFCs) make up the remainder—2-3 degrees Celsius.

Carbon dioxide contributes to warming by operating indirectly through water vapor; that is, more carbon dioxide is believed to trap more heat, which causes more evaporation, which releases more water vapor into the atmosphere, which causes more warming (the "greenhouse effect"). This is called a positive feedback. However, the additional water vapor warming produced by additional CO_2 operates logarithmically. That is, X increase of carbon dioxide produces Y amount of additional water vapor warming. But the next X of CO_2 produces only half the water vapor warming of the previous increase. And so on. Very quickly, the amount of water vapor warming produced by additional CO_2 approaches zero; as we'll see below, at 388 ppm of CO_2 we are likely already at or beyond the CO_2 "saturation" point.

Also, as the amount of water vapor increases, more cloud cover is created, which decreases global warming because low-lying, cumulous clouds radiate much of the sun's heat back into space.[11] This is called a negative feedback and on average clouds cool, rather than warm, the planet; as we all know, cloudy days are cool days.[12] However, even the IPCC admits that the effect of clouds is poorly handled in climate computer models, noting in its 2001 report: "Clouds represent a *significant source of error* in climate simulations."[13] [italics added] So, as the planet warms, perhaps due to added CO_2, more water vapor is released from the oceans into the atmosphere. More water vapor produces more clouds. More clouds tend to produce cooling. The result is a global thermostat that now and in the geological past protects the planet from "runaway" warming.

Another intriguing factor: the nasty industrial carbon emissions that, supposedly, accelerate warming also generate small particles called aerosols in the atmosphere that lead to cloud formation, and, again, cloud cover tends to promote cooling. So, ironically, the more we "clean up" our emissions, the more warming we create.[14] In short, climate is complicated: "While one agent might be leading to warming, another might counteract or enhance that warming." Also, "agents of change might interact with one another, so that the effects are not merely additive."[15] By contrast, for the IPCC the effects of more CO_2 are almost entirely additive, i.e., mainly positive feedbacks. The result is, as we will see below, a considerable overestimation of temperature increase due to carbon dioxide increase.

The sum of all these climate variables is a system that is chaotic and well beyond the power of even the most sophisticated computer models to predict with any certainty. This is even admitted in the Intergovernmental Panel on Climate Change's 2001 report, which states:

> In climate research and modeling, we should recognize that we are dealing with a coupled non-linear chaotic system, and therefore that *long-term prediction of future climate states is not possible.*[16] [italics added]

Yet, since changes in climate are, by definition, too long-term to be directly observed in the way that we can observe the weather, computer models are the main source—in fact, models are the *only* source[17]—for the alarmists' gloomy forecasts. Indeed, almost the only fact on which all sides of the climate debate can be certain is that climate is unbelievably complex. Even Gore, who dumbs down the causes of global warming to human-generated carbon dioxide, acknowledges this: "The fact that the global atmosphere operates as a complex system makes it difficult to predict the exact nature of the changes we are likely to cause."[18] And then, Gore goes ahead and predicts doomsday as if there were no difficulty at all.

Amidst all this complexity, how do we know which scenario—global doom, global boon or no effect at all—is likely to occur due to climate change? We don't know. There is no certainty, there are only speculation and hypothesis, based on inaccurate and sometimes biased computer models, masquerading as certainty. Anyone who states that anything regarding climate change is dead certain as the climate alarmists do, at least to the public, is not being honest. Or, as climatologist Judith Curry puts it, commenting on the loss of public confidence in science after the Climategate emails:

> No one really believes that the "science is settled" or that "the debate is over." Scientists and others that say this seem to want to advance a particular agenda. *There is nothing more detrimental to public trust than such statements.*[19] [italics added]

And to think—as do Al Gore and the IPCC and most climate alarmists—that reducing carbon dioxide levels alone will have the desired cooling effect on climate when so many other elements are at play makes no sense, at least to the skeptic. What we *do* know is that, based on the past million years of the planet's history, we are *fortunate* that the climate is currently warm. And we would be fortunate if the planet continued to warm rather than getting colder. If carbon dioxide acts as a warming agent, we should welcome it, not shun it.

WHY PICK ON CARBON DIOXIDE?

Carbon dioxide is a convenient choice for global-warming villain because none of the other possible, and more likely, culprits, such as fluctuating solar intensity, ocean currents, cosmic rays, water vapor, and many other natural forces, are in any way human-caused. That means we can't do anything about them. Also, the human component of carbon dioxide emissions is largely produced by industry. Many in the environmental movement, in particular, don't like industrial capitalism, so, again, carbon dioxide is a convenient villain since industrial processes emit carbon. We'll look more plausible sources of global warming (and cooling) in Chapter 5, but common sense, if nothing else, makes the IPCC's emphasis on human-caused carbon dioxide puzzling given the awesome power of the many natural warming forces, not least the sun and oceans.

However, the IPCC regards changes in solar intensity as too weak to be the cause of warming in the past few decades although, curiously, the Climate Change 07 Summary for Policymakers states: "A significant fraction of the reconstructed Northern Hemisphere inter-decadal temperature variability over [the seven centuries before 1950] is *very likely* attributable to volcanic eruptions and changes in solar irradiance."[20] [italics the author's] So, "changes in solar irradiance," allied with other natural causes, were strong enough to cause, for example, the Medieval Warm Period (850-1350 AD), then the Little Ice Age (1350-1850), then the warming from 1850 to 1950, at which point (the IPCC says) human carbon emissions became a factor. But changes in solar irradiance are apparently not strong enough to cause warming today. This view is especially curious considering that some scientists believe the sun was more active than usual in the last part of the 20th century (see Chapter 5 for details).

CARBON DIOXIDE AND PLANT GROWTH

The amount of carbon dioxide today is the highest, the IPCC says, in 650,000 years. Also the highest in that time, it says, is the rate of temperature increase in the past two centuries and particularly this century. However, as we saw in Chapter 2, the level of CO_2 has been much higher in the distant past—as much as 20 times current levels (over 7,000 ppm). In the warm era known as the Eocene, roughly 55 million to 33 million years ago, carbon dioxide was four to eight times higher than today's levels—that is, 1,600-3,000 ppm. These levels were *naturally* caused, and they harmed neither the planet nor its wildlife. Why not? Because carbon dioxide is neither a poison nor a pollutant that is "choking our atmosphere," as alarmists like Tim Flannery would have the public believe.[21] Instead, CO_2 is an essential component of life—animal life and plant life. Just as humans are "carbon-based" life forms, so, too, are plants, and the more carbon dioxide available in the atmosphere, the better plants grow. Plants are also more drought-resistant as CO_2 levels increase.

So, the levels of carbon dioxide are not, as the alarmists charge, becoming too high; the current levels of carbon dioxide are very *low* by geological standards because CO_2 levels have been falling steadily for 175 million years. As geologist Ian Plimer writes: "So depleted is the atmosphere in CO_2 that horticulturalists pump warm CO_2 into glasshouses to accelerate plant growth."[22] How much carbon dioxide do greenhouse growers put into their hothouses? On average, at least 1,000 ppm, almost three times the current level of CO_2 in the atmosphere, or an amount that the IPCC would have the public believe will lead the planet to disaster. An Ontario government website for greenhouse farmers notes:

> CO_2 increases productivity through improved plant growth and vigour. Some ways in which productivity is increased by CO_2 include earlier flowering, higher fruit yields, reduced bud abortion in roses, improved stem strength and flower size. *Growers should regard CO_2 as a nutrient.* For the majority of greenhouse crops, net photosynthesis increases as CO_2 levels increase from 340–1,000 ppm. Most crops show that ... increasing the CO_2 level to 1,000 ppm will increase the photosynthesis by about 50% over ambient CO_2 levels.[23] [italics added]

The best CO_2 level for plant growth, the Ontario website notes, is between 1,000-1,300 ppm; after that, the beneficial effects taper off. In other words, far from being "global warming pollution," as Al Gore would have us believe, or "choking our atmosphere," as Flannery would have us believe, CO_2 is essential to plant growth, and, hence, essential to life. How much does CO_2 increase plant growth? Even the IPCC acknowledges that doubled CO_2 levels produce increases of up to 33 per cent in plant growth.[24] Physicist and biologist Sherwood B. Idso, who has specialized in charting the relationship between CO_2 and plants, notes:

> A simple 330 to 660 ppm doubling of the air's CO_2 content *will raise the productivity of all plants, in the mean, by about one-third.* ... As atmospheric CO_2 concentrations more than double, plant water-use efficiencies more than double, with significant improvements occurring all the way out to CO_2 concentrations of a thousand ppm or more. [italics added]
>
> Think of what such a biological transformation will mean to the world of the future. Grasslands will flourish where deserts now lie barren. Shrubs will grow where only grasses grew before. And forests will make a dramatic comeback to reclaim many areas presently sustaining only brush and scattered shrubs.[25]

If a doubling of CO_2 did produce a 33 per cent increase in plant growth, what a boon that would be to humanity and to the planet! In other words, a warmer planet with higher levels of CO_2 would be, overall, a greener planet and not, as the alarmists charge, a devastated planet. As geologist Robert M. Carter notes: "For the past few million years, the Earth has existed in a state of relative carbon dioxide starvation compared with earlier periods. ... [Carbon dioxide's] increase in the atmosphere leads mainly to the greening of the planet."[26]

To repeat: the planet doesn't have too much carbon dioxide today, it has too little, and it appears the optimum CO_2 level for plant growth is more than 1,000 ppm, or almost three times today's levels. On the other hand, levels of CO_2 below 200 ppm stunt plant growth[27] and much lower than that would make plant, and therefore animal, life on earth impossible. CO_2 levels of less than 300 ppm have been a characteristic of glaciated times over the past million or more years. At 388 ppm, the planet's carbon-dioxide budget is much closer to the lower limit for optimal plant growth than the higher.

CARBON DIOXIDE AND TEMPERATURE

Are levels of CO_2 today radically higher than in the past 650,000 years, as the IPCC charges? Ice-core readings are the main evidence for the IPCC's claim. However, Polish scientist Zbigniew Jaworowski has suggested that due to evaporation, these ice-core readings show CO_2 levels in the geological past as lower than they actually were. That is, although the ice-core readings from the past 650,000 years may show 280 ppm, CO_2 levels in the our interglacial, or in previous interglacials that were warmer than today's, may well have been higher, and perhaps even close to or as high as today's levels of 388 ppm.[28] For his pains, Jaworowski's proposal for research on the accuracy of the ice-core samples was labeled "immoral" by his potential funders (not the IPCC in this case) and he was fired because his findings might cause doubt about the foundations of the global warming hypothesis. This is a very strange way to do science, but quite in line with the climate alarmists' approach of see-no-evil unless it's human-caused and suppress any voices that say otherwise.

Moreover, the relationship between temperature and carbon dioxide is not constant. At some times in the past, carbon dioxide levels have been high when the temperature was low, and low when the temperature was high (see Figure 2.1, above). As Carleton University paleoclimatologist Tim Patterson told a House of Commons committee in 2005:

> Through deep time *there is no meaningful correlation between carbon dioxide levels and earth's temperature*. In fact, when CO_2 levels were over 10 times higher than they are at the present time, about 450 million years ago, our planet was in the depths of the absolute coldest period in the last half billion years. On the basis of this evidence, how could anyone still believe that the recent relatively small increases in CO_2 levels ... would be the major cause of the past century's modest warming?[29] [italics aded]

To my knowledge, this is the only time Canadian parliamentarians have officially heard the skeptical side of the climate change issue.

This lack of meaningful correlation is clear in Figure 3.2, which shows

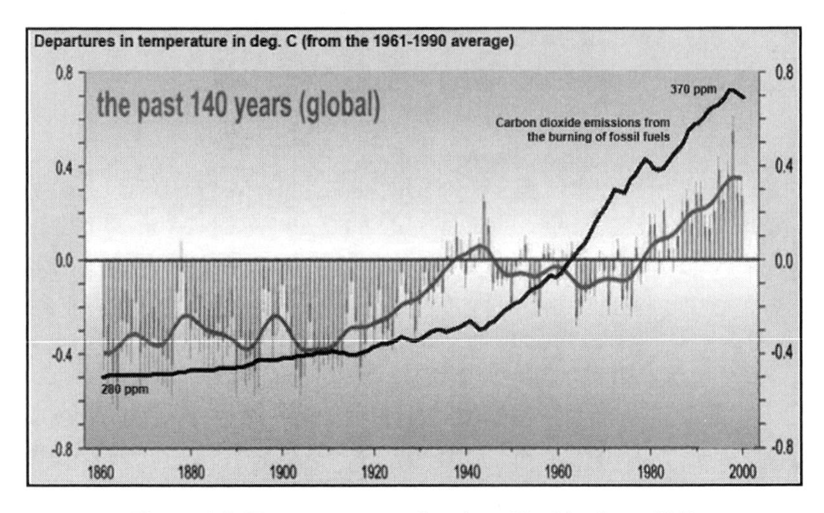

Figure 3.2: Temperature and carbon dioxide since 1860. Source: R.M. Carter.

the relationship, or, rather, lack of relationship, between carbon dioxide and temperature since 1860. The disconnect between temperature and CO_2 is even clearer when we look at a graph of carbon dioxide and temperature from 1979 to 2008 (Figure 3.3).

In these figures we see carbon dioxide steadily rising while the temperature rises and falls, making it clear that CO_2, natural or human-caused, cannot be the "principal" source of climate change over the past century. Further, temperatures have flat-lined and even fallen since the late 1990s, an event that none of the IPCC's sophisticated computer climate models predicted. In fact, almost the entire IPCC case is based on 23 years, from about 1975 to 1998, when carbon dioxide and temperature rose in tandem. Based on the past geological history of temperature and carbon dioxide, including the fact that temperature change *leads* changes in carbon dioxide levels, this late 20th-century parallel is more likely random correlation than causation.

CARBON DIOXIDE SATURATION

So, how much will the increase in carbon dioxide levels affect global temperatures, according to the IPCC? The 2007 IPCC report predicts that a doubling of carbon dioxide concentrations will increase global temperatures by

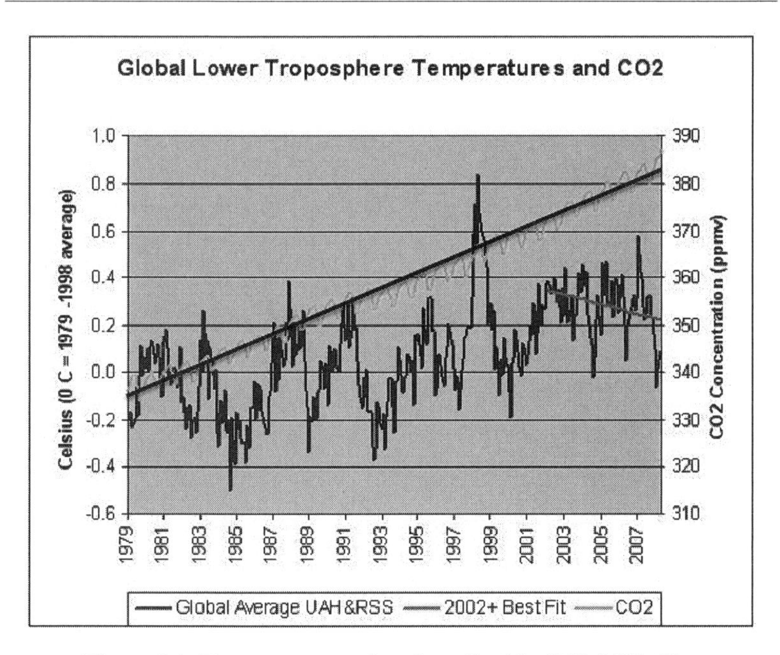

Figure 3.3: Temperature and carbon dioxide 1979-2008. The jagged line is temperature, the dark line, carbon dioxide

2 to 4.5 degrees Celsius over the next century or so. One degree is barely noticeable—the IPCC says we increased .6°C in the 20th century and who would have noticed if the IPCC hadn't been spreading fear? Indeed, IPCC chairman Rajendra Pachauri stated this explicitly: "If the IPCC wasn't there, why would anyone be worried about climate change?"[30] Why, indeed? An increase of 4.5° would be considerable, although still not as warm as in the geological past, but the IPCC considers a 4.5° increase unlikely. It estimates an increase in the range of 3°C.[31]

There's a problem with this analysis, however. As you put more CO_2 into the atmosphere, you get *less* warming effect per molecule. For example, climatologist John Bluemle notes that if you plot CO_2 and temperature increases on a graph, temperature increase "is very rapid initially and then flattens out. Doubling CO_2 from today's concentration, holding all other parameters constant, has a 'negligible' effect."[32] Even alarmist climatologist William Ruddiman has written: "Earth's temperature reacts strongly to small changes in CO_2 values at the lower end of the range (less than 200 ppm), but changes

Table 3.1: Effect of added CO2 on temperature

Added CO2 ppm	*Total CO2 in air 1900=290*	*°C temp. increase /80 ppm*	*Cumul. increase °C*
80	370	0.60	0.60
160	450	0.30	0.90
240	530	0.15	1.05
320	610	0.08	1.13
400	690	0.04	1.16
480	770	0.02	1.18
560	850	0.01	1.19
640	930	0.00	1.20
720	1010	0.00	1.20
800	1090	0.00	1.20

much less at the high end of the range (greater than 800 ppm)."[33] In other words, carbon dioxide and temperature have a logarithmic relation: small amounts of carbon dioxide may initially raise the temperature considerably, but as more CO_2 is added the temperature increase falls off. So, if 100 ppm raised the temperature 1°C, 200 ppm would raise the temperature only .5°C, 300 ppm would raise the temperature only .25°C, and so on.

Putting this into more concrete terms: The CO_2 level in 1900 was about 290 ppm. At the end of the century, in 2000, the level was 370 ppm, for an increase of 80 ppm. The IPCC reckons the global temperature to have increased .6°C in the 20th century, with half of that due to human activity, the other half due to natural causes. However, let's assume that *all* of the warming in the 20th century was due to the CO_2 increase (it wasn't, but let's take the worst-case scenario). If we also assume that the laws of physics haven't changed to accommodate the AGW hypothesis, then 80 ppm of carbon dioxide produced .6°C of warming (see Table 3.1).[34]

If this rough calculation is correct, the *next* 80 ppm of CO_2, which would put CO_2 levels at 450 ppm, would produce .3°C of warming, for a total of .9°C. The next 80 ppm would produce .15°C of warming, for a total of 1.05°C. The next 80 ppm, which is well into "oblivion" territory (610 ppm)

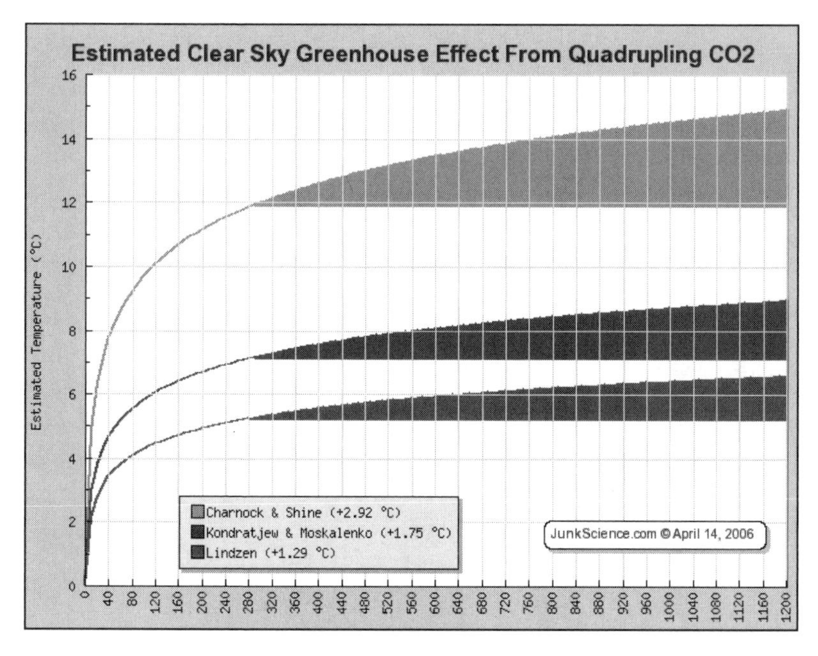

Figure 3.4: Carbon saturation effect: three estimates. With each addition of carbon dioxide, the warming effect decreases logarithmically. Shaded areas indicate a range of temperature.

according to the IPCC, would produce .08°C of warming, and if CO_2 goes up another 80 ppm to 690 ppm, the temperature will rise a mere .04°C, for a total increase of 1.16°C. And so on.

By the time CO_2 levels in the atmosphere have increased to 930 ppm, or more than double today's CO_2 level, the temperature increase per increment of CO_2 will be just slightly above zero and the total temperature increase stalls out at 1.2°C, which is hardly Thermageddon territory. Steven Malloy's Junk Science website shows three different estimates of the saturation effect for doubled CO_2 (see Figure 3.4[35]).

What the three estimates agree on is that the first 20 parts per million of carbon dioxide have the most warming effect, and that by the time the CO_2 level reaches 400 ppm, the current level, the warming effect is relatively small or even virtually nil (the shaded areas are possible ranges of temperature).

It's possible, of course, to argue with these numbers—perhaps the temperature increase per 80 ppm is a bit higher than .6°C, although it is likely less given that even the IPCC blames only half of the 20th century warming on human-caused carbon dioxide. Perhaps the logarithmic equation is something other than half the warming per doubling of CO_2. Some scientists believe the point of maximum CO_2 warming is as low as 20 ppm, others 50 ppm, while Ruddiman suggests it might be 200 ppm. We may quibble with the figures, but it's not possible to quibble with the physics. Once carbon dioxide's atmospheric band is saturated, very little extra warming occurs no matter how much CO_2 we put into the air.

That carbon dioxide gets saturated seriously weakens or even falsifies the hypothesis that higher levels of CO_2 will cause potentially disastrous temperature increases. It also explains why, despite alarmist fears, a runaway greenhouse based on carbon dioxide alone has never, as far as we can tell, occurred since complex life appeared on earth more than 600 million years ago (see Figure 3.1).[36] In the deep geological past, carbon dioxide levels have been 20 times or more higher than today's, but in the past 600 million years the earth's temperature has rarely gone above 25° Celsius, about 10°C higher than today's average global temperature. This high temperature (sometimes called "sauna earth") persisted for most of the time of the dinosaurs, from 250 million to 55 million years ago. Further, at several times in the earth's history, the planet has been in an ice age while CO_2 levels were more than 10 times today's levels. Clearly, CO_2 wasn't driving the climate then and it almost certainly isn't now, either.

At 388 ppm, we are near, at, or well beyond the CO_2 saturation point. As further evidence, a falling off of warming seems to have occurred since the late 1990s, although, of course, other warming and cooling factors (almost all natural) will continue to cause warming and cooling, as they always have and always will.

CO_2 LEVELS AND FOSSIL-FUEL DEPLETION

At some point in the next 100 years the world is expected to run out of readily available fossil fuels, particularly coal and oil, the two main fuels that produce carbon emissions. Another major fossil-fuel source, natural gas (methane), also produces carbon emissions, but emits 43 per cent less carbon per unit of energy than coal and 30 per cent less than oil.[37] The point at which oil extraction reaches its maximum and begins to decline is called Peak Oil, but the same process will create Peak Natural Gas and Peak Coal. After that, it's only a matter of time before fossil-fuel resources become difficult to find and extract, and therefore too expensive for general consumer use.

In its 2008 *Statistical Review of World Energy*, oil giant BP estimates that the oil will effectively run out in about 42 years. Natural gas is expected to last 60 years, coal 133 years.[38] In other words, at some point in the not-too-distant future, technological civilization will have either switched to non-carbon-emitting fuels like nuclear, fusion, solar, wind and tidal power, or technological civilization will have collapsed. The climate question is: if we reach the point where the planet has burned up most of the readily available fossil fuels, what would the levels of CO_2 in the atmosphere be?

If we believe the alarmists, these levels will be catastrophically high, with runaway greenhouse temperatures. In fact, as a background paper prepared for the British Government Treasury Department shows, when fossil fuels are effectively used up, CO_2 levels by 2150 would be between 562 and 728 ppm, with the latter as a worst-case scenario. The paper's author, American aerospace engineer Keith Jackson, believes the maximum temperature increase at these levels would be 2°C, with his best guess being 1.7°C.[39] Jackson concludes:

> On average, a temperature increase of 1.7 degrees will be equivalent to an average southward shift in latitude of about 140 nautical miles (300 kilometers). The effective shift will be somewhat more than that in the higher northern latitudes and a little less at more southerly latitudes. Changes within these limits are well within the experience of most people.[40]

Geologist Ian Plimer calculates: "If humans burned all the fossil fuels on Earth, the atmospheric CO_2 content would not even double,"[41] that is, it would be less than 800 ppm.

Even arch-alarmist James E. Hansen seems to agree with this analysis. A study published by NASA's Goddard Institute for Space Studies, which is run by Hansen, offers several scenarios of fossil-fuel use of which the most extreme (Business As Usual, or BAU), raises CO_2 levels to about 570 ppm by 2100 (Figure 3.5, upper left graph).[42] All other scenarios produce CO_2

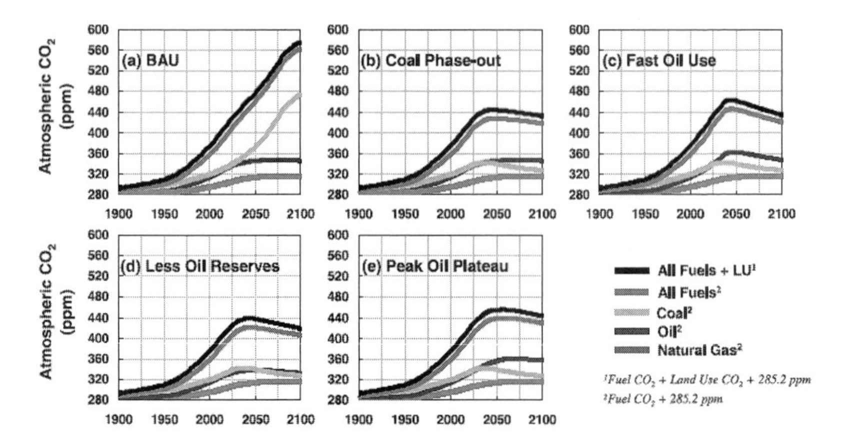

Figure 3.5: Carbon dioxide levels and fossil fuel use. Source: Goddard Institute for Space Studies

levels below 500 ppm. By 2100, under the Business As Usual estimate, most of the planet's readily available fossil fuels will be gone, so even an arch-alarmist scenario suggests that CO_2 levels will not go much higher than 570 ppm, at least from human cases.

For Hansen, a CO_2 level above 450 ppm would create serious warming (i.e., catastrophe). Anything is possible, of course, when it comes to climate, but the scary temperature projections may also result from setting the CO_2/temperature sensitivity too high. Skeptical climatologists Robert C. Balling, Jr., and Patrick Michaels note:

> It's highly debatable whether we could get to ... 1,120 ppm, even if we deliberately tried to do so. To maintain such a level for the next millennium assumes that we will still be burning fossil fuel—and at more than three times the current rate—in the year 3000.[43]

The point here is that since oil, natural gas and coal are running out, even if *all* the available fossil fuels discharge *all* their available carbon into the atmosphere over the next century, the worst that could happen is a CO_2 level of about 600-700 ppm. Given the carbon dioxide saturation effect, a 600-700 ppm level of CO_2 might produce a temperature increase of 1 to 2°C, although even this estimate may be too high. Moreover, a level of 600-700 ppm would produce a dramatic increase in the growth of many plants, keeping in mind that the ideal CO_2 level for plants appears to be in the 1,200-ppm range, and keeping in mind that for most of the time multicellular life as been on earth, CO_2 levels were higher and sometimes much higher than 700 ppm. Also, one or two degrees of warming is a small price to pay for avoiding the kind of carbon-limiting, economy-destroying measures the climate alarmists would like to see, especially when the upside is a greener, wetter planet. As even alarmist climatologist William Ruddiman acknowledges: "A warmer Earth is likely to be a wetter Earth."[44]

And, given that the fossil fuels will then be gone forever, fossil-fuel carbon emissions will virtually end, and the CO_2 component in the atmosphere will again begin, gradually, to decline, as it has for the past 150 million years. This is not a good thing since the planet will lose whatever small extra warming effect and extra fertilizing effect the higher levels of CO_2 had provided. Indeed, as the late science-fiction writer Arthur C. Clarke once speculated, it's possible, as a next glaciation looms, that "the cry in the next millennium may be 'Spare that old power station—we need more CO_2'!"[45]

OTHER HUMAN FACTORS OUTWEIGH CO_2 IN GLOBAL WARMING

Carbon dioxide plays, of course, a part in global climate change, but it appears to be a small part, with *human* carbon dioxide emissions playing an even smaller part within the carbon dioxide cycle. In the view of climatologist Roger Pielke, Sr., other human activities play a larger role in climate than our CO_2 emissions:

> Research has shown that the focus on just carbon dioxide as the dominant human climate forcing is too narrow. We have found that natural variations are still quite important, and moreover, the human influence is significant, but it involves a diverse range of first-order climate forcings, including, but not limited to the human input of CO_2. ... These other forcings, such as land use change and from atmospheric pollution aerosols, *may have a greater effect on our climate than the effects that have been claimed for CO_2.*[46] [italics added]

Pielke adds, elsewhere: "This does not mean that we should not work to limit the increase of this gas [carbon dioxide] in the atmosphere, but *it is not the dominant climate forcing* that affects society and the environment."[47] [italics added] If this is so, then the hysteria around reducing carbon dioxide emissions is unwarranted because a reduction in anthropogenic CO_2 will have a minimal effect in preventing warming.

WHAT WE KNOW

In conclusion, what do we *know* about carbon dioxide and climate, as opposed to computer-model speculation? We know the following:

- Carbon dioxide is a very small part of the atmosphere: less than .04 per cent, or 40 molecules in 100,000. It is a *trace* gas.
- The human contribution of carbon dioxide per year is even smaller: 2-6 per cent, or about 1.5 additional CO_2 molecules per 100,000 per year. The rest of the atmosphere's yearly carbon dioxide—about 95 per cent—comes from nature. Even if the relationship between carbon dioxide and temperature was as the alarmists believe—more CO_2 equals potentially runaway greenhouse temperatures—the human contribution to the yearly carbon cycle is relatively small. If our emissions are adding to warming, it is only as a small percentage of the temperature change that would occur naturally.
- We know that CO_2 levels have been steadily falling for 150 million years and that the level of CO_2 is unusually *low*, not unusually high, at this time in the planet's geological history.
- The increase in CO_2 in the atmosphere since 1750, which the IPCC sets as the start of the Industrial Revolution, is 12 molecules per 100,000, or .012 per cent, of the total atmosphere. Some of this increase—perhaps as much as half, perhaps more—will be due to natural warming since warming oceans release carbon dioxide. Logic, if nothing else, suggests that an additional .012 per cent of carbon dioxide in the atmosphere is not going to be the *principal* driver for global warming, particularly warming on the scale the IPCC predicts. As we will see in Chapter 5, massive natural forces such as the sun, ocean and wind currents, and even cosmic rays are far more likely suspects for the 20th century's warming than 12 extra molecules per 100,000 molecules of atmospheric gases.
- The relationship between carbon dioxide and temperature is complex. At some times when the planet has been in an ice age, for ex-

ample, CO_2 levels have been very high. At times when the planet was much warmer than today, CO_2 has been low. There is, in fact, little correlation between temperature and CO_2 over geological time. The strongest correlation has been over the past few hundred thousand years; in that time as temperature has gone up or down so have carbon dioxide levels, but CO_2 has gone up or down several hundred years *after* the temperature has gone up or down.

- Human activity is causing some warming, of course, but CO_2 emissions are not, at least in the view of some climatologists, the principal anthropogenic-warming factor. Land-use, mainly agriculture and cities, is probably a stronger warming factor than fossil-fuel emissions and, unless we are planning to give up farming and cities, there is little we can do about it.
- Other climate factors, most notably the sun and oceans, are and have always been stronger drivers of climate than carbon dioxide.
- A runaway greenhouse has *never* occurred based on CO_2 increases alone, at least in the past 600 million years, even when CO_2 levels were 18-20 times higher than today's.
- At a certain point, perhaps with a CO_2 level as low as 200 ppm, the atmospheric band that traps carbon dioxide gets saturated and very little additional warming occurs. With a CO_2 level of near 400 ppm, we are almost certainly at or beyond the saturation point now, which means additional CO_2 will not produce more than a degree or two Celsius of warming, if that. Humanity should be able to cope with this small increase.
- We know that at some point in the next 100 years, significant fossil-fuel emissions will end as oil, natural gas and coal are depleted and replaced with alternative forms of energy. If we can't find other forms of energy in adequate quantities, then Western technological civilization will collapse. Either way, human-caused greenhouse gases will cease to be an issue. In other words, the current run-up of CO_2 levels is a temporary phenomenon that will produce CO_2 levels, at the maximum, in the 600-700 ppm range, which is below, and sometimes way below, what CO_2 levels have

been in the past. Those higher levels did not damage the planet or its living beings, plant and animal, then, and nor will they now.

- Carbon dioxide is not "global warming pollution," as Al Gore and other alarmists would like the public to believe, nor is it "choking our atmosphere" as alarmist Tim Flannery charges. Carbon dioxide is a fertilizer. A CO_2 level of 800 ppm would be *good* for the planet's plant life and the environment, causing increases in plant growth of up to 33 per cent and plants that are more drought-resistant. That is, whether CO_2 causes additional warming or not, more CO_2 means a greener planet, not catastrophe.

In other words, carbon dioxide is not the global-warming villain the climate alarmists would have us believe. In Chapter 4, we'll look at the Intergovernmental Panel on Climate Change, the main source of alarmism over CO_2 levels.

When the number of factors coming into play in a phenomenological complex is too large, scientific method in most cases fails us. One need only think of the weather, in which case prediction even for a few days ahead is impossible. Nevertheless, no one doubts that we are confronted with a causal connection whose causal components are in the main known to us. Occurrences in this domain are beyond the reach of exact prediction because of the variety of factors in operation, not because of any lack of order in nature.

—Albert Einstein[1]

The determinants of complex processes are invariably plural and interrelated.

—Historian David S. Landes[2]

Review by experts and governments is an essential part of the IPCC process.

—IPCC Mission Statement[3]

I tried hard to balance the needs of the science and the IPCC, which were not always the same.

—Climatologist Keith Briffa[4]

Chapter 4

THE IPCC'S RUSH TO JUDGMENT

When skeptics challenge the anthropogenic global warming orthodoxy, the most common defence is that anthropogenic global warming must be true because the Intergovernmental Panel on Climate Change says so. Thousands of climate scientists say that human-caused greenhouse gases, not natural factors, are the main global warming villain, and how can thousands of climate scientists be wrong?

Actually, it's quite possible for thousands of scientists to be wrong—it's happened many times in the past. As recently as the 1950s, thousands of scientists pooh-poohed the idea that continents might move. But let's put aside past scientific errors for the moment and concentrate on what these thousands of climate scientists involved in the IPCC are studying and then what they are telling the public.

The three main greenhouse gases are carbon dioxide, methane, and nitrous oxide, but for the IPCC, the prime villain is carbon dioxide, and specifically human-caused carbon dioxide from industrial civilization. Curiously, water vapor is not on the IPCC's "most wanted" list, yet water vapor is responsible, along with clouds, for up to 90 per cent of the greenhouse effect. Water vapor is ignored, the IPCC says, because it is not a "forcing" factor but an effect (feedback) of forcing factors such as carbon dioxide. Methane, a greenhouse gas that is a 20 times more powerful than CO_2, also doesn't get a bad rap, perhaps because it's associated with agriculture while CO_2 is more a product of industry. Methane is also more than 200 times rarer in the atmosphere than carbon dioxide.

There's a reason for the IPCC's focus on greenhouse gases: it's not that human-caused greenhouse gases *are* the climate villain (although they might be), but that the IPCC's mandate from the start was to proceed *as if* human-caused greenhouse gases were the villain. The IPCC's definition of climate change is as follows:

> A statistically significant variation in either the mean state of the climate or in its variability, persisting for an extended period (typically decades or longer). Climate change may be due to natural internal processes or external forcings, or to persistent anthropogenic changes in the composition of the atmosphere or in land use.[5]

However, despite this broad definition, the IPCC is interested in only one type of climate change: human-caused. Here's the IPCC's mission statement:

> The role of the IPCC is to assess on a *comprehensive, objective, open and transparent basis* the scientific, technical and socio-economic information relevant to understanding the scientific basis of risk of *human-induced climate change*, its potential impacts and options for adaptation and mitigation.[6] [italics added]

"Natural internal processes or external forcings" are thus, by definition, excluded from the IPCC's mandate right from the start. The IPCC's mission is to investigate and promote the idea that *human activities* are the main cause of global warming. Its mandate *isn't* to explore the idea that perhaps humans aren't at fault. Yet, the IPCC promotes itself to the public as an *unbiased* and *objective* source of information on climate change when, as anyone who carefully reads its mission statement can clearly see, the organization is extremely biased toward anthropogenic warming and was from its beginnings in 1988. Unfortunately, few in the public, media or politics have read the

IPCC's mission statement and so they continue to believe the IPCC is, as it claims, an objective source of information on climate.

REVIEW BY GOVERNMENTS

The next sentence of the IPCC's mission statement reads:

> "Review by experts *and governments* is an essential part of the IPCC process."[7] [italics added]

Yet, surely, if the IPCC was taking an "objective" look at climate change from a balanced, purely scientific perspective, its findings would not be subject to "review by governments," which are political bodies with political and ideological, not scientific, agendas. No self-respecting journalist in a democratic country would allow his or her stories to be submitted to "review by governments," nor would readers in a free society willingly buy a newspaper whose stories were subject to "review by governments"—that's something we'd expect in a dictatorship. Yet, this is precisely what climate scientists agree to when they contribute to the IPCC.

This is not to say that politicians and governments shouldn't have the final say on what is done with the *results* of scientific research—in a democratic society, elected representatives must have the last word. As philosopher of science Paul Feyerabend has written:

> There must be a formal separation between state and science just as there is now a formal separation between state and church. Science may influence society but only to the extent to which any political or other pressure group is permitted to influence society. Scientists may be consulted on important projects but *the final judgment must be left to the democratically elected consulting bodies.* ... Organs of the state should never hesitate to reject the judgment of scientists when they have reason for doing so.[8] [italics added]

That is, in a free society, elected representatives do not interfere with what scientists publish as their *results*, but politicians have the final say in how those results are *used*. In the case of the IPCC, however, and astonishingly, politicians have a say *in what scientists publish as their results*. The IPCC, by allowing "review by governments," by allowing politicians and governments to decide what will or won't be presented as its scientific findings, has eliminated the formal separation of state and science that is essential for honest science. Instead, the IPCC is revealed as a *political* enterprise and, as always, those who pay the piper (politicians) get to call the tune. In making this devil's political bargain, the IPCC and its scientists forfeited their scientific credibility right from the start.

SEEKING 'CONSENSUS'

Nor would a truly objective scientific body be striving for "consensus" in its reports as the IPCC does, although we expect consensus from social and political think-tanks—the Fraser Institute, for example, isn't going to put out a document calling for the nationalization of the Canadian oil industry. By contrast, the 1990 IPCC report said of dissenters: "Whilst every attempt was made by the Lead Authors to incorporate [dissenting] comments, in some cases these formed a minority opinion which *could not be reconciled with the larger consensus*"[9] [italics added]. This striving for consensus meant that the IPCC was not interested, right from the start, in giving legitimacy or a forum to views that didn't fit its narrow mandate of finding anthropogenic causes to blame for warming.

The IPCC's position here is puzzling. The purpose of scientific research isn't, surely, to reach a "consensus" but to find the truth, or as close to the truth as possible, whether everyone agrees or not. As IPCC contributor and alarmist climatologist Mike Hulme has written:

> Reaching consensus about climate change, recognising that these statements emerge from *processes of deliberation and discussion rather than from pure observation, experimentation and falsification*, can therefore be an uncomfortable thing for scientists and public alike. Scientists need to be prepared to argue about their "considered opinions", to embrace consensus but *without closing down argument or suggesting that matters are settled.*[10] [italics added]

Yet reaching a consensus and closing down argument was the IPCC's mission when it was started in 1988 by the World Meteorological Organization and the United Nations Environment Program. What consensus? Why, the one approved by the politicians who bankrolled the process, hence, "review by governments." That is, the IPCC has to meet a *political* (ideological) as well as a scientific agenda, another fact that is, perhaps, not generally known to the public. We'll discuss in Chapter 13 what the political agenda in creating the IPCC might have been, but climatologist Judith Curry makes the political goal clear when she observes:

> The IPCC itself doesn't recommend policies or whatever; they just do an assessment of the science. But it's sort of framed in the context of the UNFCCC [the United Nations Framework Convention on Climate Change]. That's who they work for, basically. The UNFCCC has *a particular policy agenda*—Kyoto, Copenhagen, cap-and-trade, and all that—so *the questions that they pose at the IPCC have been framed in terms of the UNFCCC agenda.* That's caused a narrowing of the kind of things the IPCC focuses on. *It's not a policy-free assessment of the science.* That actually torques the science in certain directions, because a lot of people are doing research specifically targeted at issues of relevance to the IPCC. Scientists want to see their papers quoted in the IPCC report.[11] [italics added]

This need to find consensus to meet a *political* agenda also explains why unlike, say, Supreme Court rulings, the IPCC's documents don't have minority reports. Think about it: you gather the work of 2,000 scientists. How likely is it that *none* of them—not one!—will disagree, perhaps even passionately, with the final report? Does this make sense? And yet, the dissidents may well have valid criticisms of the consensus science, especially if, as Hulme notes, that consensus is *not* based on "pure observation, experimentation and falsification," i.e., on empirical evidence, but on what are in effect informed opinions. If the IPCC was truly interested in an open process of scientific discovery, it would be publishing the criticisms of its majority reports as, if nothing else, appendices. That way, the public would get the full spectrum of scientific opinion on the possible causes of global warming rather than only one side—the side "reviewed" and approved by politicians.

One final point: undoubtedly, global warming will bring problems, perhaps severe problems for certain areas on the globe. But, surely, warming would also have *benefits*, would it not? It's an ill wind that blows nobody good, after all. But this isn't the IPCC's view. Its 2007 report presents almost nothing but gloomy forecasts, from more hurricanes to rising oceans to droughts. That there might be any benefits to, say, "fewer cold days and nights over most land areas"[12] is downplayed, although these benefits might include better health (cold weather kills far more people than warm weather[13]), longer growing seasons, more rainfall, and the ability to farm, raise animals and extract natural resources in vast areas now under a near-permanent cover of snow and ice. For example, a recent report by the U.S. Geological Survey stated that 13 per cent of the world's undiscovered oil might be found in the Arctic.[14] But, it's the Summary for Policymakers that undergoes "review by governments," so this gloomy outlook isn't, perhaps, surprising since gloomy is what the politicians who are bankrolling the IPCC want.

This criticism doesn't mean the IPCC's findings are wrong; they may be quite correct. Nor does it mean that the scientists involved in the IPCC aren't being as objective and honest as possible in their evaluations of the data (although, as the Climategate emails showed, this is not true of several leading IPCC contributors and possibly many more we don't know about). It just means that right or wrong, the IPCC's position on global warming was fixed before the full evidence was in, in part due to political pressure. To reach a firm conclusion before seeing all the evidence and under political direction is not what most people, including most scientists outside the alarmist "consensus," would consider good science. Or, as novelist Michael Crichton has noted, correctly: "The IPCC is a political organization, not a scientific one."[15]

THE CASE AGAINST CARBON DIOXIDE: NO PROOF REQUIRED

That said, let's look at the IPCC's politically loaded charges against human-caused carbon dioxide.

Most readers are probably familiar with Al Gore's accusations. For Gore, human beings are the "principal" cause of global warming. By "human beings" Gore means human-caused greenhouse gases, and particularly carbon dioxide. For Gore, the actions of the villainous carbon dioxide are, moreover, "so dangerous as to warrant immediate action,"[16] although, logically, the danger of carbon dioxide is what he needs to prove, and hasn't, before "immediate action" makes sense. He calls CO_2 "global warming pollution."[17] Nor is there any need, in Gore's mind, for a trial (in scientific terms, confirming empirical data): the accused is so obviously guilty that, in Gore's words, "We must act boldly, decisively, comprehensively, and quickly, even before we know every last detail about the crisis."[18] First, Gore creates a straw man when he argues that skeptics are requiring "every last detail" about the so-called "crisis"; skeptics just want to see enough empirical evidence to make a "crisis" probable. So far, that evidence does not exist. But then, evidence doesn't matter to Gore since he completely discounts the possibility that other, possibly natural factors might be the "principal" cause of warming, with human carbon dioxide emissions in a small contributing role. Yet this small role seems increasingly likely given that, at the time of writing (2010), the planet had not warmed for at least a decade *despite* steadily increasing carbon dioxide levels. Clearly, as we will see in more detail in Chapter 5, other natural factors are more important in determining climate than carbon dioxide, and far more important than the relatively small contribution made by humans.

In support of Gore's approach, North Carolina economist and alarmist Andrew Brod wrote: "Fortunately, people finally seem to understand the fallacy of requiring proof"[19]—proof that humans are the prime cause of global warming, that is. Christine Stewart, environment minister in the Jean Chrétien Liberal government, went so far as to say she didn't care a whit about proof: "*No matter if the science is all phony*, there are collateral environmental benefits."[20] [italics added] This repudiation of scientific honesty came from a senior Canadian government environment minister! A senior American government official and a major force behind the Kyoto Accord, Timothy Wirth, said: "We have got to ride the global-warming issue. *Even if the theory of global warming is wrong*, we will be doing the right thing in terms of economic policy and environmental policy."[21] [italics added] And Richard Benedick, a U.S. deputy assistant Secretary of State, stated in 1989: "A global warming treaty must be implemented *even if there is no scientific evidence to back the [enhanced] greenhouse effect*."[22] [italics added] In other words, the IPCC's job is to provide scientific cover for what some politicians want to do anyway.

So, yes, heaven forbid that we should require scientific proof before we take drastic, very expensive, and probably useless action to curb carbon emissions and stop global warming. Yet, if governments are determined to reduce carbon emissions for economic and "collateral" environmental reasons, why pollute climate science by dragging in the red herring of global warming? Why not sell these economic and environmental policies on their own economic and environmental merits? And why would scientists with integrity go along with this charade? We'll discuss some possible reasons in Chapters 11 and 12.

The IPCC is more circumspect than Gore in the wording of its charges against CO_2, but the message is the same. The IPCC's 1995 report concluded, rather weakly: "The balance of the evidence suggests a discernible human influence on global climate." The 2007 report is more emphatic, arguing that "very strong evidence" of human influence through carbon emissions has been detected.[23] Interestingly, though, the original final draft of the full 1995 report contained the following sentences:

- None of the studies cited above has shown clear evidence that we can attribute the observed [climate] changes to the specific cause of increases in greenhouse gases.
- No study to date has positively attributed all or part [of the climate change observed to date] to anthropogenic causes.
- Any claims of positive detection of significant climate change are likely to remain controversial until uncertainties in the total natural variability of the climate system are reduced.[24]

However, these key sentences *were taken out* and replaced with a reference to "discernible human influence."

S. Fred Singer, professor emeritus in environmental science at the University of Virginia, notes that IPCC officials said they'd made the revisions "to ensure that it [the body of the report] conformed to a 'policymakers' summary' of the full report. ... Their [the IPCC's] claim raises the obvious question: Should not a summary conform to the underlying scientific report rather than vice-versa?" Singer and others have suggested "a possible distortion of science for political purposes."[25] The alarmist journal *Nature*'s explanation of the change was: "Some phrases that might have been (mis)interpreted as undermining these [IPCC] conclusions ... have disappeared."[26] The IPCC's response? Just a "normal" and "accepted" practice.[27]

If so, it is a practice that has clearly made its judgment regardless of the evidence: as far as human-caused global warming was concerned, the IPCC fix was in from the start. The IPCC has its culprit. No other suspects need apply. And if the IPCC has to torture the evidence to get a conviction, well, it's in a good (political) cause. But just to be sure what happened is clear: for the 1995 report, the IPCC's scientists found *no evidence* that global warming was primarily caused by humans. The IPCC's political Central Committee, the bureaucrats and politicians who put together the Summary for Policymakers, wrote that humans were causing warming, and then deleted passages in the main report that said otherwise.[28] It's a clear case of the tail (politics) wagging the dog (science) in an ideological cause.

LIES, DAMN LIES, AND COMPUTER MODELS

As we saw in Chapter 3, at a certain level of CO_2 the atmospheric band that traps warmth gets saturated and little extra warming occurs. However, the IPCC prefers to discount the CO_2 saturation effect, and so the IPCC's 2001 report predicted a temperature increase of 1.4 to 5.8 degrees Celsius over the next century due to increased carbon dioxide. In the 2007 report, the highest projected temperature is lower (4.5° C), and the lowest is higher (2° C). However, *all* of the IPCC's computer projections for temperature increases in the future have been *too high* when checked against recent real-world data (Figure 4.1[29]).

Bjorn Lomborg, the skeptical environmentalist, writes: "The original computer models used in the 1990 IPCC report and well into the nineties just didn't match up with the data: they predicted way too much warming from CO_2 and other greenhouse gases."[30] This predictive failure continues to be true in the 2007 report. Meteorologist Roy W. Spencer writes in 2009: "Computerized climate models tracked by the IPCC—*all 23 of them*—predict too much warming for our future."[31] [italics added] Christopher Monckton, a British environmental journalist, has calculated that the IPCC's estimates for CO_2 warming sensitivity are at least *three times* too high.[32] The IPCC's too-high CO_2 sensitivity will, naturally, lead to too-high predictions of warming.

Overall, the uncertainty surrounding carbon dioxide's effect on temperature is larger than the public is led to believe which, again, the IPCC is willing to admit in its better moments. In its 2007 report it says:

> Models still show *significant errors*. Although these are generally greater at smaller scales, *important large-scale problems also remain*. ... The ultimate source of most such errors is that *many important small-scale processes cannot be represented explicitly in models*, and

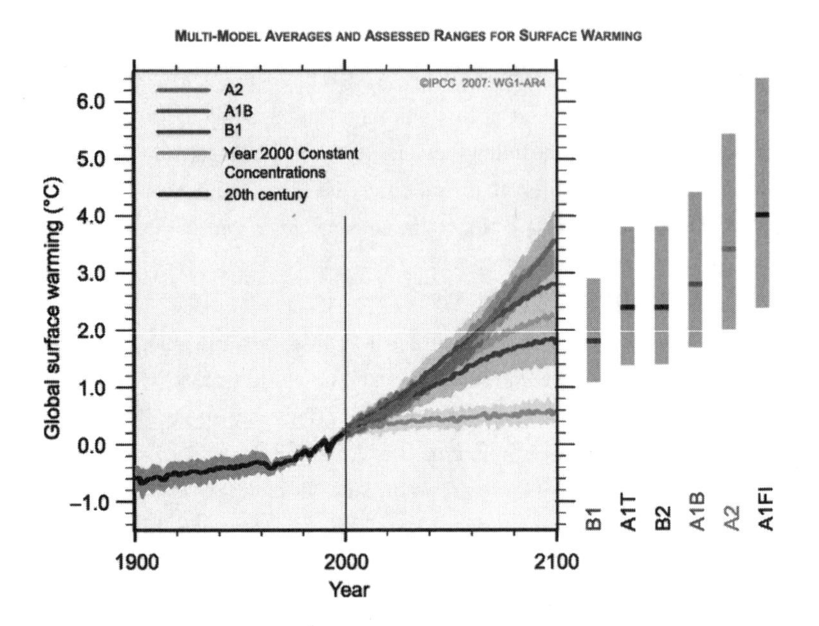

Figure 4.1: IPCC 2007 scenario temperature projections. The bottom line shows the flat temperature if CO2 emissions had stayed at year 2000 levels. This line could also represent the lack of warming since at least 2001.

> so must be included in approximate form as they interact with larger-scale features. This is partly due to limitations in computing power, but also results from *limitations in scientific understanding* or in the availability of detailed observations of some physical processes.
>
> *Significant uncertainties, in particular, are associated with the representation of clouds*, and in the resulting cloud responses to climate change. Consequently, *models continue to display a substantial range of global temperature change* in response to specified greenhouse gas.[33] [italics added]

So, taking into account these possible "limitations in scientific understanding," "significant errors," "large-scale problems" and "uncertainties," the IPCC's predictions are deliberately wide (a "substantial range of global temperature change"). In fact, what the IPCC offers is a scattergun of computer-generated scenarios based on different variables such as expected carbon emissions, efforts to control emissions, economic development, and many others.

In this respect, it's worth considering words of Albert Einstein on science's ability to predict the future. Einstein wrote:

> When the number of factors coming into play in a phenomenological complex is too large, *scientific method in most cases fails us.* One need only think of the weather, in which case prediction even for a few days ahead is impossible. Nevertheless, no one doubts that we are confronted with a causal connection whose causal components are in the main known to us. Occurrences in this domain *are beyond the reach of exact prediction because of the variety of factors in operation*, not because of any lack of order in nature.[34] [italics added]

In other words, the weather, because so many factors come into play, is not predictable. Nor is climate predictable for the same reason, although consensus climate scientists seem to think they know better than the 20th century's premier scientific genius. The reason why we can't with confidence predict either weather or climate is simple: both are, as the IPCC acknowledges, chaotic, non-linear processes. Chaos theory teaches that small changes in initial conditions can create large-scale changes later on; the flap of a butterfly's wings in Peking can cause a tornado in Chicago. Similarly, small errors in initial conditions can accumulate into large-scale errors in long-range climate predictions. It is not possible for the IPCC's climate models to fully encompass all the initial conditions, such as clouds, as the IPCC admits. Therefore, the longer the range of the climate forecast, the more *inaccurate* it is likely to be. And yet, a prominent climate alarmist, Vicky Pope of the British Meteorologist Office's Hadley Centre for Climate Prediction and Research, claims: "In many ways we know more about what will happen in the 2050s than next year."[35] Actually, for the reasons stated above, we don't. Or, we'll have to wait 40 years to find out, making Pope's prediction meaningless now.

But, wait! The IPCC says it isn't making predictions at all. In *Climate Change 92* the IPCC warns: "Scenarios are not predictions of the future and should not be used as such." The Special Report on Emissions Scenarios 2000 states: "Scenarios are images of the future or alternative futures. *They are neither predictions nor forecasts*." [italics added] From *Climate Change*

2001: Mitigation: "The possibility that any single emissions path will occur as described in the scenario is *highly uncertain*" and "there is no objective way to assign likelihood to any of the scenarios."[36] [italics added] Also from *Climate Change 2001: Advancing Our Understanding*:

> In climate research and modeling, we should recognize that we are dealing with a coupled non-linear chaotic system, and therefore that *the long-term prediction of future climate states is not possible*. The most we can expect to achieve is the prediction of the probability distribution of the system's future possible states by the generation of ensembles of model solutions.[37] [italics added]

Kevin Trenberth, an IPCC author, head of the U.S. National Centre for Atmospheric Research, and a prominent figure in the CRU "Climategate" emails, also makes it crystal clear that the IPCC does not make predictions and that, until it has that ability, it is not possible to say that global warming science is "done," as so many in the alarmist consensus claim (i.e., "the science is settled"). Trenberth writes:

> In fact *there are no predictions by IPCC at all*. And there never have been. The IPCC instead proffers "what if" projections of future climate that correspond to certain emissions scenarios. ... There is no estimate, even probabilistically, as to the likelihood of any emissions scenario and no best guess.
>
> Even if there were [estimates], the projections are based on model results that provide differences of the future climate relative to that today. None of the models used by IPCC are initialized to the observed state and *none of the climate states in the models correspond even remotely to the current observed climate*. ... [Therefore,] *the science is not done because we do not have reliable or regional predictions of climate*.[38] [italics added]

Yet, as Vincent Gray, an IPCC editor who is critical of its computer models, notes:

> Despite these assertions *the scenarios are widely presented as predictions and forecasts*, not only by the media, politicians and "activists" of various colours, but often by the scientists who are responsible for them in their public appearances.[39] [italics added]

In other words, by the time the IPCC's reports get into its "government-reviewed" Summary for Policymakers stage, these hesitant not-forecasts/not-predictions/just "what ifs" become scary, "settled" certainties. Thus, the IPCC's claim that it "doesn't predict" and only offers "what if" scenarios is, in practice, a distinction without a difference. Nor is it likely that any computer climate model ever will achieve reliable predictive ability (the computer modeler term is "robustness"), at least not for many years. This, again, is more or less what the IPCC says in its better moments—and then glosses over in presenting its findings to the politicians, media and public as urgent calls for action based on what it claims are robust (in IPCC terminology, "very likely") predictions.

And so, the IPCC continues to vigorously defend its computer models, including the models' ability to predict climate (even though it says elsewhere its models do not predict):

> Despite such uncertainties ... models are unanimous in their *prediction* of substantial climate warming under greenhouse gas increases, and this warming is of a magnitude consistent with independent estimates derived from other sources, such as from observed climate changes and past climate reconstructions.[40] [italics added]

And the IPCC claims its 2007 models are considerably more accurate than those used in its 2001 report:

> In summary, confidence in models comes from their physical basis, and their skill in representing observed climate and past climate changes. Models have proven to be extremely important tools for simulating and understanding climate, and there is considerable confidence that they are able to provide *credible quantitative estimates of future climate change* [most people would call these "predictions"], particularly at larger scales.
>
> Models continue to have significant limitations, such as in their representation of clouds, which lead to uncertainties in the magnitude and timing, as well as regional details, of *predicted* climate change. Nevertheless, over several decades of model development, they have consistently provided a robust and unambiguous picture of significant climate warming in response to increasing greenhouse gases.[41] [italics added]

So, the models *are* predictions after all—the IPCC itself says so on one page, while denying it is offering predictions on another page. The IPCC reports also use shifting terminologies that acknowledge the deficiencies of its models on the one hand while, on the other, asserting their reliability. For example, in its 2007 report the IPCC talks of "significant limitations" to its models. Yet, *on the same page* (!), the IPCC acknowledges that its computer models have "significant errors" (as opposed to "significant limitations") and that "important large-scale problems also remain." The term "prediction," meanwhile, has morphed into "quantitative estimates" which, again, is a distinction without a difference.

A computer expert assesses the accuracy of climate computer models as follows:

> When evaluating model reliability … if [a model] has correctly predicted major climate changes over and over again, that is pretty good evidence that its predictions should be taken seriously. There are plenty of studies that show what is called "hindcasting," in which a model is built on the data for, say, 1900-1950, and is then used to "predict" the climate for 1950-1980. Unfortunately, it is notoriously common for simulation models in many fields to fit such holdout samples in historical data well, but then fail to predict the future accurately. … *No global climate model has ever demonstrated that it can reliably predict the climate over multiple years or decades—never.*[42] [italics added]

THE IPCC'S 'HOCKEY STICK'

Another problem with the IPCC computer models is subjectivity. You don't just input a bunch of data into a computer and wait for the result. The data has to be selected, "adjusted" and manipulated—which is what the computer modelers do when they add parameters—and this selection is, necessarily, a subjective process. Thus, if you're looking or hoping for a particular result in your computer modeling, it's easy to find it—you just tweak your inputted data and parameters to get the result you want. These adjustments are strongly affected by confirmation bias, which is the human tendency to filter out or ignore data that doesn't fit our biases and include data that does. Part of confirmation bias is not wanting to get too far from what *other* scientists believe. As astrophysicist Lowell Wood notes: "Everybody turns their knobs [the model parameters] so they aren't getting the outlier, because the outlying model is going to have difficulty getting funded."[43] Of course, scientists shouldn't get caught in confirmation bias, and know they shouldn't, and undoubtedly try not to be caught. But some confirmation bias is inevitable with computer models, and sometimes the bias overwhelms the data.

The best-known and most blatant example of subjectivity leading to confirmation bias in climate science is the "hockey stick" temperature graph that the IPCC highlighted in its 2001 report (see Figure 4.2).[44] The "hockey stick" presented a fairly flat average temperature gradient (the shaft, shown as a solid line in Figure 4.2) for the past 1,000 years, with a sudden leap up after 1900 (the blade). This graph was the basis of the IPCC's 2001 claim that the current climate was the warmest in a thousand years.

A more sophisticated and critical analysis of the model's parameters by mathematician and retired Canadian businessman Steve McIntyre and Canadian economist Mike McKitrick revealed numerous flaws. In particular, the hockey's "stick" isn't that straight after all in that it doesn't show the Me-

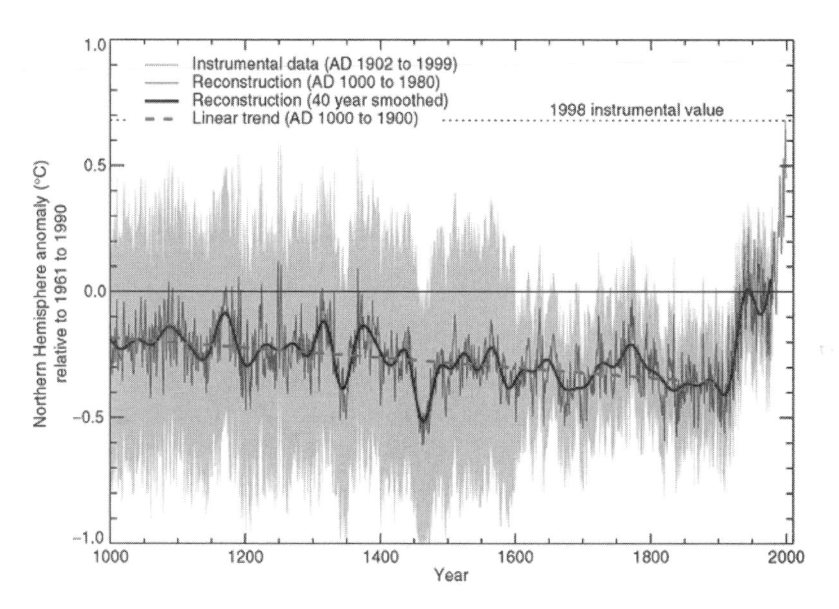

Figure 4.2: The IPCC 2001 'hockey stick' graph. The shaded areas are ranges of variation in the data.

dieval Warm Period and Little Ice Age.[45] The IPCC backed down and said in its 2007 report: "Average Northern Hemisphere temperatures during the second half of the 20th century were *very likely* higher than during any other 50-year period in the last 500 years." But rather than give up completely on that 1,000-year figure, the IPCC adds: "… and *likely* the highest in at least the past 1,300 years."[46] But 500 years is not a significant length of time on which to be making long-term, catastrophic climate predictions, especially when 500 years ago the planet was in its coldest period (the Little Ice Age) in 10,000 years. If the global temperature was warmer than now a thousand years ago without human carbon emissions, and 1-3°C warmer than now 7,000 years ago during the Holocene Optimum,[47] then why couldn't the late 20th-century warming also be just another natural cycle?

To make matters worse, although the hockey stick appeared at least seven times in the IPCC's 2001 report, and the Michael Mann, et al., paper that produced it went through a peer-review process, *nobody in the peer-review process or at the IPCC bothered to check the math that led to the hockey stick.*[48] Even though Mann's paper wiped out two major features of the climate of the past thousand years, the Medieval Warm Period and the Little Ice Age, none of the paper's peer reviewers thought this omission worthy of additional fact-checking. In other words, confirmation bias had settled in and the hockey stick graph gave the IPCC the result it expected and wanted. The 1,000-year temperature graph in the 1990 and 1995 reports, based on the work of climatology pioneer H.H. Lamb, showed the Medieval Warm Period, Little Ice Age, and warming for our time that was not at all unusual (see Figure 4.3).

And so, until McIntyre actually re-did the numbers, the IPCC was happy to promote the hockey stick without double-checking its accuracy. In other words, the IPCC failed to do due diligence, just as politicians, the media and the public fail at due diligence when they accept the IPCC's reports without questioning or criticism. Why the lack of due diligence by the IPCC in the case of the hockey stick? Because the hockey stick confirmed the IPCC's political and scientific bias.

EVALUATION BIAS

Computer models normally receive various real-world tests called valida-

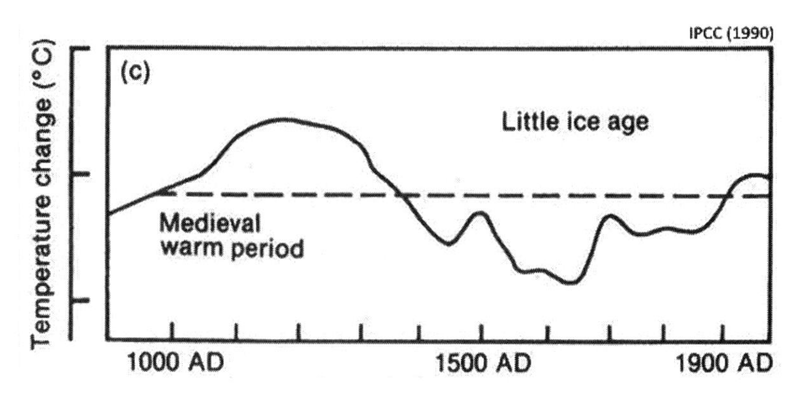

Figure 4.3: 1,000-year temperature graph from 1990 and 1995 IPCC report

tions. The model's results are compared to real-world data and "validated," or not. However, as IPCC contributor Vincent Gray observes, "no computer climate model has ever been validated."[49] Indeed, these climate models can't be validated since what they are predicting what won't happen for years into the future, although, of course, models can be tested backwards ("hindcasting") against previous climate data. Here is what the IPCC says about its evaluation method in Chapter 8 of Climate Change 2001:

> We fully recognize that many of the evaluation statements we make contain a *degree of subjective scientific perception* and may contain much "community or "personal" knowledge (Polanyi, 1958). For example, the very choice of model variables and model processes that are investigated are often based upon *the subjective judgment and experience* of the modeling community.[50] [italics added]

IPCC climatologist Mike Hulme notes:

> All of this means that climate scientists frequently have to reach their conclusions on the basis of the partial, and sometimes poorly tested, evidence and models available to them. And when their paymasters— elected (or non-elected) politicians—ask them for advice, as in the case of the IPCC, *opinion and belief become essential for interpreting facts and evidence.* Or rather, incomplete evidence and models have to be worked on *using opinions and beliefs* to reach considered judgments about what may be true.[51]

It is this "subjective judgment and experience" and reliance on "opinions and beliefs" rather than testable, falsifiable empirical evidence that leaves climate modelers open to confirmation bias. Also worth noting: the 2007 IPCC models are still "evaluated" rather than "validated," which means there is still a large *subjective* element in deciding what data and parameters the models will use and therefore what results they will produce. In other words, however good the IPCC thinks its predictions are, its models should always be regarded with healthy skepticism because of their strong subjective element—advice that the IPCC itself acknowledges. And then ignores as it makes its scary predictions (or non-predictions).

'2,000 SCIENTISTS' CAN'T BE WRONG

One of the IPCC's most common claims is that it is the product of the "consensus" of more than 2,000 scientists. And how can more than 2,000 scientists be wrong? The only problem is that this IPCC claim is, as so often with climate alarmists, misleading and exaggerated if not outright false. In discussing this massive "consensus," Nigel Lawson, a former *Spectator* editor and Britain's Chancellor of the Exchequer from 1983-89, writes:

> It is sometimes claimed … that the scientific account published in the reports of the [IPCC] … represents the unanimous view of some 2,500 scientists. In fact, the physical science section of its most recent report [2007] was written by 51 named authors (and subsequently edited by representatives of member governments and the UN). The other scientists engaged in the process were involved as "reviewers" and the like, and many of these have made clear their disagreement with important aspects of the IPCC account.
>
> Then there is the even larger number of reputable climate (and allied) scientists not involved in the IPCC process, literally hundreds of whom have, at one time or another, made public their disagreement with (often fundamental) aspects of the conventional wisdom. Finally, there are large numbers of dissenting climate scientists who have chosen not to stand up and be counted, *for fear that to do so would damage either their career prospects or their chances of securing research grants*. All that can be said with confidence is that the dissenting minority of reputable climate (and allied) scientists is a sizeable one.[52] [italics added]

TAKING ON THE IPCC

The IPCC is a huge undertaking, backed by millions of dollars of government money, presenting the work of literally thousands of dedicated scientists in many fields. And this formidable institution says that human activities are the principal cause of what the IPCC sees as accelerated and unnatural global warming. To challenge this citadel of scientific truth, to contradict the opinions of so many eminent men and women, is a formidable and daunting task. Furthermore, many of these eminent men and women have nothing but contempt for those who don't share their vision of a planet devastated by warming—even though the planet has been much warmer in the past and it and its creatures survived quite well.

The IPCC could be quite correct. Global warming may be a huge threat to humankind and the planet, given our growing numbers, and the disaster may be all our fault. But, in science, there is no such thing as being absolutely, totally, completely right—there is, or should be, room for differing theories that are listened to with respect, tested, and rejected or incorporated based on the evidence.

However, although most of the IPCC's scientists would probably agree with historian David S. Landes that "the determinants of complex processes are invariably plural and interrelated"[53]—and climate processes are about as complex as they come—the IPCC has its villain, human-caused carbon dioxide, and will not seriously consider any other culprit. Indeed, those who suggest other culprits have been labeled "immoral," or, as we will see in the next chapter, "irresponsible," and even "dangerous." This is language more suited to religious dogma than the open-minded, "objective" scientific discourse that the IPCC ought to be pursuing, especially with a subject as complicated as climate. Instead, the IPCC in practice has become a kind of Holy See, with its reports and models a scientific Bible that must not be questioned. And the IPCC has become a purveyor of fear to the public, politicians and media. AGW believers use the IPCC findings to say we must act on global warming *right now* or terrible things will happen to us and our world.

How do we know terrible things will happen to our world? Because—and this is the *only* reason—the IPCC's computer models, that is, a bunch of mathematical formulas, say so. There may well be a disaster in our future, but there is *no* objective, empirical evidence pointing to *human-caused* disaster, because it's not possible to take data from the future and because it is extremely difficult to discern the distinctly human "fingerprint" in climate data. For example, are humans responsible for the Arctic melting (although the northern icecap seemed to be returning to "normal" in 2009-2010)? Or is this melting just a normal consequence of being in an interglacial? The latter seems far more likely given that this interglacial has already melted millions of tons of ice in the past 15,000 years and has already gone on longer than most previous interglacials.

Physicist Freeman Dyson has commented on the IPCC's weaknesses and tendency to promote fear over rational discourse:

> I have seen that happen in many fields. You sit in front of a computer screen for 10 years and *you start to think of your model as being real.* It is also true that the whole livelihood of all these people depends on people being scared. Really, just psychologically, it would be very difficult for them to come out and say, "Don't worry, there isn't a problem." It's sort of natural, since their whole life depends on it being a problem. I don't say that they're dishonest. But I think it's just a normal human reaction. It's true of the military also. They always magnify the threat. Not because they are dishonest; they really believe that there is a threat and it is their job to take care of it. I think it's the same as the climate community, that *they do in a way have a tremendous vested interest in the problem being taken more seriously than it is.*[54] [italics added]

In other words, in their focus on models, scientists, including climate scientists, can be like TV or radio weatherpersons who report rain because that's what their teleprompter (i.e., computer model) says when, if they just stuck their heads out the window, they'd discover the weather is actually sunny.

WHAT WE KNOW

What do we *know* about the Intergovernmental Panel on Climate Change and its approach to carbon dioxide?

- We know that the IPCC's origins were *political* and *ideological* rather than scientific, which is why the IPCC's conclusions are subject to "review by governments," something that no true scientist should be willing to accept. Few citizens in a democracy would uncritically accept the reports of newspapers subject to "review by governments." Surely we should be equally skeptical about the IPCC's government-vetted reports.

- We know that the IPCC's politically motivated mission is to find *human-caused* warming; natural causes are examined but dismissed because it's not the IPCC's mandate to find non-human causes of warming.
- We know, because the IPCC itself says so, that its computer models cannot make reliable predictions for a system as complicated as climate. And we know that the IPCC goes ahead and makes predictions (which it disguises as "what if" projections, i.e., forecasts, i.e. "credible quantitative estimates of future climate change," i.e., predictions) anyway.
- We know that the planet's temperature is *less sensitive* to CO_2 than the IPCC models claim. How do we know? Because the actual temperatures in the 21st century have consistently been well *below* the IPCC's estimates and the planet may even be cooling. If CO_2 was as powerful as the IPCC claims, warming might have slowed a bit, but it would not have stopped and even going into cooling.
- We know that the IPCC's condemnation of carbon dioxide is based *entirely* on computer models—there is no empirical evidence whatsoever to indicate that *human* CO_2 emissions are driving the climate, nor is there any empirical evidence whatsoever that higher CO_2 levels are dangerous, especially since CO_2/temperature sensitivity is almost certainly lower than the IPCC assumes. In other words, just because the planet is warming does not mean that humans are the principal cause of that warming.
- We know that, as happened with the "hockey-stick," computer models can fall victim to the biases of their creators.
- We know that the IPCC, based on its mission-statement bias toward human-caused warming, did not do due diligence on the hockey-stick model that was its poster boy in the 2001 report.

So, overall, is human-caused carbon dioxide guilty of being the *principal* driver of global warming, as the IPCC climate police charge? Or is it innocent? Based on the evidence so far—and all the evidence is far from in—the *scientific* answer should be the Scottish verdict when a jury remained unsure: "Not proved."

POSTSCRIPT ON 'CLIMATEGATE'

In late 2009 and early 2010, several revelations publicly and officially confirmed what this chapter and the book overall have been arguing: that the IPCC has primarily a *political* rather scientific agenda, as if more proof was needed (see hockey stick, above).

The major revelation was, as noted earlier and as we will see in more detail in later chapters, more than a thousand hacked or surreptitiously released emails from East Anglia University's Climatic Research Unit (CRU). These emails showed top IPCC scientists manipulating data to get the result the IPCC wanted (the scandal's signature phrase is "hide the decline"), trying to conceal data from critics, and rigging the peer-review process in favor of pro-AGW journal papers and against skeptical papers. The emails became known as "Climategate."

Then, in January 2010, the IPCC admitted that its 2007 report included a claim that the Himalaya's glaciers would melt by the year 2035 even though this claim had not been peer reviewed, as IPCC rules require, and was flat-out ridiculous. The figure came from the World Wildlife Fund, hardly an unbiased source. The IPCC report reads: "Glaciers in the Himalaya are receding faster than in any other part of the world and, if the present rate continues, the likelihood of them disappearing by the year 2035 and perhaps sooner is very high if the Earth keeps warming at the current rate. Its total area will likely shrink from the present 500,000 to 100,000 km^2 by the year 2035 (WWF, 2005)."[55] In January 2010, but only after this gaffe was revealed publicly, the IPCC included a correction. Worse: some IPCC authors knew the claim was absurd and the IPCC had even been warned before publication that the claim was speculation, not science.[56] Why, then, was this wildly inflated figure included in the IPCC report? Dr. Murari Lal, whose 2035 speculation appeared in the WWF document and who, even knowing the claim was false, helped write the chapter that included this comment, told Britain's *Daily Mail*:

> It related to several countries in this region and their water sources.

> We thought that if we can highlight it, *it will impact policy-makers and politicians and encourage them to take some concrete action.* It had importance for the region, so we thought we should put it in.[57] [italics added]

In other words, although scientifically incorrect and even absurd, the phony glacier prediction was included to put pressure on politicians and the public. That is, *political* considerations were allowed to corrupt the *science.* In blunter terms: the IPCC deliberately lied to support its political case.

On the heels of the glacier fiasco other errors have emerged, including the IPCC's inclusion of a non-peer-reviewed claim that world had "suffered rapidly rising costs due to extreme weather-related events since the 1970s" such as hurricanes, floods and wildfires.[58] These rising costs were, the IPCC implied, due to global warming. In fact, as the IPCC authors could not help but know, the world's weather events have been no worse than usual over the past few decades; the higher costs are due to more people living in areas prone to extreme weather, not global warming as the IPCC wanted politicians, the media and the public to believe. Indeed, the paper on which the IPCC based its claim was eventually peer-reviewed in 2008, but the paper concluded: "We find *insufficient evidence* to claim a statistical relationship between global temperature increase and catastrophe losses." [59] [italics added]

The 2007 IPCC report claimed that, due to global warming, more of the world's population would be "water-stressed." In fact, the opposite is true—*fewer* people will be water-stressed, which is what you would expect if, thanks to warming, there is more water vapor in the atmosphere. The IPCC claimed that global warming would cause growing hunger, without noting that growing hunger would occur in any event if the population increases, as the UN expects, to nine billion over the next 40 years. Always false has been the IPCC's claim that warming will cause an increase in malaria. In fact, malaria is a disease of poverty, not climate. Developed nations, even in zones where malaria was once endemic, such as the U.S. capital, Washington, D.C., have virtually eliminated malaria. The IPCC also asserted that large parts of Africa would be affected by drought if warming continued, a claim that is not scientifically supported; that large parts of Holland would be submerged by sea level rise, not true; that 40 per cent of the

Amazon would be blighted if warming rose 2°C, completely wrong—and these are just a few of the IPCC's "gates." The IPCC was warned about many of these errors or sins of omission before publication but, because the false information bolstered its message, chose to ignore the warnings.[60] Finally, a Toronto blogger, Donna Laframboise, found dozens of references to World Wildlife Fund-supplied information in the 2007 IPCC report, none of it peer-reviewed as the IPCC procedures require, but included because the WWF's alarmist political agenda agrees with the IPCC's.[61]

How many other errors in science, included because they agree with the IPCC's alarmist political bias, are in its reports? These errors, now publicly acknowledged by the IPCC, show clearly that the IPCC cannot be trusted, and never could be trusted, as an objective, purely scientific source of climate information. It is a creature of politics, not science.

When one thinks of the alternative hypotheses [to increased carbon dioxide causing warming]—that climate is controlled by volcanic eruptions, oceans, clouds, phytoplankton, vegetation, the sun, and aerosols—placing immense bets on a single basis for predictions is unwise.

—**Aaron Wildavsky**[1]

Among the thousands of human generations, ours may be the first that was ever frightened by warming.

—**Henrick Svensmark**[2]

Chapter 5

THE *REAL* GLOBAL WARMING SUSPECTS

As we saw in Chapter 2, the sun seems the most likely candidate for "principal" cause status when it comes to warming the climate. The IPCC does not agree. Its scientists argue that changes in the sun's radiation aren't strong enough to be driving the climate over the last half of the 20th century, although the IPCC is willing to accept that natural factors caused the 19th and 20th century warming before that.[3] Other scientists, however, consider the sun a significant factor. Richard Willson, a researcher with the NASA's Goddard Institute for Space Studies, believes the sun has increased in brightness by 0.05 per cent per decade since the late 1970s, when good satellite data became available, and has probably been increasing in brightness since the late 19th century. He writes:

> This trend is important because, if sustained over many decades, it could cause significant climate change. ... Historical records of solar activity indicate that solar radiation has been increasing since the late 19th century. If a trend comparable to the one found in this study persisted during the 20th century it would have provided a significant component of the global warming that the Intergovernmental Panel on Climate Change report claims to have occurred over the last 100 years.[4]

A 1997 study suggested that 40 per cent of warming up to then was caused by increased solar activity over the previous three decades.[5] In 2004, the Max Planck Institute in Germany announced that the sun's activity was the highest in a thousand years.[6] And Carleton University paleo-climatologist Tim Patterson notes:

> My own research shows that, on all time scales, there is a very good correlation between the Earth's temperature and natural celestial phenomena, such as changes in the brightness of the sun. The fact that the sun is now brighter than it has been in 8,000 years should have a major impact on climate.[7]

You'd think so, and even the IPCC's scientists acknowledge that the Milankovitch cycles, which create relatively small changes in the amount of sunlight the earth receives, have triggered a series of glacials and interglacials capable of creating and then melting billions of tons of glacial ice, some of it up to three kilometers high. Curiously, while the IPCC doesn't consider small changes in solar radiation capable of driving the current climate, it believes very small changes in the amount of atmospheric carbon dioxide have this ability. This refusal to acknowledge the role of the sun in recent climate change appears to be another example of ideological and political bias at work.

THE SUNSPOT CYCLE

Sunspots are a visible marker of fluctuating solar activity. Scientists used to believe the sun's energy was unchanging—a "solar constant." We now know this isn't the case. During periods with much sunspot activity the sun radiates more heat; therefore, the earth warms. There is much evidence in the recent historical record directly relating sunspot activity, which operates on an 11-year cycle, to warming and cooling (see Figure 5.1[8]).

For example, there was very little sunspot activity between about 1645 and 1715 (this time is called the Maunder Minimum), which corresponded to the coldest part of the Little Ice Age (1350-1850) and caused enormous hardship for much of humanity.

Sunspots may also explain, in a way that carbon dioxide levels cannot, the curious fact that global warming accelerated from the mid-1800s to 1940 *before* human carbon emissions had reached major proportions. In other words, *most* of the warming in the 20th century occurred *before* carbon dioxide emissions were a major factor. But then temperatures fell slightly

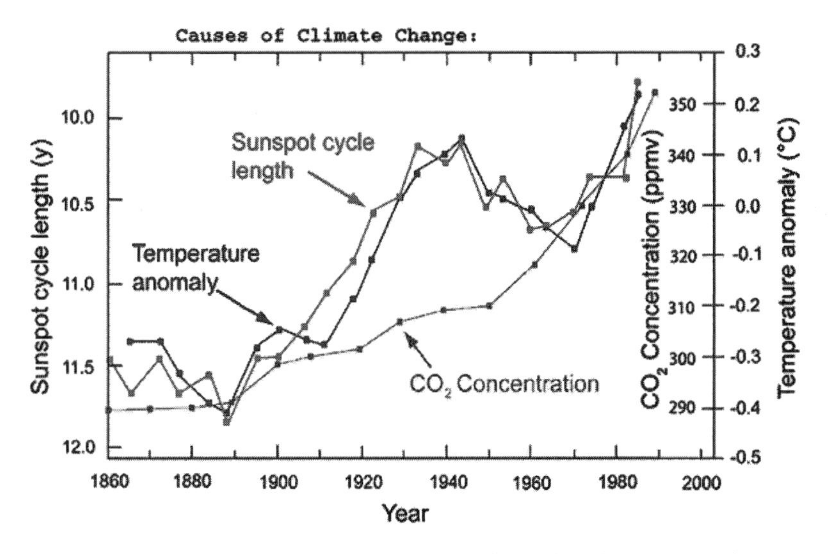

Figure 5.1: Sunspots and temperature since 1860. Sunspots and temperature seem to be closely linked

from 1940 to 1975, which is puzzling since the post-war period saw boom times for industrial growth and therefore increased carbon emissions. Then temperatures accelerated again in the late 1970s. Clearly, something other than carbon dioxide is at work here; if carbon emissions have been steadily increasing since the Industrial Revolution of the 1800s, as they have, how could there have been a *decrease* in the global temperature after 1940 (this decrease is shown in Figure 5.1)?[9]

Astronomer John W. Eddy, of the High Altitude Observatory in Boulder, Colorado, offers a possible explanation for this warming that doesn't involve CO_2:

> The present day frequency of sunspots and auroras is probably unusual, and … since the 17th century the activity of the Sun has risen steadily to a very high level—*a level perhaps unequaled (over the past million years)*.[10] [italics added]

In other words, our planet's supposedly accelerated warming in the 20th century was probably due not to a higher level of human industrial activity but to a higher level of sunspot activity, or just solar activity—after all, the sun's intensity has slowly but inexorably increased by an estimated 30 per cent since the earth was formed 4.5 billion years ago.[11] This increase in solar intensity might also explain why increased warming has been detected on Mars and other planets in our solar system.[12]

Diminishing sunspot activity might also explain the flat-lining of global temperature since the late 1990s and the plunge in global temperature in 2007-2010. Phil Chapman, a geophysicist and former NASA astronaut, writes:

> The sunspot number follows a cycle of somewhat variable length, averaging 11 years. The most recent minimum was in March last year [2007]. The new cycle, No. 24, was supposed to start soon after that, with a gradual build-up in sunspot numbers. It didn't happen. … Pray that there will be many more, and soon. The reason this matters is that *there is a close correlation between variations in the sunspot cycle and Earth's climate.* The previous time a cycle was delayed like this was in the Dalton Minimum, an especially cold period that lasted several decades from 1790.[13] [italics added]

As of early 2010, the normal sunspot activity had not resumed.

Of course, alarmist climate scientists are unwilling to accept that this close connection between sunspots and climate could be affecting climate in the past 30 years or so. The University of Victoria's Andrew Weaver points out that while the sunspot-climate correlation worked quite well until 1980, temperature outstrips sunspots after that date; to this extent, he argues, Figure 5.1 is incorrect after 1980. Therefore, although sunspots and climate have very strongly correlated previously, for Weaver the sun has ceased to be a player and the sunspot-climate hypothesis is simply another product of the nefarious "denial industry."[14] The only problem with Weaver's criticism is that the planet is currently not warming, hasn't since the late 1990s, and may be cooling, at a time when, as we've seen above, sunspot activity is at the lowest in a century and perhaps the lowest since the 18th century. In other words, sunspot activity and global temperature *are,* currently, correlated. It's also worth noting that Weaver's climate models show that, if it wasn't for human activities, the planet would be cooling right now.[15] In other words, Weaver argues, in its *natural* state the earth would be cooling. Curiously, cooling seemed to be happening in the first decade of the 21st century, giving more evidence that human carbon emissions are not as powerful when it comes to climate as the alarmists would have us believe.

A 1,500-YEAR CYCLE

In their book *Unstoppable Global Warming: Every 1,500 Years,* climatologist Fred Singer and environmental writer Dennis Avery describe a 1,500-year climate cycle, based on fluctuations in the sun, that seems more plausible than human-caused CO_2 in explaining global warming, now and in the 15,000 years since the most recent interglacial began. This 1,500-year cycle was discovered in the early 1980s by climate researchers Willi Dansgaard and Hans Oeschger based on pioneering studies of Greenland ice-core samples going back 250,000 years. Dansgaard and Oeschger expected to see the Milankovitch climate cycles, and did, but they also found a "smaller, moderate, more persistent temperature cycle" with an average length, they estimated, of 2,550 years. Later research reduced this time to 1,500 years, +/- 500 years. This cycle is now called the Dansgaard-Oeschger cycle. One aspect of this *natural* cycle is *accelerated warming* followed by cooling—the pattern we may be following at the moment. That is, there was apparently accelerated warming in the late 20th century, followed by the 21st century cooling. During this cycle the temperature rises and falls by about 4° C, which is enough to create warm and cold spells within our interglacial, and the Dansgaard-Oeschger cycles seem to be triggered by fluctuations in the sun's radiation.[16]

This idea of an alternating warm/cold cycle fits well with what we know of the ups and downs of climate over the 15,000 years since our Holocene interglacial began (see Figure 5.2[17]).

In the last six thousand years there have been several warming and cooling episodes, including the Minoan Warming of 3,500 years ago, then a cold spell from 600 to 200 BC, then the so-called Roman Warming (200 BC-600 AD), a cold spell until about 800 AD, then the Medieval Warm Period (800-1350), then the Little Ice Age (1350-1850), and finally our own warming time starting in about 1850. All previous cycles occurred *without significant human influence* and are "an embarrassment for those who, in recent years, wished to play down the natural variations in climate that occurred before the Industrial Revolution."[18]

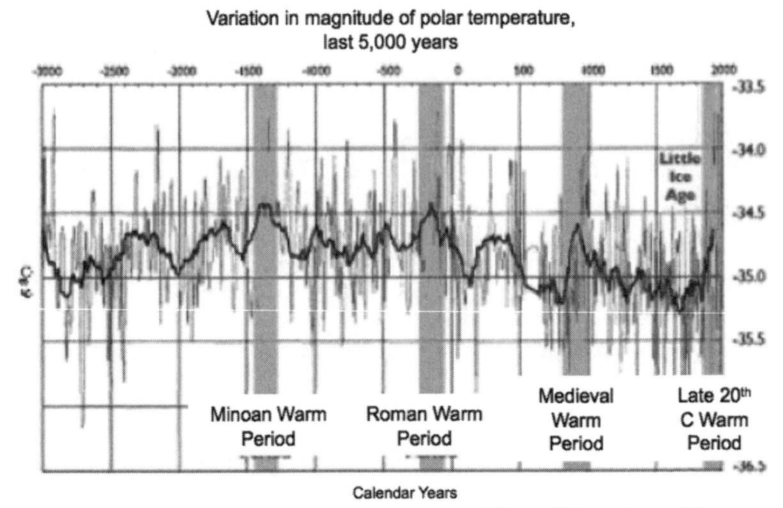

Source: Grootes, P. M. (et al), 'Comparison of oxygen isotope records from the GISP2 and GRIP Greenland ice cores', Nature, 366, 1993, pp. 552-4.

Figure 5.2: The 1,500-year climate cycle.

If this 1,500-year (+/- 500 year) observation is valid, then we are currently going through a warm part of the cycle and the planet will eventually cool again, and warm again, and so on. Moreover, the current warming fits the pattern of previous warm cycles. As Singer and Avery write:

> The [IPCC's] Greenhouse Theory has … a warming trend that started rapidly and has now turned erratic. *That's exactly the pattern of the natural, solar-driven warmings in previous 1,500-year cycles.* The [current] warming began too early and too suddenly for manmade CO_2 to be a likely candidate as its driving force. There is no overarching polar warming trend in either the Arctic or the Antarctic, as the Greenhouse Theory says it should be. There is little warming in the lower atmosphere, which the Greenhouse Theory says should be warming faster than the Earth's surface.[19] [italics added]

However, these 1,500-year cycles don't fit into the IPCC's belief that the current warming is primarily human-caused, and so the Dansgaard-Oeschger cycles are for practical purposes ignored and its proponents sometimes ostracized as "deniers" or in the pay of the oil companies.

SUN PLUS COSMIC RAYS

As we've seen, the earth's climate is strongly affected by sunspot activity—the more sunspots, the warmer the earth's temperature; the fewer the sunspots, the cooler the earth's temperature. Indeed, the correlation is remarkable, at least up to 1980 and perhaps up to today (see Figure 5.1). And, as the sunspot hypothesis would predict, at a time (2008-2010) when sunspot activity has reached an almost century-long low the earth stopped warming and even started cooling, with the coldest winter of the 21st century so far occurring in 2008-2009.[20] Solar fluctuation, then, seems like a plausible explanation for both the current, 21st-century cooling and the previous warming of the late 20th century when sunspots were more frequent.

Nevertheless, the IPCC cannot see—and does not want to see—a mechanism by which changes in sunspot activity could be translated into temperature changes, given that these solar changes are relatively small.[21] However, climatologist Henrik Svensmark has suggested the sun's fluctuations may be boosted or dampened by another possible climate factor: cosmic rays affecting cloud formation. According to Svensmark, clouds promote global cooling, but clouds don't just appear magically in the sky; they form around tiny suspended particles. These particles can be carbon soot, for example, but they can also be cosmic ray particles that come at us from outer space.

Here's how Svensmark believes the cycle works: When the sun is active (hotter, therefore more sunspots), it sends out charged particles called the solar wind. This wind pushes aside some of the cosmic radiation also coming at us from space. With less cosmic radiation getting through there are fewer charged particles in the atmosphere, therefore fewer clouds and a warmer temperature because more sunlight reaches the earth's surface. When the sun's intensity wanes as part of its 11-year cycle and the solar wind eases, more cosmic rays enter the atmosphere. More clouds form around these particles and the temperature cools because the sun's rays are reflected back into space by the tops of the clouds (cloudy days are cooler days). Svensmark and other researchers believe they have found a strong correlation between sunspot activity, cosmic particle intensity, and global temperatures:

> The temperature records show the Earth gradually warming by about 0.6 degrees Celsius overall during the 20th century. About half of the warming occurred before 1945, when the sun was becoming more

> active and cosmic rays were diminishing. ... A period of pronounced cooling intervened in the 1960s and early 1970s, which accorded very well with a temporary weakening of the Sun's magnetic activity and a rise in the cosmic rays. After 1975, the upward trend in solar activity resumed for a while, the cosmic rays fell again, and the warming of the world resumed. That was the period when growing concern about carbon dioxide culminated in the creation of the Intergovernmental Panel on Climate Change.[22]

One would think the climate researchers on the IPCC would be interested in the cosmic ray theory since it provides a plausible and testable explanation for solar activity as it affects the climate. If nothing else, the theory is surely worth serious investigation before we spend billions of dollars trying to fix a problem—human-caused carbon emissions—that might not be the problem at all when it comes to warming. After all, in the IPCC's Climate Change 2001, Summary for Policymakers, we read: "Mechanisms for the amplification of solar effects on climate have been proposed, *but currently lack a rigorous theoretical or observational basis.*"[23] [italics added]

Supporting and funding research on the cosmic ray theory might provide this "rigorous theoretical and observational basis." If nothing else, further research could show that Svensmark's theory was false and the process could move on. The IPCC not only wasn't interested in further research on this idea, it was actively hostile. The late IPCC chairman Bert Bolin denounced Svensmark's theory as "extremely naïve and irresponsible," and another scientist at a conference called it "dangerous."[24] These comments recall the charge, noted in Chapter 3, that Zbigniew Jaworowski's research questioning the validity of ice-core CO_2 readings was "immoral." These comments are reminiscent more of religion than science. Once again, alarmist climate science has made up its mind as to the culprit in global warming, isn't interested in any other suspects, and has even made it as difficult as possible for Svensmark to fund his research.[25]

GALAXY QUEST

As part of the research on cosmic rays comes another intriguing possibility: that our position in the galaxy has an effect on climate.

The earth is on the edge of our galaxy, the Milky Way galaxy, and makes a circuit around it every 225 million years or so that is called the "cosmic year." The galaxy has several arms of stars that spiral out of it like scythes. Israeli astrophysicist Nur Shaviv has suggested that every 135 million years, as the planet enters one of the more populated arms of the galaxy, it receives an unusually intense bombardment of cosmic rays, which in turn leads to greater cloud formation, and hence to cooling. When we are not in a spiral arm and there are fewer rays, the planet is warmer. These effects may occur for tens of millions of years. Shaviv and others believe this cycle fits earth's hothouse and icehouse climate cycles over the last several hundred million years.[26] The IPCC is not eager to seriously consider this possibility, either.

WIND AND OCEAN CURRENTS

There isn't space here to look at all the other possible natural global warming suspects the IPCC prefers to dismiss in favor of one greenhouse gas, but two need to be mentioned: wind and ocean currents.

There are several wind and ocean current systems around the planet and they strongly affect climate. One ocean current is the North Atlantic Oscillation (NAO), which determines the winds for Europe. When the winds from the west are strong, Europe has cool summers and mild winters. When the westerly winds are weak, Europe has more extreme temperatures—hot summers and severe winters. Three other ocean current oscillations are the North Pacific Oscillation, the Pacific Decadal Oscillation, and the El Nino/ Southern Oscillation. Research by mathematician Anastasios Tsonis and his team indicates that while these oscillations have their own separate patterns, at times the patterns act in concert. When they are acting together, major climatic effects can occur and these effects are completely independent of human activity. Tsonis believes the see-saw effect of these overlapping oscillations has contributed to the ups and downs of global warming in our century:

> The first synchronization of the 20th century occurred around 1910. For almost three years, the coupling strength increased and then in late 1912 or early 1913, the synchronous state among the oscillations vanished. That coincided with a sharp increase in global temperatures.[27]

Tsonis believes one synchronization produced the cold temperatures starting in the 1940s, and another had the opposite effect in the mid-1970s, when the planet got warmer again. His model predicts a cooling spell in 2027 and a warming in 2065. He concludes: "The model shows that the climate changes are intrinsic to the natural climate system. Man-made activities doubtless also influence the climate, but not through the profound synchronizations that have foretold the climate in the past and will foretell it in the future."

Ocean currents are, next to the sun, the most likely suspect in driving climate change. The world's oceans are linked by currents, called the "Global Ocean Conveyor Belt," one of which is the Gulf Stream that brings warmed water to the North Atlantic and keeps Europe from going into the deep freeze. This current begins in the Gulf of Mexico and evaporation due to tropical warmth means the current starts its journey in a salty state. As the current proceeds northward it meets colder water that is less salty. Because saltier water is denser than the fresher water the current sinks, pulling its water after it like a conveyor belt. Thanks to this conveyer belt, warm water from the tropics moves north and Europe remains livable despite being quite far north. Some scientists fear an influx of fresh water from melting Arctic ice caps, triggered by global warming, will make the current less salty, and therefore less dense, thereby slowing or stopping the belt and drastically affecting the climate. This failure of the ocean conveyor belt is believed to have occurred 12,800 and 5,000 years ago and caused considerable cooling. However, it is difficult to believe that the relatively small rise in temperature we are experiencing now would have such a drastic effect on the oceans. The Younger Dryas cooling of 12,800 years ago was caused by a geophysical event of gigantic proportions—either the draining of North America's Lake Agassiz, which slowed the ocean conveyor belt, or an asteroid strike, or both; nothing like that is on the horizon at the moment.[28]

An abrupt change in climate is always, of course, possible due to changes in the ocean currents, but research reveals that a) abrupt coolings and warmings like those of 12,800 and 5,000 years ago have often occurred *naturally* and b) there are many possible causes for climate change, oceans being one of them, and almost all of these changes are natural. For example, the Pacific Decadal Oscillation (PDO) may be partly responsible for the current non-warming trend. Figure 5.3 shows the cycles from warm to cool currents, and how they correlate remarkably well with the ups and downs of climate since

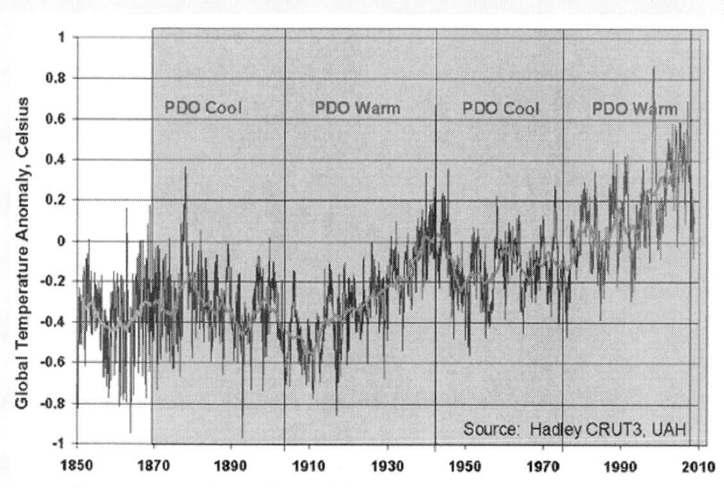

Figure 5.3: Pacific Decadal Oscillation and temperature since 1850. As with sunspots, the correlation with climate is remarkable. Source: Hadley Centre

1900. If the PDO is going into a cooling, as seems to be the case, then, along with the decline of sunspots, we can expect a planetary cooling that may last, some researchers believe, as long as "several decades"[29] before warming resumes. And if the cooling continues much longer than that, we could be in for another Little Ice Age. As consensus climatologist Judith Curry notes: "These oscillations and how they influence global temperature haven't received enough attention, and it's an important part of how we interpret 20th-century climate records."[30]

ABRUPT CLIMATE CHANGE

The process by which climate warms or cools quite suddenly is called "abrupt climate change" and abrupt climate change undermines the claim of climate alarmists that the warming since the mid-20th century is entirely due to human activity. For example, Andrew Weaver has written: "The scientific community has a very solid understanding of what is causing global warming: It is overwhelmingly because of the combustion of fossil fuels."[31] In other words, Weaver and the IPCC believe that if it wasn't for fossil-fuel emis-

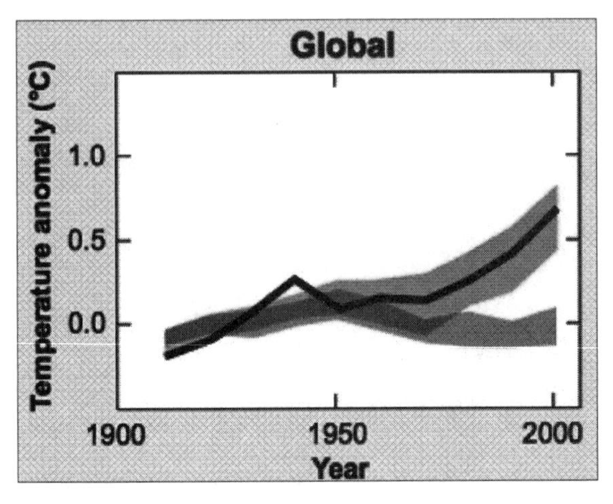

Figure 5.4: IPCC graph showing "natural" cooling (falling line) compared to human-caused warming (rising line). Source: IPCC 2007

sions, the planet would be getting colder. Or, as Weaver puts it: "Since the turn of the century the overall simulated [computer model] response shows very little warming. .... Since 1979, natural forcing has actually acted to cool the planet."[32] Figure 5.4 is a graph from the 2007 IPCC report showing the "natural" cooling that would be occurring without human carbon emissions (the falling line), compared to what the climate models show as the human-caused warming (the rising line).[33]

Leaving aside the question of how accurate these computer climate models are (and, as we saw in Chapter 4, they aren't very accurate), Weaver and the IPCC appear to be arguing that a cooling planet is preferable to a warming planet. Yet, a cold climate is almost always more difficult for the earth's human population than a warm climate. If fossil fuels are having the effect Weaver and the IPCC claim, we should be grateful to carbon emissions for keeping the planet from plunging into another cold spell like the Little Ice Age.

However, carbon emissions are, as of 2010, *not* having this warming effect. Since at least 2002 the climate has been cooling, even if slightly, just as the IPCC predicts would happen *without human influence*. Apparently fossil fuels aren't causing the warming the models predict. Putting this another way: it's likely the climate isn't nearly as sensitive to carbon dioxide "forcing" as the IPCC claims. But then, the IPCC's models have always, with unfailing consistency, overestimated the warming effect of greenhouse gases.

Weaver and other alarmist climate scientists worry about "dangerous" carbon dioxide levels, although this is puzzling since they know perfectly well that carbon dioxide levels have been much higher than today's for most of earth's geological past without harm to animal or plant life. And, since humans hadn't evolved yet, those high CO_2 levels were not anthropogenic. Alarmists also worry about the *speed* of the changes to climate brought about by the rapid increase in carbon dioxide levels. As Warrick et al. write: "Quickened rates of change could exceed the capacity of natural and human systems to adapt without undue disruption or cost."[34] What cost? Weaver believes that, should the planet's temperature go above 3.3° C higher than today's, the result will be "the sixth and perhaps greatest extinction" in which "between 40 per cent and 70 per cent of the world's species" will perish.[35]

Alarmist climate science would have the public believe that a sudden temperature change of this speed and magnitude would be humanity's fault. In fact, ice-core studies show that abrupt climate change is quite common in earth's history, including within the past 15,000 years since the end of the last glaciation. The Younger Dryas cooling, a mere 12,800 years ago, was very sudden, with temperatures dropping an estimated 5° Celsius back into glacial conditions within a decade.[36] The rise in temperature 800 years later (11,600 years ago) was equally rapid: 8° C in a decade.[37] Human carbon emissions had nothing to do with these sudden fluctuations.

That abrupt climate is far from unusual—and therefore quite natural—is reflected in the title of a book by the U.S. National Research Council: *Abrupt Climate Change: Inevitable Surprises*. The authors write:

> Recent scientific evidence shows that major and widespread climate changes have occurred with startling speed. For example, roughly half of the north Atlantic warming since the last ice age was achieved in only a decade. ... similar events, *including local warmings as large as 16°C*, occurred repeatedly during the slide into and climb out of the last ice age.[38] [italics added]

And, yes, these sudden rises and dips have had "adverse effects on societies," in part because the coolings created droughts and warming and cooling

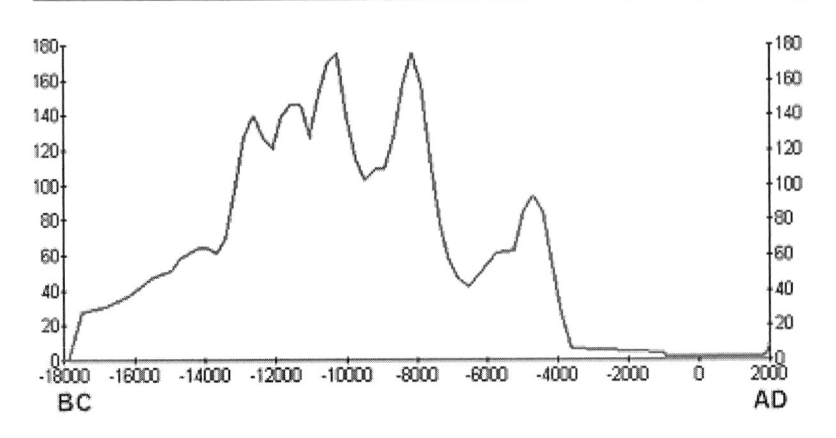

Figure 5.5: Rate of sea level rise over the past 20,000 years. Source: IPCC 2001

caused ocean levels to rise and fall quite rapidly. The lore of B.C.'s Haida Indians recalls that, as the planet was warming thousands of years ago, the water levels rose so quickly that seaside villages had to move every generation or two.[39] Figure 5.5 shows the rate of increase and decrease of sea levels over the past 18,000 years.

The rate of sea-level rise was very high as the ice melted; in the last 3,000 years, sea level rise has been about two millimeters a year and certainly nothing to worry about. It's also worth noting that sea levels at the end of the most recent glaciation, about 15,000 years ago, were 120 metres (almost 400 feet) *lower* than today's levels. In other words, after the glaciation ended, sea levels rose *400 feet* in 15,000 years, dwarfing anything that could happen in the next few centuries or millennia. And, as we saw in Chapter 2, sea levels during the most recent interglacial, the Eemian 125,000 years ago, were 4-6 metres (14 to 20 feet) higher than today's levels—again, all quite naturally; no human carbon emissions could be blamed. In fact, as we saw in Chapter 2, sea levels were higher than today's for all of the interglacials of the past 500,000 years, and perhaps earlier than that as well.[40]

From 1961-2003, the IPCC says, sea levels rose about 2 mm per year, or 20 cm (eight inches) per century.[41] With the increased warming it predicts, the 2007 IPCC report estimates that sea levels could rise in the 21st century from 18 cm (seven inches) to 59 cm (23 inches), with the latter as a worst-case scenario. The lower levels are hardly cause for alarm given that the seas have been rising about this amount for the past few centuries anyway, and have risen (and fallen) much more quickly in the past whether humans were here or not. That said, an acknowledged expert on sea levels, Nils-Axel Morner, argues that sea levels rose only one mm per year from 1850-1940, and from 1992-2002 have not risen at all.[42]

So, abrupt climate change is most definitely not something to be desired from the human perspective. But, as the geological record shows, abrupt climate change is also an act of nature that is utterly beyond our control. And the past geological record shows very clearly that it is highly unlikely, as some alarmist climate scientists tell the public, that temperature and sea-level increases could accelerate to "unprecedented rates."[43] Will the temperature warm faster than 8ºC in a decade, as happened at the end of the Younger Dryas cooling? This is highly unlikely under current conditions. And if it did happen, the cause would be, as in the past, abrupt, natural, and therefore "inevitable" climate cycles.

One of the curious facts about abrupt climate change is that these relatively sudden rises and falls of temperature occur even within the depths of the glacial periods, so that glaciations are not uniformly cold. Figure 5.6[44] shows some of the ups and downs during the most recent glacial, just before our Holocene interglacial. The sharp points at the graph's peaks and valleys show that the change from cold to warmer and vice-versa was sudden, sometimes occurring over decades, not centuries or millennia, and with no human input whatsoever.

As science writer John D. Cox notes in his very readable book *Climate Crash: Abrupt Climate Change and What It Means for Our Future*, the glacial data "illustrates a climate pattern that is very different from the slumbering stability that researchers traditionally envisioned for ice ages."[45] We see the same sharp peaks and valleys, by the way, in the changes from one glacial episode to the next (see Figure 2.2 in Chapter 2). What we don't see are gradual, rolling curves of change; when a major climate shift occurs, it seems to occur quickly.

Meanwhile, the IPCC estimates the temperature rise in the 20th century was about .6 degrees Celsius—hardly grounds for alarm considering the speed of, say, the Younger Dryas cooling and warming. Or, if the current rate

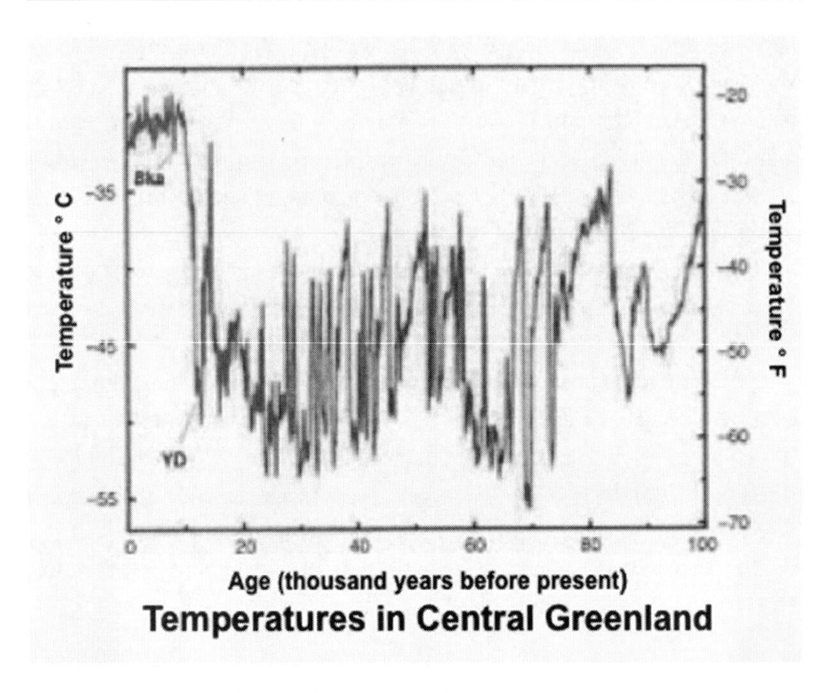

Figure 5.6: Abrupt climate changes *within* the most recent glacial

of rise is as alarming and as "unprecedented" as the alarmist climate scientists would have us believe, then perhaps it is because we are at the start of one of these abrupt climate roller-coaster rides that occur *naturally*, whether humans are here or not. The natural explanation works as well as a human-caused one once we accept the fact of abrupt climate change.

But aren't we tempting fate by adding more CO_2 to the atmosphere? Cox notes, "Climate scientists now realize that just as a moving hand is more likely to throw a switch than a still one, anything that changes the system runs the risk of provoking abrupt climate change."[46] Therefore, the alarmists believe, we should not do anything (such as add carbon dioxide to the atmosphere) that might provoke this abrupt change that could have "adverse effects" on human society. However, Cox's book is one of those unusual cases in which the author's even-handedness actually works against the case he wishes to make. Cox argues that increased anthropogenic carbon dioxide could trip the climate switch, so we should curb carbon emissions. What he ends up demonstrating, although unintentionally, is that the planet goes into warm and cold spells far more often than was previously believed and will continue to do so *no matter what we do or don't do.* If we increase carbon dioxide levels, we may trigger an abrupt climate change. On the other hand, even if we decrease carbon dioxide levels, the planet may go into an abrupt warming or cooling anyway due to other triggering factors—factors that are natural. In other words, *any* climate factor could trip the switch, or keep it from tripping. And as abrupt climate surprises are "inevitable," they *are* going to occur. Therefore, one could argue that we might as well carry on as best we can. The alternative is to go back to living in caves while walking on eggshells lest we anger the climate gods. Like every other creature, we affect the earth — that, too, is "inevitable." Our job is to cope with these inevitable, and ultimately natural, changes as they occur.

Once abrupt climate change is understood, the whole issue of trying to curb carbon dioxide to stop anthropogenic warming is seen as irrelevant — the planet will warm or cool as natural forces (not human carbon "forcings") dictate. This doesn't mean we shouldn't seek more efficient, less carbon-emitting forms of energy; it just means that there's no urgency to do so as far as *climate* is concerned. An understanding of abrupt climate change also gives the lie to the alarmist climate science claim that, if humans weren't interfering, the climate of the 20th century would have been stable, without the warmings and coolings that have occurred. The geological record makes it abundantly clear that a stable climate doesn't exist. That means that, whether we were emitting carbon dioxide or not, the 20th century's climate would have fluctuated, as will the 21st's.

HUMAN ACTIVITIES CAUSING CLIMATE CHANGE

None of this is to say that humans are completely blameless when it comes to climate change. For example, as we saw in Chapter 3, climatologist Roger Pielke, Sr., is not a "denier," in the sense of denying that humans affect the climate. But he argues, quite persuasively, that carbon emissions are only a small part of the picture when it comes to human impacts on climate. He writes:

> Humans are significantly altering the global climate, but in a variety of diverse ways beyond the radiative effect of carbon dioxide. The

> IPCC assessments have been too conservative in recognizing the importance of these human climate forcings as they alter regional and global climate.[47]

In particular, Pielke suggests that regional land-use patterns such as cities, which create the "urban heat island" effect, have more of an impact on climate than our carbon emissions, yet these other forcings are not adequately addressed by the IPCC. Nor can these land-use factors be eliminated except in small ways—it's unlikely human beings will decide not to live in cities or stop practicing agriculture, another human activity that causes warming.[48] Therefore, there is little we can do about these anthropogenic warming activities, and curbing carbon emissions, at great cost, does not touch the heart of the problem (assuming warming is a problem).

WHAT WE KNOW

It wasn't so long ago—only the 1970s—that some scientists feared the planet was getting colder. They had good reason. After almost 100 years of warming temperatures, in the early 1940s the planet became cold again. It remained cold until the mid-1970s, by which time the panic over global cooling was well under way, at least in the popular media and among some scientists (for a chapter on the history of the 1970s global cooling scare that was not included in this book, see paulmacrae.com). Then warmth returned. After a while, the scientists decided it was getting too warm too quickly. Some of them decided the warming would continue to "unprecedented" heights unless we did something, because clearly we were at fault. Even the IPCC admits humans weren't at fault for any of the coolings or warmings before the 1950s, but we were at fault for this one.

And so, in 1988, the IPCC was born based on the belief that human beings had caused the late 20th century warming, on the faith that we could stop the warming, and on the conviction that climate cycles had ceased being cyclical, so that the current warmth would be, on average, permanent and ever-increasing. And, based in these beliefs, alarmist climate science created the Doctrine of Certainty: warming is caused by human greenhouse gas emissions and no other culprits will be seriously considered.

However, as this chapter has tried to show, there are many causes for global warming and cooling; humanity is only one cause and far from the most powerful. Do humans have an influence on climate? Of course. Are we the "principal" influence? Unlikely. Indeed, to believe this makes no sense. For, despite what we're told by the IPCC and its followers—"the science is settled," "the evidence is overwhelming"—the scientific case against carbon dioxide is far from conclusive; other, non-human influences on climate are far more likely to be at fault. At the very least, other culprits should be far more carefully considered than the IPCC is prepared to do, and the alarmists' Doctrine of Certainty should be abandoned in favor of more scientific investigation not just of skeptical claims but of the IPCC's alarmist claims as well. Instead, the IPCC and the global warming pessimists have adopted what philosopher of science Karl Popper has called a "closed" rather than "open" model of science, which we'll look at in greater detail in Chapter 7. The result is bad science, leading to bad public policy, to the point where, as noted by Henrik Svensmark and Nigel Calder: "Among the thousands of human generations, ours may be the first that was ever frightened by warming."[49]

In general we look for a new law by the following process. First we guess it. Then we compute the consequences of the guess to see what would be implied if this law that we guessed is right. Then we compare the result of the computation to nature, with experiment or experience, compare it directly with observation, to see if it works. If it disagrees with experiment it is wrong. In that simple statement is the key to science.

— **Richard Feynman**[1]

Under what conditions would I admit that my theory is untenable?

—**Karl Popper**[2]

How is it possible for a theory, which is false in its component parts, to be true as a whole?

—**Jean-François Revel**[3]

Chapter 6

AL GORE: GOOD SCIENCE, BAD SCIENCE, AND BOGUS SCIENCE

In *The Logic of Scientific Discovery*, philosopher of science Karl Popper sets out the criterion for empirical science not as the test of truth but the test of falsehood. That is, a theory doesn't stand or fall based on whether it is true—for science, it is often impossible to say something is incontrovertibly, eternally true—but on whether the theory can avoid being proved false. Therefore, the true scientific question is not, "Under what conditions is my theory true?" but "Under what conditions would I admit that my theory is untenable?"[4] That is, although scientific tests may *refute* a theory, they can never establish it as a certainty because, Popper argues, the whole function of the "true scientific attitude" is to undermine the Doctrine of Certainty. This attitude, for Popper, "is utterly different from the dogmatic attitude which constantly claimed to find 'verification' for its favorite theories."[5]

So, for example, Darwin's theory of natural selection is a "theory" because there is a possibility that it may—or that some part of it may—be proved false. So far, it has stood up. A theorist who has not passed the test of experience, so far, anyway, is Paul Ehrlich, author of *The Population Bomb* in 1968 and many other doomsday books. To date, *all* of Ehrlich's predictions of disaster (e.g., mass famines by the mid-1980s) have flunked the experience test, thereby casting into doubt his overall theory that the planet can't survive industrial civilization. There is more on Ehrlich's and other alarmists' failed prophecies in Chapter 8.

With the falsification criterion of science in mind, let's examine Al Gore's 1992 book *Earth in the Balance: Ecology and the Human Spirit* and his 2006 film *An Inconvenient Truth*, which has also come out as a book, to see how well his theories stand up to the "falsifiability" test. He claims his truths are "inconvenient," which they are, but are they also "true"?

WHY DOES CLIMATE CHANGE? GORE DOESN'T SAY

Earth in the Balance has a chapter entitled "Climate and Civilization: A Short History" that is quite good as far as it goes, which isn't very far. Gore keeps his focus on the last 13,000 years, when the last glacial ended, the ice melted, and civilization developed, and shows some of the ups and downs of climate change in that time. But he has a problem: human agricultural and industrial activity weren't advanced enough to have caused any of the climate ups and downs before the Industrial Revolution. In 13,000 years the climate warmed, sometimes 2 or 3 degrees Celsius beyond today's levels (e.g., the Holocene Optimum), but the planet also cooled, sometimes abruptly and considerably, thousands of years before humans had developed industrialization. How did these warming and cooling episodes occur if they weren't caused by human industrial activity?

Gore's best guess for a climate culprit in *Earth in the Balance* is volcanoes,[6] a theory that doesn't have much scientific support in that volcanoes can cause short-term climate changes but rarely long-term changes unless the vulcanism is extensive.[7] As a text on paleoclimatology puts it: "Volcanic events, even very large ones, often do not have a global-scale effect."[8] The Little Ice Age lasted 500 years, from about 1350-1850—no volcanic eruption in the 14th century was powerful enough to have caused a climate change of that length. And never mind what happened after the glaciers began to melt: Gore also doesn't tell us what first *caused* the glaciers to thaw 13,000 years ago. Nor can vulcanism explain why the planet has gone into a series of glacials every 100,000 years over the past million years, and every

41,000 years for a million or so years before that. By confining himself to the climate ups and downs of our Holocene interglacial, Gore tries to show that today's climate is a human-caused extreme. Unfortunately for Gore, the climate data doesn't back him up. Human-caused carbon dioxide levels could not, at least until the last century or two, have caused the various warmings and coolings of the past 13,000 years, including one warming, the Holocene Optimum of about 7,000 years ago, when temperatures were 1-3 degrees Celsius higher than today's. So Gore's book doesn't try to explain these changes; instead, it ignores the reasons for them.

GLACIALS AND INTERGLACIALS NOT EXPLAINED

In his film *An Inconvenient Truth,* Gore highlights the series of glacials and interglacials over the past 650,000 years using a huge and very impressive line graph that compares temperatures and carbon dioxide levels over that time—it's the visual centrepiece of the movie. But, again, he doesn't explain *why* the planet has had so many cooling and warming cycles over that long stretch of time. Or, rather, he tries to, but he gets the scientific facts completely backward.

In his film, Gore notes that changes in carbon dioxide levels in the past 650,000 years have coincided with changes in temperature. He then implies, without stating it directly, that the fluctuations in carbon dioxide levels *caused* the fluctuations in temperature. But if this is so, then he needs to explain why the levels of carbon dioxide increased and decreased in the first place. He doesn't, perhaps because he knows that the scientific reality is the other way around—temperature changes cause CO_2 changes. In other words, the scientific evidence does not support his hypothesis so he ignores the science.

Ice-core records show that in the current interglacial and at least the two previous ones, changes in carbon dioxide levels lagged 400-1,000 years behind changes of global temperature.[9] In other words, the planet's temperature goes up or down (for whatever reasons, human or natural), and several hundred years *later* carbon dioxide levels also go up or down. Why? Mainly because as the planet warms, over time the oceans release carbon dioxide; when the planet cools, the oceans slowly re-absorb carbon dioxide.[10] That is, the colder the ocean, the more CO_2 it can hold, a scientific fact we all witness whenever we open a can of cold soda pop. As the pop warms, it releases carbon dioxide. So, carbon dioxide and temperature levels are related, but not in the way Gore claims. He is either aware of this scientific fact, in which case he is deliberately deceiving his audiences, or he isn't aware of this scientific fact, in which case he is incompetent.

Also, Gore's long and impressive line graph of temperature and carbon dioxide levels appears to have the two processes—temperature and CO_2 levels—happening simultaneously, rather than one causing the other. Yet, surely, if one process causes the other, there must be a time lag. On the other hand, if the two processes were simultaneous—they aren't, but if they were—then clearly some third mechanism must be operating on both carbon dioxide and temperature.

This major Gore error was pointed out in my May 22, 2007, *Times Colonist* column that also endorsed a skeptical British TV documentary entitled *The Great Global Warming Swindle.*[11] The column was challenged in a letter to the editor from Martin Golder, chair of the Sierra Club of British Columbia. Golder wrote: "This key point to the movie [*Swindle's* assertion that temperature precedes CO_2 levels] was examined and discredited by the normal scientific process of peer review before the movie [*Swindle*] was even released."[12] On the same page, a letter to the editor from Andrew Weaver took the opposite position: "Yes, temperature changes came before carbon dioxide changes in the recent glacial record. Temperature change was driven by small changes in the Earth's orbit [i.e., the Milankovitch cycles]."[13] Golder says one thing, Weaver says the exact opposite. So much for consensus although, in this case, Weaver is right.

John Houghton, a lead author for the Intergovernmental Panel on Climate Change, is highly critical of *The Global Warming Swindle*. But he also explicitly declares Gore wrong on the relationship of carbon dioxide and temperature:

> That carbon dioxide content and temperature correlate so closely during the last ice age is not evidence of carbon dioxide driving the temperature but *rather the other way round* . ... I often show [Gore's] diagram in my lectures on climate change but always make the point that *it gives no proof of global warming due to increased carbon dioxide.*[14] [italics added]

In other words, Houghton teaches the *exact opposite* of what Gore claims in his movie. But Golder's letter also shows how tenaciously the environmental movement clings to its belief that CO_2 is driving climate change in the face of scientific evidence to the contrary. Why alarmists cling to this belief so fiercely we will examine in Chapters 11 and 12, which deal with the religious and ideological influences on global warming alarmism.

Here are more scientific facts that Gore doesn't include in his argument because, unscientifically and unethically, he prefers to ignore data that doesn't fit his theory.

PAST TEMPERATURES WERE WARMER THAN TODAY'S

First, as we've already seen, today's global temperatures are still 1-3°C below the Holocene Optimum of 7,000 years ago. Human beings did not cause the Holocene Optimum spike in temperature. Nor did human beings cause the Medieval Warm Period from about 800 AD to 1350 AD, when global temperatures were up to a degree Celsius warmer than today's. Gore is aware of the Medieval Warm Period. In *Earth in the Balance* he writes: "Thus began the global climate shift known as the medieval warm epoch."[15] This was when the Vikings colonized Greenland and Iceland, although, as usual, Gore focuses on the negative—the Medieval Warm Period may also have caused the collapse of the Mayan civilization through drought.[16] Still, not even Gore is suggesting that human-caused carbon dioxide emissions triggered the MWP. And Gore only indirectly acknowledges that this warm spell was generally a *good* time for humanity. For example, of an earlier warming period called the Roman Warming (roughly 200 BC–600 AD and also a bit warmer than today), he writes:

> This same period of relative warming cleared the Alpine passes separating Italy from the rest of Europe and corresponded to the awakening of Rome's *imperial ambition*. Moreover, the simultaneous clearing of mountain passes in Asia led to the *expansion of Chinese civilization* and the opening of the Silk Route.[17] [italics added]

(Note that, for Gore, Rome has "imperial ambition" while Chinese civilization merely undergoes "expansion," apparently free of any imperial ambitions. But, then, for Gore, industrialized Western civilization can do little right.) As for the Medieval Warm Period, anthropologist Brian Fagan notes:

"The generally stable weather of the Medieval Warm Period was an *unqualified blessing* for the rural poor and small farmers"[18] [italics added].

The Vikings had to abandon Greenland when temperatures dropped again in the 14th century to begin what is known as the Little Ice Age, which lasted until about 1850. Gore is aware of the Little Ice Age, of course, and even notes that life for humanity became much more difficult during this period, including several disastrous famines and the Black Death, which killed about a third of the population of Europe. The irony here—an irony that is lost on Gore—is that the *warm* Medieval period was a time of growth and prosperity for humanity; it was the *falling* temperatures that triggered humanity's problems. Similarly, the Roman Warming saw the rise of the great civilizations of Rome and China but, as the planet cooled, Europe entered a Dark Age. In fact, as geographer Harm de Blij notes: "*Global cooling is a far greater long-range threat than global warming* at a time when humanity's numbers approach seven billion on this small planet."[19] [italics added] For some reason—again, perhaps because it destroys his argument—Gore alludes to the problems of global cooling but doesn't draw the logical conclusion: that warming is preferable to cooling.

CAUSE OF ICE AGES LEFT UNANSWERED

Earth in the Balance leaves many logical questions unanswered. For example, Gore observes: "Approximately 17,000 years ago ... global temperatures ... shot upward to relatively constant levels, which have persisted more or less for the last several thousand years."[20] What caused global temperatures to shoot up? Gore doesn't say, but it couldn't have been human-caused carbon dioxide—humankind hadn't even invented agriculture 17,000 years ago. Was the cause of this warming a sudden natural increase in carbon dioxide? If so, Gore doesn't explain what caused this carbon dioxide increase. Similarly, talking about the onset of earth's descent into several ice ages, including our current, two-million-year ice age of glacials and interglacials, Gore only tells us they took place "after average global temperatures dropped by only a few degrees."[21] But what *caused* temperatures to drop? In the movie Gore attributes falling temperatures to falling carbon dioxide levels. If so, what caused the carbon dioxide levels to drop? He doesn't say.

In *Earth in the Balance*, the only explanation for rising and falling temperatures and CO_2 levels that Gore is willing to offer is that these changes may be "tied to variations in solar intensity."[22] This is a rather important piece of information—most climate scientists, including even global-warming alarmists, assign a much more prominent role than Gore to the sun, and particularly the Milankovitch cycles, in creating the current ice age. But Gore's comment is buried in a caption under a graph and isn't part of his main text, i.e., Gore ignores the role of the sun as much as possible; he doesn't even mention the Milankovitch cycles. Yet, as we saw in Chapter 2, the Milankovich cycles are very likely the *main* initiating force for the glacials and interglacials of the past two million years. Even Weaver agrees with this, as noted in his letter to the *Times Colonist*: "Temperature change was driven by small changes in the Earth's orbit."[23] No wonder Gore buries this extremely inconvenient solar fact rather than address it—it utterly destroys his case that carbon dioxide fluctuations are the drivers of temperature change, at least for most of the past 650,000 years. And if carbon dioxide didn't drive climate over this long period of time, why would we assume carbon dioxide drives the climate now?

And yet, despite this huge scientific error and many others, detailed below, Gore's film remains largely uncriticized by alarmist climate scientists, in part because they believe he is getting the "right" point of view across. For example, University of Colorado climatologist Kevin Vranes praises Gore for "getting the message out," while wondering whether his film was "overselling our certainty about knowing the future." Weaver gushed that he was "fortunate to attend" a talk by Gore in Victoria in 2007, and that Gore's speech was "passionate, inspirational and empowering."[24] Gore's Victoria talk was, if it followed the script of the movie, scientifically misleading, exaggerated and/or inaccurate, but scientific accuracy doesn't seem to matter to alarmists as long as the anthropogenic warming theory is supported. Similarly, James E. Hansen, director of NASA's Goddard Institute for Space Studies and a key figure in the global warming alarmism industry, acknowledges Gore's approach may have "imperfections" and "technical flaws," but "he has the bottom line right."[25]

Does he? I've been in an academic environment since 2002, when I left my job as an editorial writer at the Victoria *Times Colonist* to get an MA and then teach writing at the University of Victoria. I am certain that Gore's book, which ignores data that contradict his thesis and contains gross errors of fact (not just "technical flaws"), would not have passed muster as an undergraduate paper, much less as one of the bibles of global warming alarmism today. Gore's scientific supporters like Hansen seem happy to forgive these errors, but they wouldn't be so generous if Gore was one of their students or colleagues rather than a former politician with the right "message."

WRONG ON MORE SEVERE AND FREQUENT HURRICANES

As one of Gore's "technical flaws," Hansen notes that Gore has his facts wrong in blaming increased storm frequency and intensity, such as Hurricane Katrina that devastated New Orleans in 2005, on global warming. On the contrary, notes the *New York Times* article that quotes Hansen: "This past Atlantic season [2006-2007] produced fewer hurricanes than forecasters predicted (five versus nine), and none that hit the United States." Hansen concludes: "We [scientists] need to be more careful in describing the hurricane story than he [Gore] is."[26] The IPCC has also not been careful about falsely claiming more hurricanes, and extreme weather in general, due to global warming. For example, one of the world's leading experts on hurricanes, Christopher Landsea of the Atlantic Oceanographic and Meteorological Laboratory, quit the IPCC in protest over its misrepresentation of the facts,[27] and the Climategate scandal has revealed numerous other errors in the IPCC's reports. Yet, surely anyone trying to influence public policy as much as Gore does—with costs, if his recommendations are accepted, that will total in the billions of dollars—shouldn't just be "more careful" but *scrupulously* careful about the facts. Let's look at a few more cases where Gore not only isn't careful, but is actually either misleading, exaggerating, or just plain wrong.

WRONG ON '100 PER CENT' CONSENSUS

One of the most striking "truths" in *An Inconvenient Truth* is Gore's contention that of 928 published studies on global warming over the past decade, not one—zero—disagreed with the "consensus" view that global warming is "real," that "human beings are the principal cause," and that "its consequences are so dangerous as to warrant immediate action."[28] Note that Gore's "consensus" actually embraces three "truths," otherwise known as hypotheses, not one: first, that warming is real; second, that it is principally human-caused; and third, that is an impending catastrophe. This three-part hypothesis is a very broad net that would include many skeptics who acknowledge the planet is warming or has warmed but dispute (as this book does) that humans are the *principal* cause of warming and that the results of warming must inevitably be disastrous. Even Dr. Patrick Michaels, a leading global warming skeptic, begins his book *Meltdown* with the words: "Global warming is real and human beings have something to do with it."[29] In the real world of science, "100 per cent" consensus would be a miracle and perhaps even unprecedented, especially in a field like climate science that is barely four decades old: total agreement among thousands of scientists! How could anyone dare to question that! Unfortunately, as with so much of the film, Gore's claim is false and Gore gets the consensus he claims through a statistical sleight of hand.

The research Gore refers to appeared in *Science* in 2004 and looked at the abstracts (summaries) of 928 articles found on the ISI (Institute for Scientific Information) database between 1993 and 2003, using the search terms "global climate change." The researcher, Naomi Oreskes, states:

> The 928 papers were divided into six categories: explicit endorsement of the consensus position, evaluation of impacts, mitigation proposals, methods, paleoclimate analysis, and rejection of the consensus position. Of all the papers, 75% fell into the first three categories, either explicitly *or implicitly* accepting the consensus view; 25% dealt with methods or paleoclimate, *taking no position* on current anthropogenic climate change. Remarkably, none of the papers disagreed with the consensus position. Admittedly, authors evaluating impacts, developing methods, or studying paleoclimatic change might believe that current climate change is natural. However, *none of these papers argued that point.*[30] [italics added]

In other words, Oreskes and Gore take silence as consent: if a scientist doesn't explicitly reject the consensus, then he/she accepts the consensus. In marketing this is called "negative option billing" and it got Canada's cable companies in trouble—if you didn't explicitly refuse new channel services, you got them and paid for them. The cable companies were forced to change to a more honest system. Oreskes' study pulled the same dodge to create the phony "100 per cent consensus" that Gore refers to.

When anthropologist Benny Peiser of Liverpool's John Moore University checked Oreskes' data, he found that the abstracts of only 13 of the 928 papers (one per cent) explicitly endorsed the "consensus," while 322 (29 per cent) implicitly endorsed it. Peiser also found 34 abstracts that explicitly rejected the idea that humans are the main cause of global warming, 44 that dealt with *natural* factors in warming, and 470 (42 per cent) pulled in by the keywords "global climate change" that had no reference to human activities causing warming at all.[31] In other words, of the 928 papers, only about a third explicitly or implicitly endorsed the Gore "consensus"—a far cry from the scientific unanimity that Gore would have us believe exists. Here, for example, is what the abstract of one paper, cited by Oreskes as supporting Gore's "consensus," says about the effect of global warming (human-caused or otherwise) on tropical cyclones:

> The authors conclude in their abstract that, even though the possibility of some minor indirect effects of global warming on TC [tropical cyclone] frequency and intensity cannot be excluded, they must effectively be "swamped" by large natural variability.[32]

For the authors of this paper, it isn't humans causing increased cyclone activity. Another paper's abstract states: "The implications of global climate change are enormous. However, there are major questions concerning whether climate change is occurring."[33] In other words, this paper even questions (as any good scientist should) the "consensus" that global warming is occurring at all.[34] Peiser concludes:

> This is not to deny that there is a majority of publications that, although they do not empirically test or confirm the view of anthropogenic climate change, go along with it by applying models based on its basic assumptions. Yet, it is beyond doubt that a sound and unbiased analysis of the full ISI databank will find hundreds of papers (many written by the world's leading experts in the field) that have raised serious reservations and outright rejection of the concept of a "scientific consensus on climate change." The truth is that there is no such thing! In light of the data presented above (evidence that can be easily verified), *Science* should withdraw Oreskes' study and

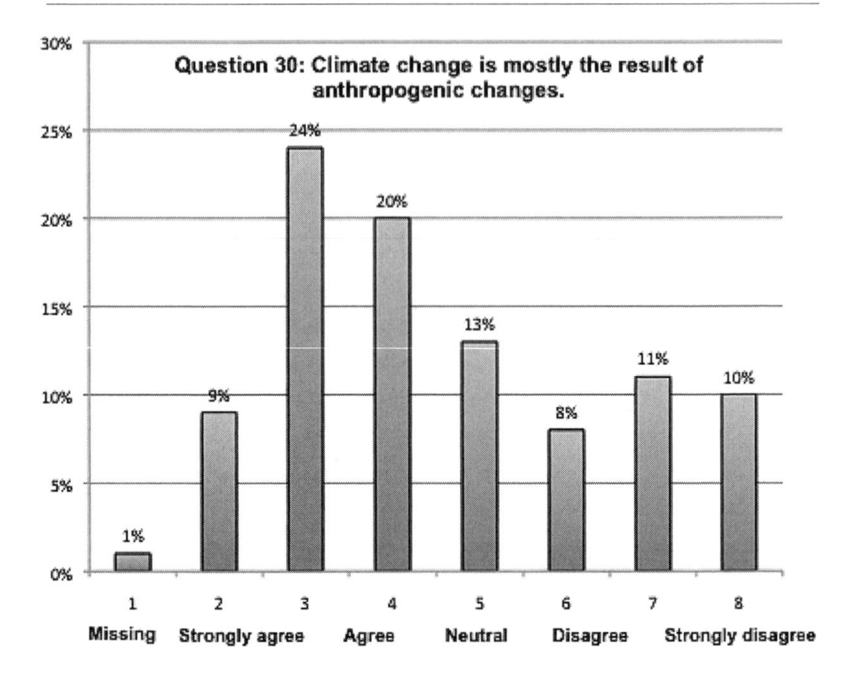

Figure 6.1: Question 30 in 2003 Bray and Storch poll on climate change. 42% of respondents disagreed or were neutral on whether humans are the main cause of global warming.

> its results in order to prevent any further damage to the integrity of science.[35]

However, Peiser notes that *Science* refused to publish his letter criticizing the original article. As we'll see in greater detail in Chapter 10, refusing to publish material that doesn't support the "consensus" is one way that alarmist climate science maintains the illusion of a consensus.

In 2003, Dennis Bray and Hans von Storch, of Germany's Institute of Coastal Research, conducted an online survey of 530 climate scientists in 27 countries on their perceptions of climate science, building on a similar survey the pair had done in 1996. Among the survey statements is: "Climate change is mostly the result of anthropogenic causes," which is Gore's hypothesis and for which, in *An Inconvenient Truth*, he claims "100 per cent" agreement among climate scientists. The survey of 530 climate scientists returned the result shown in Figure 6.1.[36]

The vast majority of the scientists polled—81 per cent—agreed or strongly agreed that "we can say for certain global warming is a process already underway," keeping in mind that, in 2003, it still wasn't clear that warming had stopped. However, on the issue of whether "climate change is mostly the result of anthropogenic causes," Gore's claim, only 53 per cent "strongly agreed" or "agreed"; 13 per cent were neutral and 29 per cent either "disagreed" or "strongly disagreed." In other words, in 2003, a full 42 per cent of 530 climate scientists *disagreed* with the position that humans are principally to blame for global warming.[37]

Gore's *An Inconvenient Truth*, both movie and book, came out in 2006. Gore must have been, or should have been, aware of the Bray-Storch poll. And yet he continued to claim "100 per cent" agreement among climate scientists that humans are principally at fault for global warming. In other words, he was promoting what he must have known to be an outright falsehood, although this is just one Gore falsehood among many. It is a wonder that so many people, including environmentalists, alarmist climate scientists and, worst of all, the Nobel Prize committee, have been so blinded by global warming ideology that they continue to take Gore seriously even when his "truths" are so blatantly untrue. Perhaps feminist literary and social critic Camille Paglia best summed up Gore's absurd claims:

> When I tried to watch Al Gore's *An Inconvenient Truth* on cable TV recently, I wasn't able to get past the first 10 minutes. I was snorting with disgust at its manipulations and distortions and laughing at Gore's lugubrious sentimentality, which was painfully revelatory of his indecisive, self-thwarting character. When Gore told a congressional hearing last month that there is a universal consensus among scientists about global warming—which is blatantly untrue—he forfeited his own credibility.[38]

WRONG ON MASS EXTINCTIONS

Gore's film has a computer-generated polar bear drowning because, thanks to global warming, it doesn't have enough ice to cling to. In fact, as Gore would know if he'd done any research that didn't fit his preconceived ideas, global polar bear populations are not endangered, at least not at the moment. For example, Dr. Mitch Taylor, manager of wildlife for the northern

Canadian territory of Nunavut, reports that polar bear populations in his jurisdiction are increasing, but "good news about polar bear populations does not seem to be welcomed."[39] He adds: "No evidence exists that suggests that both bears and the conservation systems that regulate them will not adapt and respond to the new conditions. Polar bears have persisted through many similar climate cycles." Indeed, polar bears are actually brown (grizzly) bears that evolved to meet Arctic conditions somewhere between 250,000 and 100,000 years ago. In other words, they are relatively recent arrivals on the planet (as are we), the products of evolution, and they may have to evolve or perish again in the future (as may we). That's natural selection. For now, the bears are coping very well, just as they coped with the last interglacial 125,000 years ago, which was warmer than our interglacial so far, and perhaps the one before that. The famous photograph of two polar bears apparently stranded on a melting ice floe, i.e., victims of global warming, is now recognized as misleading—the photo was taken in August when the ice usually melts anyway. And, polar bears are very strong swimmers.

Gore isn't just worried about polar bears: in *An Inconvenient Truth* he argues that "human civilization ... is destroying as many as half of all the living species on earth during the lifetimes of persons now living."[40] His reference? A theologian, Thomas Berry, writing in a book published by the Sierra Club, which is hardly the most unbiased source for scientific information on the environment. A decade earlier, in *Earth in the Balance*, Gore asserted that human activities are causing 100 extinctions a day, or almost 40,000 species a year.[41] In fact, as Julian Simon, Bjorn Lomborg, and many others have pointed out, the charge that technological civilization is destroying species by the tens of thousands is a complete myth, based on equations from climate models rather than hard data.

Lomborg notes in *The Skeptical Environmentalist* that there's no doubt human beings cause extinctions—prehistoric human hunters either caused or speeded the extinctions of all the large mammals in North and South America, the moa (huge flightless birds) in New Zealand, and many other creatures. But how many species are we eliminating now? Lomborg suggests, based on a lot of statistical evidence, that the planet may be losing perhaps a species a year to human activity.[42] For many environmentalists, this is one lost species too many, but it is still a far cry from the eco-Holocaust that Gore and others wish to alarm us with. It's possible to quibble with Lomborg's figures, of course, in part because, as Lomborg himself notes, we don't actually know how many species there are on earth, and because all statistical assertions are open to question. But think about it: does it seem reasonable, as Gore suggests, to believe that species are vanishing at a rate of 100 a day, 40,000 a year? Wouldn't we notice, especially in an era when the numbers of endangered spotted owls and condors are considered of international interest? And, why aren't alarmists able to name these 40,000 extinct species? There's a simple reason: because these "extinct" species are the product of mathematical calculations, not real, flesh-and-blood creatures.

The figure of 40,000 extinct species a year was an estimate by environmentalist Norman Myers in 1989, but with no actual research data to back up that estimate.[43] Myers has since retracted this figure, noting: "The estimate of 40,000 extinctions per year was strictly a first-cut assessment, preliminary and exploratory, and advanced primarily *to get the issue of extinction onto scientific and political agendas.*"[44] [italics added] Note the use of exaggeration to get the public's, the media's and the politicians' attention; we'll see other examples of misleading scare tactics by *leading* climatologists in Chapter 8. Myers says his recent figures have been "more documented, modified, refined, and generally substantiated estimates." A later, more "moderate" Myers' estimate was 50 species extinctions a day or about 18,000 extinctions a year, which still seems high and is, again, the product of a mathematical calculation, not a census of actual missing species.[45] In his latest estimate, however, Myers is up to his old hyperbolic tricks. In the next century, he expects half of the world's 10 million species (that is Myers' estimate; in reality, nobody actually knows how many species the earth has) will become extinct, creating what he calls a "sixth mass extinction."[46] That's 50,000 species a year, or about 140 species a day.

For those curious about how many species have *actually* gone extinct in the real world, rather than species that have theoretically gone extinct within an alarmist computer program or as a result of a wild guess, an organization called the International Union for Conservation of Nature and Natural Resources (IUCN) publishes online a list of species that are *known* to have gone extinct since 1500. The total, as of 2009: 717 extinct species, or fewer than one and a half species a year in more than 500 years. The

2009 list includes 76 mammal species, 134 birds, and 60 insects. The largest number, 257, are snails and slugs.[47] Again, while nobody wants a species to go extinct, less than two extinct species a year is hardly the environmental holocaust that alarmists like Gore wish the public to believe. As philosopher of science Jeffrey Foss notes:

> Since the scientific *data* do not support 6th X (the Sixth Extinction), it is hardly surprising that the 6thX scientific literature bases it on models rather than observation. Scientific theories are supposed to be based on evidence.[48]

But, then the entire global warming hypothesis is based on models rather than observation and evidence, so perhaps it isn't surprising that the alarmist extinction figures are equally imaginary.

The reality is that even if human beings didn't exist at all, extinctions would occur: the United Nations Environment Program estimates that more than 95 per cent of all species that have ever lived on earth have vanished, the vast majority long, long before humans appeared on the scene.[49] Furthermore, the figure of tens of thousands of extinctions a year is based on the assumption that there are millions of species that we aren't aware of. Therefore, if we believe we are losing, say, one per cent a year of species due to human activity, this one per cent is calculated not on the basis of the total number of known species but on the basis of the total number of estimated but so far *unknown* species. In other words, 40,000 species lost a year is an imaginary number based on pure speculation, run through mathematical models with little evidential backing, although, of course, the makers of the alarmist models claim otherwise.[50]

BRITISH COURT RULING FINDS MOST OF GORE'S MOVIE INACCURATE

Gore's film is so riddled with errors and exaggerations that even a British judge could easily spot them. In a judgment released in October 2007, the judge found nine of Gore's "truths" scientifically invalid—that is, they contradicted the IPCC findings (which themselves should be regarded with some skepticism, as we saw in Chapter 4). The judge forbade British schools from presenting *An Inconvenient Truth* as scientific truth; instead, if schools did show it, teachers had to clearly identify the film as "presenting one-sided views about political issues" rather than as being scientifically based. What had Gore gotten *scientifically* wrong, according to the judge?

- Sea levels won't go up seven metres in the immediate future due to melting of the Arctic and Antarctic ice caps, as Gore says. Even the IPCC estimates that sea levels might only go up perhaps half a metre in the next century. Also, the Antarctic ice cap is actually *increasing*, which would either *lower* sea levels or keep them steady.
- The Pacific atoll islands aren't in danger of sinking under the ocean, and whole island populations have not been evacuated, as the film says.
- Very few climate scientists believe that global warming will shut down the Gulf Stream and cause another ice age.
- Gore's huge temperature-CO_2 wall graph has cause (temperature) and effect (CO_2 levels) reversed.
- Global warming is not threatening polar bears or causing them to drown. Only four have drowned that we know of, but that was due to a storm, not global warming, and all but two of 17 world polar bear populations are either stable or increasing.
- The glaciers on Mt. Kilimanjaro in Africa aren't melting because of global warming. The Kilimanjaro glacier has actually been melting since the 1800s, so industrial carbon emissions can't be the reason, and most scientists think surrounding deforestation is causing the problem.
- The drying up of Lake Chad was likely due to reasons other than global warming.
- The world's coral reefs are not bleaching due to global warming.
- Global warming is not the reason for major weather events like Hurricane Katrina.[51]

That is, virtually the entire movie is scientifically misleading, exaggerated, or wrong. A judge could find these obvious errors—and there are many, many more[52]—but, for some reason, consensus climate scientists like James Hansen and Andrew Weaver do not think the errors are important because Gore is getting out the "message." Not only do IPCC climate scientists approve of Gore's astonishing number of errors for an ostensibly "scientific" film, but, shamefully for science, they were willing to accept a Nobel prize with him in 2007.

WRONG ON ALMOST EVERYTHING

In his 1999 book *Hoodwinking the Nation*, the late economist Julian Simon, who would be classified as a skeptic on human-caused global warming and most other environmental scares, had several criticisms of *Earth in the Balance*. The gist of Simon's criticism was that "just about every assertion in the book points in the wrong direction—suggesting that conditions are getting worse rather than getting better, which they are." Furthermore, "the advisors Gore leans heavily on—Paul Ehrlich and Lester Brown—have been proven wrong in every one of the forecasts they have made in the past two decades, a truly astonishing record of consistency. Yet it is still their agenda that Gore puts forth, almost as if he is writing from handouts of the environmental movement." Indeed, says Simon, "the book is filled with this sort of environmental gossip, backed by no sources, and contradicted by solid data."[53] In short, Gore's work does not meet any standard of scientific or academic accuracy—and yet, *An Inconvenient Truth* is being shown in Canadian schools as an ostensibly "scientific" educational tool.

COMBINING SCIENCE AND RELIGION: BAD IDEA

Chapter 10 will look in detail at Gore's and the alarmists' attitude toward science itself, but it's worth highlighting one comment from *Earth in the Balance* that attacks Francis Bacon (1561-1626), one of the pioneers of the scientific method. Gore writes:

> His [Bacon's] moral confusion—the confusion at the heart of much of modern science—came from the assumption, echoing Plato, that human intellect could safely analyze and understand the natural world without reference to any moral principles defining our relationship and duties to both God and God's creation.[54]

Bacon made respectable the method of induction, that is, testing theories based on observation and experimentation rather than, as had mostly been the case since Plato and Aristotle, and as carried on by the medieval religious thinkers, building theories deductively, that is, making scientific assertions based on accepted principles or dogma without reference to actual experiments or facts. For example, based on a deduction that women were inferior to men, Aristotle asserted that men had more teeth than women, an anatomical detail he could have disproved with a simple physical examination.[55] Without Bacon (or someone like him), we could still be living in that medieval mind-set, relying on religious, deductively based remedies for our physical, social and mental ills rather than the experimentally based benefits of modern science, medicine and technology. Thus, the great Victorian thinker Thomas Macaulay (1800-1859) described Bacon's aim as "the multiplying of human enjoyments and the mitigating of human sufferings."[56] As for Bacon "echoing" Plato, Macaulay noted:

> To make men perfect was no part of Bacon's plan. His humble aim was to make imperfect men comfortable. ... In Plato's opinion, man was made for philosophy; in Bacon's opinion philosophy was made for man; it was a means to an end; and that end was to increase the pleasures and to mitigate the pains of millions who are not and cannot be philosophers. ... The aim of the Platonic philosophy was to raise us far above vulgar wants. The aim of the Baconian philosophy was to supply our vulgar wants.[57]

Of God, Bacon wrote in *Novum Organum*, his treatise on the scientific approach to solving problems:

> If the matter be truly considered, natural philosophy [the 17th century term for what today is called science] is, after the word of God, at once the surest medicine against superstition and the most approved nourishment for faith, and therefore she [science] is rightly given to religion as her most faithful handmaid, since the one displays the will of God, the other his power.[58]

In other words, Bacon believed that the development of scientific knowledge and power was a way of glorifying the God who'd made human beings capable of reason. Gore's God, on the other hand, would have us abandon our reason in favor of the untested, irrational, and mythic religious influences on science that Bacon and those who followed him were trying to overcome. In short, as a Creationist (he writes of "our relationship and duties to both God and God's creation"), Gore wants to take us back a thousand years to a time when science and religion were one. If Gore had any grasp of history, he'd know this is a very bad idea. Bacon certainly thought so:

> It is not surprising that the growth of natural philosophy [science] is checked when religion, the thing which has most power over men's minds, has by the simpleness of incautious zeal of certain persons been drawn to take part against her.[59]

That is, Gore is using science (asserting that his views are scientifically valid) to attack science (the experimental method, as free as possible from religious or other bias) to promote a religious (probably born-again Christian) agenda. Gore's religious agenda, by the way, is explicit in the title of his first book, *Earth in the Balance: Ecology and the Human Spirit.*

GORE'S CARBON-SPEWING LIFESTYLE

Finally, let's look at Gore's *personal* response to the enormous threat he perceives if the planet warms. We need, he says in *Earth in the Balance*, a complete change in our attitudes toward the earth, towards science, and toward ourselves. Indeed, he waxes quite romantic on the topic, in the sense that the European Romantics of the late 18th and early 19th century, like Jean-Jacques Rousseau, believed (understandably but wrongly) that returning to a closer relationship with Nature would cure our psychological and material ills. Gore writes:

> I believe our civilization is, in effect, addicted to the consumption of the earth itself. The addictive relationship distracts us from the pain of what we have lost: a direct experience of our connection to the vividness, vibrancy, and aliveness of the rest of the natural world.[60]

He calls this a "spiritual loss," with the implication that if only we consumed less and were closer to nature, we'd be happier and more spiritually grounded; this closeness would assuage the "hunger for authenticity" that haunts us. "Thus, the essence of denial is the inner need of addicts not to allow themselves to perceive a connection between their addictive behavior and its destructive consequences."

In the book version of *An Inconvenient Truth*, Gore helpfully provides five pages of tips on how to "Save Energy at Home."[61] In short, he is an advocate of the anti-technological ideology that would have us all living with much, much less, to the point of what the developed world would call poverty. British journalist George Monbiot's ultra-alarmist book *Heat* is explicit about this. Monbiot quotes a veteran British environmentalist as stating that, if Britain cut back its carbon emissions to the level the climate extremists want, the standard of living there would be equivalent to "a very poor third-world country."[62]

Naturally, Gore himself doesn't live like the inhabitant of a poor third-world country, any more than the movie and rock stars who proclaim themselves as Green while crossing the world in their private jets. Gore's Nashville mansion reportedly uses more than 20 times as much electricity as the average American household.[63] Apparently Gore is willing to forgive his own, personal "addictive behavior and its destructive consequences," and clearly the environmental crisis isn't serious enough for him to make more than token sacrifices to his own lifestyle: he buys "green credits"—from a company he owns, which is convenient and profitable—to atone for his profligate electricity use. Meanwhile, he is amassing millions of dollars selling "green" products and services—a classic case of making a killing by creating an artificial demand and then filling it.[64]

Gore's reluctance to reduce his lifestyle isn't surprising, because most of us don't want to live any closer to Nature than we have to, thanks. Indeed, if humanity was so "authentic" and spiritually fulfilled in ages past, it's worth asking why our ancestors worked so hard over thousands of years to free themselves from the tyranny of nature. Even a cursory knowledge of the "authentic," "spiritual" lifestyles of our ancestors—lifestyles that included crushing poverty, famine, early death, illnesses that are today easily cured, and many other evils all caused by dependence on nature rather than control of it—should persuade most of us that we are glad we live now, not then. Gore's Romantic view of Nature is a myth that, like much of what Gore says and writes, sounds plausible, but can't stand up to the most perfunctory analysis.

WHAT WE KNOW

Chapter 11 will have more on the relationship between seeing the world through mythical, romanticized, religious eyes, as Gore does, and the scientific/rational vision that Gore deplores. For now, let's see how well Gore's book and film do in meeting Karl Popper's "falsifiability" test, recognizing that it's impossible to prove a theory totally true based on the data—there may always be exceptions we don't know about. We might also recall physicist Richard Feynman's test for the validity of an hypothesis in the chapter epigraph: if an hypothesis isn't backed up by experimental data, it should be discarded. Therefore, although we can't be sure a theory is *true*, it is possible under Popper's philosophy to find a theory or parts of it *false*, or probably false (we don't want to fall into the alarmist error of the Doctrine of Certainty).

Gore's record from the book and the film:

- That carbon dioxide has driven temperature in the past 650,000 years? False.
- *Cause* of warmings and coolings over the past 650,000 years if it isn't CO_2? He doesn't say.
- Overwhelming, 100 per cent consensus on human causes of global warming? Utterly false.
- Global warming causing extinction of polar bears? So far, false.
- Global warming causing more extreme weather such as hurricanes? False.
- Extinction of 40,000 species a year? These extinctions exist only in computer models (i.e., false).
- Global warming accelerating loss of glaciers on Mt. Kilimanjaro? False.
- Oceans rising several metres in the foreseeable future? Even the 2007 IPCC report estimates a sea-level rise of at most 59 centimetres (23 inches) over the next century, so almost certainly false.
- Inundated Pacific islands with evacuated populations? False.

Plus many other exaggerations or outright falsehoods too numerous to list here. In fact, Gore's work is an astounding record of scientific ineptitude and dishonesty, to the point where there is little in Gore's book or movie that can stand up to scientific and sometimes just logical scrutiny, in part because Gore was writing very much from the gut rather than from actual research—exactly the kind of mainly deductive approach that Francis Bacon warned against. That is, Gore has a belief and then finds the data that supports his belief, rather than objectively examining the data before publishing. And yet, incredibly, Gore's book and film are a major source of the Canadian, American and European public's information and therefore views on global warming.

This is not to say that Gore's basic hypothesis—global warming is caused by human activity and needs immediate correction—is wrong. Gore's theory may turn out to be correct. It's just that Gore presents no valid *scientific evidence* to prove he is right. But if he is right, if he has this valid scientific evidence, why not present it in his movies and books? Why mislead, exaggerate and lie if the truth is as obvious, if the evidence is as overwhelming as Gore clearly believes? Even if Gore is 100 per cent sincere, his *hypothesis* as he presents it—the "inconvenient truths"—cannot pass the falsifiability test. His so-called facts are, most of the time, based on opinion, belief, ideology, myth, religion, morality, or anecdotal evidence rather than scientific data, as the British judge noted.

Millions of people have seen Gore's film and thousands have read his book and been convinced that human beings are the main cause of global warming and that we need to curb technological civilization. Yet Gore's argument is based not on science but myth, ideology and conjecture, i.e., it is pseudoscience that is, again, supported by alarmist climate scientists. The best advice to readers and viewers of Gore's "truths"? Don't reject his arguments out of hand but be critical. Think for yourself. Do some reading of your own. Ask yourself: If Gore's theory is so strong, why are the facts supporting it so flimsy? Or, as the late French thinker Jean-François Revel put it, "How is it possible for a theory, which is false in its component parts, to be true as a whole?"[65]

How, indeed? Based on the evidence from his own books and movie, Gore's hypothesis—that humans are the principal cause of climate change and the result will be disaster—is shown to be clearly wrong. In Popper's terms, it has been falsified.

Global warming has little to do with science and everything to do with politics. Those scientists who endorse the theory command the lion's share of government-funded research grants. Since the global warming prediction emerged in the late 1980s, climate science funding has gone through the roof. Scientists know, however, that they won't get funded unless their research confirms global warming. Too many enormous reputations would go down the plug otherwise; too many political agendas depend on the theory. So global warming has become big business.

—Columnist Melanie Phillips[1]

It is always flattering to belong to the inner circle of the initiated, and to possess the unusual power of predicting the course of history.

— Karl Popper[2]

Chapter 7

'OPEN' SCIENCE AND ITS ENEMIES: THE GLOBAL WARMING PARADIGM

When a climate scientist or other specialist is invited to contribute to the United Nations' Intergovernmental Panel on Climate Change, it is a singular honor. That individual is being recognized as one of the top scientists in his or her field by other top scientists. It's a bit like joining a very exclusive club. But, as with every other exclusive guild, there is a price to be paid in joining the IPCC—you must accept the rules of the club. Challenge the rules and you risk becoming an outsider, perhaps even a pariah. The basic rule of the IPCC club is that global warming is primarily human-caused. The IPCC is not "open" (in the sense that philosopher of science Karl Popper uses the term, as discussed below) to other possible theories of climate change. Anthropogenic climate change is the rule under which the IPCC and the global warming alarmist movement in general operates and that they defend against all competing theories.

This chapter will argue that the global warming alarmist movement, both scientific and lay (i.e., the various environmental groups), is operating on a "closed" model of science. In other words, alarmist climate science is defending what science historian Thomas Kuhn, in *The Structure of Scientific*

Revolutions, called a paradigm, a fixed position, rather than opening itself up to being proved false by offering testable hypotheses (as described in Chapter 6). Or, if a testable hypothesis is offered, the paradigmists won't abandon the hypothesis if it is proved false (there will be more on falsified alarmist hypotheses in Chapter 10). In short, the debate over the causes of global warming is a battle between two views of science. One, preferred by Karl Popper, is "open" to competing hypotheses; the other, as described by Kuhn, is basically "closed" to all but its own hypothesis.

In *The Structure of Scientific Revolutions*, which is reportedly one of the most cited sources in all of scientific literature, Kuhn popularized the term "paradigm" and the idea of "paradigm shifts." A paradigm is a way of seeing the world, "a common theoretical and methodological framework within which meaningful scientific problems can be posed and solved."[3] Paradigms are essential for collecting and organizing information; we can't do without them, either in science or in daily life, because the alternative is chaos. Therefore, a paradigm is the filter of *belief* through which we organize (select, evaluate and criticize) what we believe are the "facts." This is illustrated by a sign outside a church that read: "The atheist believes what he sees; the Christian sees what he believes."[4] But we all, ultimately, see what we believe, which is why two scientists (or two anyones, for that matter) can reach completely different conclusions from the same set of facts. As Albert Einstein noted, "For science can only ascertain what *is*, but not what *should be*, and outside of its domain value judgments of all kinds remain necessary."[5] [italics Einstein's]

In other words, since paradigms are rooted, as Kuhn says, in a "theoretical and methodological *belief*," then, like the Christian described by the sign outside the church, we literally see what the paradigm predisposes us to see. However, if we can't do without paradigms, we can and should be critical of what a paradigm encourages us to see, rather than blindly accepting what we see as "settled" and "certain" fact. This ability to be critical is the difference between the dogmatist and the thoughtful believer, outright skeptic, or scientist.

That scientific paradigms are ultimately underpinned not by scientific data but by beliefs is crucial to one of the central arguments of this book—that global warming alarmism is based more on belief (political, environmental and religious ideology) than on scientific data. How do we know? In part, because there are no scientific "facts" about the future, only guesses based on beliefs (really, computerized guesses) about what the future might hold. These beliefs may turn out to be true, or they may end up being false, but in the present we *cannot know* how our predictions will turn out. And yet, alarmist climate science does make this claim to certain knowledge: it states absolutely, at least to the public, that the future *will be* disastrous. The alarmist prescriptions—what they believe we *should* do—are ultimately based on this predicted future, which may be highly informed by present data but is still, in the end, the product of an informed guess, not scientific facts. It would be equally valid to say of the climate skeptics like myself that we are skeptical because of our underlying belief structures, because skeptics don't *know* what the future will bring, either.

As Bertrand Russell put it, "Truth consists in some form of correspondence between belief and fact."[6] The key question when it comes to climate change is this: which of the two belief systems about climate change, alarmism or skepticism, is closer to what might be regarded as scientific fact? For, as Kuhn writes: "To be accepted as a paradigm, a theory must seem better than its competitors," although "it need not, and in fact never does, explain all the facts with which it can be confronted."[7] Therefore, "any description must be partial."[8] This means that "research [is] a strenuous and devoted attempt to force nature into the conceptual boxes supplied by professional education,"[9] but, inevitably, some facts must be left out. In the end, we (all of us, not just scientists) fall back on an underlying belief system (the "conceptual box") that, we firmly believe or hope, fills in as accurately as possible what we don't know and perhaps can never know. For example, some of our ancestors thought the stars were the campfires of hunters who had died; they had no scientific explanation for those lights in the sky and the mind abhors a vacuum. Today we know that stars are balls of burning gases. The hunter had one story (paradigm) to explain the stars, the scientist has another. And the scientist believes (rightly) that the burning-gas paradigm explains more of the facts about stars than the campfire theory, however appealing the image of the dead warming themselves in the heavens may be. One theory is scientific, the other is religious or mythical, and each paradigm—science and religion/myth—has its place.

MIXING RELIGION AND SCIENCE

That said, the history of Western civilization has shown that, at least from the scientific perspective, it's not a good idea to mix science and religion. Indeed, the historic task of science in Western civilization has been to separate itself as much as possible from religious and mythical belief to attain a more "objective" view of the world. The process of separating science and religious belief began in earnest with Francis Bacon (1561-1626), whose work *Advancement in Learning* "amounted to a declaration of independence on the part of scientific rationalism, which sought emancipation from myth and declared that it [science] alone could give human beings access to truth."[10] We will deal with the pros and cons of this rationalist, scientific attitude vis-à-vis religion in Chapter 11. We've already seen Al Gore's negative assessment of Bacon and scientific rationalism.

The classic example of a "paradigm shift" in the Western world is, of course, the move from the religious dogma that the sun revolves around the earth to the scientific fact that the earth revolves around the sun. The Catholic Church of the Middle Ages was so threatened by the Copernican (sun-centred) world view that it sometimes imprisoned and even executed heliocentric believers. This persecution occurred because the Church made the mistake of identifying a *spiritual* belief with what is properly the domain of *materialist* science, the laws governing physical bodies. Therefore, when the science changed, Catholic belief was also under threat and the church fought back. Darwin's discovery of natural selection was a major paradigm shift in the 19th century, as was Einstein's relativity theory in the 20th.

PARADIGMS AS SCIENTIFIC 'DOGMA'

In a paradigm shift, Kuhn identifies two processes. One is the tenacious clinging to an old paradigm by its believers—the Medieval Church in the Copernican example—because "no part of the aim of normal science is to call forth new sorts of phenomena; indeed, those that will not fit the box *are often not seen at all*. Nor do scientists normally aim to invent new theories, and *they are often intolerant of those invented by others*."[11] [italics added] We will look at some examples of this intolerance in the alarmist warming paradigm later in this chapter. The second process begins after the new paradigm is accepted by most scientists in a discipline because the new theory explains the facts better than the previous theory. At this point, those who "cling to one or another of the older views ... are simply read out of the profession, which thereafter ignores their work."[12]

So, a paradigm operates as a kind of *scientific* dogma, a "drastically restricted vision ... born from confidence in a paradigm."[13] And this scientific dogma is not a bad thing in and of itself. Popper states that "there is no such thing as an unprejudiced observation," and that "there can be no critical phase without a preceding dogmatic phase, a phase in which something—an expectation, a regularity of behavior—is formed, so that error elimination can begin to work on it."[14] That is, once created, these scientific dogmas should be regularly and rigorously tested and rejected if falsified; if not falsified, they should be provisionally (but only provisionally) accepted. But note: the *scientific* paradigm is different from a *religious* paradigm because the scientific paradigm has, or ought to have, a built-in mechanism that makes it self-critical in a way that religious belief is not so that the scientific paradigm is discarded "whenever [it] ceases to function effectively."[15] In other words, as science works within the paradigm, science also aims to continuously challenge its own dogmas, and this critical attitude should eventually make a scientist aware when he or she is clinging to a paradigm that no longer works. A scientific paradigm should remain, to some extent, always "open" to challenge. This, at least, is the ideal.

However, in practice, as British sociologist Steve Fuller notes in *Kuhn vs. Popper: The Struggle for the Soul of Science*, too much attachment to a paradigm may also lead to dogmatic, unjustified rejection of other paradigms because they operate from a different belief system. Thus, the dominant paradigm is "closed" and literally blind to other possibilities. For Kuhn, this was largely a necessary blindness; he felt that without it no real scientific work would be done because there would be no basis on which to choose what problems to study.[16] In other words, as Fuller rightly observes: "It is easy to forget that *both science and religion are preoccupied with justifying beliefs.*"[17] [italics added] A paradigm, then, ultimately rests not on data but on a belief system.

'OPEN' AND 'CLOSED' PARADIGMS

For Popper the "closed" nature of the paradigm that Kuhn describes is not desirable. In *The Open Society and Its Enemies,* Popper describes a closed society as resembling "a herd or a tribe in being a semi-organic unit whose members are held together by semi-biological ties—kinship, living together, sharing common efforts, common dangers, common joys and common distress."[18] There is little personal responsibility in the closed society—all moral decisions are based on tribal tradition, expressed in taboos, and the individual's only responsibility is to follow the tradition. In short, the tribe is an example of groupthink, which we will discuss in more detail in Chapter 12. While this lack of personal responsibility reduces freedom, it has one huge psychological advantage—"certainty."[19]

By contrast, what does an "open" society look like, for Popper? "The open society is one whose members, like the citizens of classical Athens, treat openness to criticism and change as a personal ethic and a civic duty."[20] Popper believes that "[Western] civilization has not yet recovered from the shock of its birth—the transition from the tribal or 'closed society,' with its submission to magical forces, to the 'open society' which sets free the critical powers of man.[21] ... The transition from the closed to the open society can be described as one of the deepest revolutions through which mankind has passed."[22] In this transition from closed society to open society is, really, the birth of Western culture, civilization, and science.

Similarly, what does an open *scientific* society look like? It, too, is open to criticism and change as a professional ethic and a scientific duty—indeed, this welcoming of criticism is at the heart of the scientific method and ethos. Popper describes the right scientific attitude toward an hypothesis as follows:

> If you are interested in the problem which I tried to solve by my tentative assertion, you may help me by criticizing it as severely as you can; and if you can design some experimental test which you think might refute my assertion, *I shall gladly, and to the best of my powers, help you to refute it.*[23] [italics added]

Nor should the open scientific process try to crystallize its findings into the kind of certainty that Popper identified as a characteristic of totalitarian, "closed" societies, and that Christopher Essex and Ross McKitrick have called the Doctrine of Certainty. To repeat a comment by Bertrand Russell that appeared in Chapter 1: "Dogma, regarded as unquestionably true ... is incompatible with the scientific spirit, which *refuses to accept matters of fact without evidence*, and also holds that *complete certainty is hardly ever attainable*."[24] [italics added]

One aspect of an "open" scientific paradigm (if this isn't a contradiction in terms) would be the freedom of its members to explore non-paradigmatic avenues of research without the risk of being expelled from the club. This freedom is based on Popper's template of Athenian democracy in which citizens were *expected* to speak their minds, even if their views differed from the majority—not to speak up was to fail in one's duty as a citizen. This failure, this putting of self-interest ahead of the public interest, the Greeks called *stasis*. *Stasis* "undermines the ideal that citizens, unconcerned about the material consequences of their speech, will be free to think openly about what is in their city-state's best interests."[25] A modern translation of *stasis* could be "corruption," although today we use the word "stasis" to indicate immobility or stagnation. Therefore, an "open" scientific paradigm would allow its members to pursue avenues outside the paradigm without the risk of having their careers and incomes destroyed. As Fuller notes:

> The ideas at stake would be sufficiently detached from the decider's personal circumstances that neither secular power nor financial advantage is bound to them. Only then can the ideas be considered solely on their merits. The alternative is the scientific equivalent of *stasis*.[26]

ALARMIST CLIMATE SCIENCE AS A 'CLOSED' PARADIGM

What has all of this to do with climate change? It is this: the alarmist "consensus" paradigm of climate change follows in almost every respect Popper's description of a "closed" society in *stasis*. As we've seen, induction into the Intergovernmental Panel on Climate Change involves accepting the club rules, including the belief that humans are the principal cause of the late 20th-century global warming and that we are heading into a catastrophe. For most scientists contributing to the IPCC reports, this agreement probably isn't a problem—they were chosen because they already believe in the alarmist paradigm. They may or may not be aware that they are also boarding a scientific "bandwagon," as defined by climatologist Donald R. Prothero:

> One of the striking sociological features of science is a well-known

> political and social activity called "jumping on the bandwagon"—latching on to a trendy idea when it is rolling to get a free ride. Politicians do it when a particular candidate, party, or ideology has irresistible momentum to put them in power. Scientists do it when a particular theory or idea gives one a better change of getting a grant, publishing a paper, or landing a book contract.[27]

In other words, if as a climate scientist you *don't* jump onto the anthropogenic global warming bandwagon, many important career doors will be closed to you. You may be ostracized (another useful Greek term) by your colleagues as a "denier" and may, indeed, find it difficult to get a grant, publish a paper, or land a book contract. For example, meteorologist and physicist Hugh Ellsaesser writes:

> Among non-climate modelers—which includes a good portion of the meteorological profession—there are many who doubt the consensus view on climatic effects of CO_2 but their opinions rarely appear in print. The ones who did get their doubts in print ... have been severely attacked by the establishment.[28]

In academia, as I know from personal experience, the art of getting research funding consists in telling the funders exactly what they want to hear. At universities, workshops are held to coach grant-seeking professors and graduate students on how to push just the right buttons with funders, which are usually governments or government-funded bodies. One of those buttons right now is global warming. Proposals for climate research grants that don't support the human-caused global warming hypothesis are likely to be rejected; as a result, the careers of climate scientists and would-be scientists who don't accept the alarmist paradigm can be ruined or stillborn.

Global warming alarmists like Al Gore are fond of pointing out that the vast majority (100 per cent for Gore) of published, peer-reviewed papers support anthropogenic warming. But which came first, the chicken or the egg? Are there more papers on the anthropogenic global warming side because the theory is so obviously true that the vast majority of scientists agree? Or are there more papers on this side because the paradigm—the club—says that if you don't agree with the AGW paradigm, you won't get published? Based on the evidence, the latter seems more probable—play by the club rules or you will find publication very difficult and will therefore academically perish. And, indeed, this attempt to stifle "denier" publications is exactly what we find in the Climatic Research Unit, "Climategate" emails revealed to the public in November 2009. For example, in one email, IPCC author Phil Jones, then head of the CRU, writes of two papers by skeptics: "I can't see either of these papers being in the next IPCC report. Kevin [Trenberth] and I will keep them out somehow—even if we have to redefine what the peer-review literature is."[29] In another email, climatologist Tom Wigley discusses proposals by IPCC author Mike Hulme to get the editorial board of the peer-reviewed scientific journal *Climate Research* to resign because the journal has been publishing articles by skeptics.[30]

CREATING 'PROBLEMS' TO GET RESEARCH GRANTS

There is another factor at work in scientists' search for research money: if there's no problem, there's no need for research, and therefore no need for the research grants that are the lifeblood of a scientific career. As Essex and McKitrick observe, it is very much in the financial interest of climate scientists to agree that there is a *serious* problem—human-caused climate change—because the more serious the problem, the more money that will come their way. For example, Essex and McKitrick note that the Canadian government offered $50-million a year for studies that "support early actions to reduce greenhouse gas emissions" and that backed the Kyoto Protocol's call for emission reductions. Research that suggested greenhouse gas emissions *aren't* a major problem clearly won't qualify for a grant. "With the intended outcome so firmly fixed ahead of time, why bother with research, other than to feed back onto itself?"[31]

Furthermore, having gathered and then spent all this money on researching a problem, a scientist is naturally reluctant to report that there really isn't a problem after all. First, the scientist might look silly. Second, it's *always* possible to find something that could be done better. Third, the scientist's funding would dry up (no problem, no funding). And so paper after paper comes out saying there is a major climate-change problem. It's not that scientists are dishonest; they're just working within the paradigm, and the paradigm—the alarmist Doctrine of Certainty—says humans are causing climate change, it's going to be a disaster, and we need lots of research (and therefore research money) to study the problem. This doesn't mean there isn't a real problem around global warming—we don't have enough evidence to make a judgment one way or the other. It means that whether global warming is a problem or not, there is a paradigmatic imperative (the bandwagon) to identify a researchable climate-change problem, which will then get funding—why would governments or foundations fund a topic that isn't seen as a problem?—which will then produce scientific papers that, not surprisingly, discover a problem that requires more research, needing more funding, and so on.

Even this self-serving cycle need not be a major issue if the normal scientific checks and balances on this research were in operation, i.e., scientists and lay critics who were not afraid to speak out when they saw the process going bad, such as veering into certainty when none exists or into fear-mongering not justified by the facts. However, when it comes to the global warming paradigm, as we've seen and will see further below, those who dare to criticize are instead marginalized, derided, insulted and dismissed as "deniers." In other words, on the topic of climate change the normal rules of civilized scientific discourse have been suspended and corrupted, and we are left with publicly funded scientific *stasis*. For example, Reginald Newell, a meteorologist at the Massachusetts Institute of Technology, in 1989 "lost National Science Foundation funding for data analyses that were failing to show net warming over the past century. Reviewers suggested that his results were dangerous to humanity."[32] Similarly, Oregon deputy state climatologist Mark Albright was fired from his voluntary position by the state governor for disputing, with facts, the contention that Oregon's mountain snowpack had declined 50 per cent. Unfortunately for Albright, his facts contradicted the state's official position that the snowpack was melting at an unprecedented rate thanks to global warming.[33] Not even solid empirical evidence, apparently, will save you when faced with the power of the global warming alarmist paradigm.

Chapters 3 and 4 mentioned several scientists whose funding had been reduced or who even lost their jobs because of their "immoral" and "irresponsible" position that perhaps global warming might not be entirely human-caused, or at least that other, natural factors should be more seriously explored. Similarly, in May 2007, NASA's chief administrator, Michael

Griffin, made the following statement during an interview with the U.S.'s National Public Radio:

> I have no doubt that a trend of global warming exists. I am not sure that it is fair to say that it is a problem we must wrestle with. To assume that it is a problem is to assume that the state of Earth's climate today is the optimal climate, the best climate that we could have or ever have had and that we need to take steps to make sure that it doesn't change. I guess I would ask which human beings—where and when—are to be accorded the privilege of deciding that this particular climate that we have right here today, right now is the best climate for all other human beings. I think that's a rather arrogant position for people to take.[34]

To any objective observer, Griffin's comment makes perfect sense. Is our current climate the optimum, the ideal, the Eden that we should seek to preserve at any cost? After all, as we saw in Chapter 2, the Eocene era (55-33 million years ago), when temperatures and CO_2 levels were much higher than today's, is frequently described as a "paradise." And, yes, who is to decide the perfect climate? But Griffin's words were immediately denounced as "incredibly arrogant and ignorant" by James Hansen, head of NASA's Goddard Institute for Space Studies and thus Griffin's employee.[35] In other words, for Hansen, it is wrong *to even ask these questions*! If this is Hansen's idea of science, then he is working within a "closed" paradigm rather than one that leaves itself open to challenge, criticism, debate, and argument. For Hansen, the issue is settled; let there be no more discussion. Surely, if there is "an arrogant position for people to take," it is Hansen's, not Griffin's.

WHO GETS THE FUNDING?

If a scientist is foolish enough to question the alarmist climate paradigm, he or she will likely be accused of being funded by the large oil companies to spread false science about global warming. This is the charge journalist Ross Gelbspan makes again and again in his 1997 book *The Heat Is On: The High-Stakes Battle Over Earth's Threatened Climate.* Gelbspan basically accuses any scientist not agreeing with the "consensus" of being either blindly ideological or bought off by corporate interests. Ironically, he also accuses the skeptics of *ad hominem* arguments against their opponents. Yet,

Gelbspan also quotes a union lobbyist as noting: "Everything stems from the assumption that the earth is getting warmer and the cause are [greenhouse gases]. ... But when I read opposing [skeptical] articles ... they're actually ... more persuasive than the others [the alarmists']."[36] To the unbiased observer (i.e., someone who isn't benefiting from the anthropogenic global warming paradigm), the skeptical position usually makes more sense than the alarmist position. For Gelbspan, on the other hand, anyone who doesn't buy into the consensus is a conniver or a dupe. Again, a more elitist and arrogant (i.e., "closed") attitude is hard to imagine.

In its Aug. 13, 2007, issue, *Newsweek* magazine revisited the charge of corporate funding to promote false science with a cover headline that read: "Global Warming Is A Hoax: Or so claim well-funded naysayers who still reject the overwhelming evidence of climate change."[37] Again, the alarmists' blanket charge against anyone who dares to challenge their science (i.e., paradigm) is that they are in the direct pay of large corporations like the oil and gas companies and therefore basing their ideas on ideology and personal gain rather than science. Yet, in the next issue of *Newsweek*, one of the magazine's senior columnists, Robert J. Samuelson, took the extraordinary step of criticizing the article, writing: "*Newsweek* implied, for example, that Exxon Mobil used a think tank to pay academics to criticize global warming science. Actually, this accusation was long ago discredited, and *Newsweek* shouldn't have lent it respectability." Samuelson also noted:

> We in the news business often enlist in moral crusades. Global warming is among the latest. Unfortunately, self-righteous indignation can undermine good journalism. Last week's *Newsweek* cover story on global warming is a sobering reminder. It's an object lesson of how viewing the world as "good guys vs. bad guys" can lead to a vast oversimplification of a messy story. Global warming has clearly occurred; the hard question is what to do about it.[38]

What Samuelson attacked head-on was the *moralistic* attitude of the alarmist paradigm toward global warming, arguing:

> As we debate [climate change], journalists should resist the temptation to portray global warming as a morality tale—as *Newsweek* did—in which anyone who questions its gravity or proposed solu-

tions may be ridiculed as a fool, a crank or an industry stooge. *Dissent is, or should be, the lifeblood of a free society.*[39] [italics added]

The irony here is that the vast majority of the funding for climate research goes to the alarmist believers: more than $5-billion a year in the United States alone.[40] The funding from the evil oil companies and industrialists? ExxonSecrets, an organization dedicated to revealing funding by ExxonMobil to climate "deniers," reports that Exxonmobil funded climate "denier" organizations to the tune of almost $23 million from 1998-2005.[41] That's an average of less than $3 million a year, going to, ExxonSecrets says, 41 organizations, for an average per institution of $73,000 a year. Let's compare these figures. Money to warming alarmists, $5-billion a year. Money (at least from Exxon) to warming skeptics: $3-million a year. If the skeptical climate scientists were as cynical and money-grubbing as Gelbspan and *Newsweek* present them, they'd get on the alarmist gravy train, because that's where the real money is. For example, Greenpeace itself, which published the ExxonSecrets figures, has gotten almost $2-billion—yes, $2-billion—in donations from 1994 to 2005 thanks to the often unsubstantiated environmental fears it deliberately creates.[42] Clearly, again, the big money is in climate alarmism, not climate skepticism. Also worth noting: many alarmist climate scientists, too, get oil and industry money for their research. For example, alarmist climate scientist Walter Broecker notes in his book *Fixing Climate* that he got money from Exxon for research into ice ages.[43] Apparently Broecker didn't fear this "dirty" money would bias his findings. Similarly, the head of the IPCC, Rajendra Pachauri, was from 1999-2002 on the board of directors of Indian Oil.[44] Since corporate directors normally get a rather generous stipend, Pachauri was also funded by Big Oil even if, as he says, he gave his stipend to charity.

AGW PARADIGM VS. BJORN LOMBORG

Scientists should be beyond this moralistic demonizing, which is based on the Doctrine of Certainty. In practice, when it comes to global warming, too often they do not resist this temptation. The shameful treatment of Danish statistician Bjorn Lomborg is one of them most blatant examples of the attempt by the alarmist "consensus" paradigm to silence critics on moral or political rather than scientific grounds.

Lomborg describes himself as a left-wing Greenpeace member who believed the end-of-the-world rhetoric the environmental alarmist movement puts out. One day he read an interview with economist Julian Simon that challenged the doomsday theories and suggested that by almost all measures Western society was becoming better, not worse (for more on this environmental improvement in developed nations see Chapter 9). Skeptical but intrigued, Lomborg decided to check out the accuracy of Simon's unorthodox ideas for himself. He and a group of his students compared the statistics put out by the United Nations and other official organizations with what the environmentalists were saying. Here's what they discovered:

> Honestly, we expected to show that most of Simon's talk was simple, American right-wing propaganda. And yes, not everything he said was correct, but—contrary to our expectations—it turned out that a surprisingly large amount of his points stood up to scrutiny and conflicted with what we believed ourselves to know. The air in the developed world is becoming less, not more, polluted; people in the developing countries are not starving more, but less, and so on.[45]

The result of this open-minded research was a thick, comprehensive, but very readable book called *The Skeptical Environmentalist: Measuring the Real State of the World.* Based on statistical evidence, most of it from IPCC and UN sources, Lomborg's book finds false many if not most environmentalist and climate alarmist charges, such as massive species extinctions and deforestation.

Lomborg says he expected a lively debate when his book came out; what he got was implacable hostility:

> To begin with, I was surprised that the only reaction from many environmental groups was the gut reaction of complete denial. Sure, this had also been my initial response, but I would have thought as the debate progressed that refusal would give place to reflection on the massive amounts of supportive data I had presented, and lead to a genuine reevaluation of our approach to the environment.[46]

Needless to say, this didn't happen. Instead, Lomborg was labeled a traitor to both the environmental cause and to science itself, to the point where *Scientific American* ran a special 11-page section of four articles attacking Lomborg's optimistic ideas,[47] and then allowed him only a one-page reply in the magazine.[48] Even more incredibly, several alarmist scientists persuaded the Danish Committees on Scientific Dishonesty to review Lomborg's work. The Danish committee came up with a finding in 2003 that criticized *Skeptical Environmentalist* for "systematic one-sidedness in the choice of data and line of argument," but without giving any actual examples of "one-sidedness" and without examining the truth or falsehood of Lomborg's arguments.[49] One wonders how Lomborg could be found guilty of falsehood without a definition of truth, and why "one-sidedness" is a scientific sin for a skeptic but not for an alarmist.

The Danish committee's decision made little sense—no one reading Lomborg's book can fail to note that he bends over backwards to be fair to those who don't share his views[50]—and the finding was widely criticized by scientists and lay observers alike. For example, an international group of scientists published an assessment of the Danish panel's condemnation of Lomborg, writing:

> None of Lomborg's criticisms on the quoted major exaggerations [alleged by those who had complained to the Danish committee] were directly and effectively challenged. The reply of the opponents is largely restricted to areas where Lomborg presented, in their opinion, *a too optimistic view of future environmental developments*. In other words, to examples where Lomborg made suspected exaggerations himself on the opposite side. Herewith the opponents developed a secondary motivation which is presented clearly in the complaint lodged to the [Danish committee] by one of the major accusers (K. Fog) *as the danger that politicians may become induced to take environmental problems insufficiently seriously.*[51] [italics added]

In other words, Lomborg's critics were using a scientific body to attack more than his analysis of alarmist science, an analysis that they were unable to refute in part because it is *their* science as well, coming as it does from recognized bodies like the IPCC and United Nations. They were also attacking the potential *political* effect of Lomborg's non-alarmist views—an undermining of politicians', the media's and the public's belief in the possibly catastrophic possibilities of global warming. The assault, then, is on Lomborg's optimism rather than his science—which is exactly what we'd expect when a paradigm feels itself threatened but lacks counter-evidence. Don't attack the science, attack the messenger.

By taking Lomborg to the Danish panel, and as we saw in Chapter 1, the alarmists showed very clearly *they don't want politicians, the media or the public to hear any evidence except that which they, the alarmists, provide*, since only this information has (for them) the correct spin. Therefore, the alarmists are attacking Lomborg's *interpretation* of the facts, but trying to make it appear to the public and politicians that they are attacking his *facts* (i.e., that he has his facts wrong when he doesn't). If this isn't a perversion of science—*stasis*—then nothing is. Sadly, as we have seen and will see further, this moralistic attitude is common among supporters of the paradigm of human-caused global warming.[52]

Another of Lomborg's alleged sins, in the eyes of his accusers, is he is not himself a climate or environmental scientist, which he cheerfully admits:

> I have let experts review the chapters of this book, but I am not myself an expert as regards environmental problems. My aim has rather been to give a description of the approaches to the problems, as the experts themselves have presented them. ... The key idea is that we ought not to let the environmental organizations, business lobbyists or the media be alone in presenting truths and priorities. Rather we should strive for a careful democratic check on the environmental debate, by knowing the real state of the world—having knowledge of the most important facts and connections in the essential areas of our world.[53]

Lomborg is a statistician, using data compiled by the consensus climate scientists themselves to dispute some of their conclusions, although, as Lomborg is at pains to emphasize, he does not dispute the view that the planet is warming:

> Global warming, though its size and future projections are rather unrealistically pessimistic, is almost certainty taking place, but the typical cure of early and radical fossil fuel cutbacks is way worse than the original affliction, and moreover its [global warming's] total impact will not present a devastating problem for our future.[54]

Thus, Lomborg's argument is with the alarmist *conclusions*—the evaluations—drawn from the data, not the data itself.

The charge that Lomborg isn't a member of the climate science club also reflects the "closed" scientific paradigm of consensus climate science, in line with Karl Popper's view that academia today rarely accepts "that a non-scientist might criticize science for failing to abide by its own publicly avowed standards."[55] *We* are the experts, the alarmist paradigm says. Criticism shall only come from other experts—unless, of course, the non-expert agrees with the club, as Al Gore does. Then they'll gladly even accept a Nobel Prize with him. While the club is willing to debate issues *within* the paradigm, it does not welcome criticism *of* the paradigm.

For example, one of the scientists who attacked Lomborg in *Scientific American* was Stanford University biologist and climatologist Stephen H. Schneider, a long-time climate alarmist—in 1976, when the global climate was cooler, he published a book entitled *The Genesis Strategy* that warned about a new ice age. Schneider now writes books sounding the alarm about warming. After the *Scientific American* articles appeared, in a sometimes acrimonious debate with Lomborg about global warming on Australian radio, Schneider made pretty clear the elitist nature of his discipline, but more importantly, what a public-relations game the "consensus" really is:

> Well, for those of us who in my case have spent about three decades working with thousands of scientists and policy analysts and others, trying to figure out something about ... the future that we face, not just environmentally but also a whole range of other issues that we call sustainable environmental development, *we end up with a maddening degree of uncertainty*. We end up with scenarios which, if we're lucky, give us mild outcomes and we end up with scenarios that, if we're unlucky, give us catastrophic outcomes.
>
> We *fight amongst ourselves bitterly* about the relative likelihoods of these, and *have virtually no agreement* and now all of a sudden I see in *The Skeptical Environmentalist* the subtitle: "Measuring the Real State of the World," and the person who's a non-contributor to the debate has selected largely out of context the happier news. He's very confident that we're not going to get the more serious outcomes, a confidence that's not based on any significant analysis by him, or any properly balanced citation from the literature.[56] [italics added]

In his *Scientific American* assault on Lomborg, Schneider says about the same thing:

> For three decades, I have been debating alternative solutions for sustainable development with thousands of fellow scientists and policy analysts—exchanges carried out in myriad articles and formal meetings. Despite all that, I readily confess a lingering frustration: *uncertainties so infuse the issue of climate change that it is still impossible to rule out either mild or catastrophic outcomes* let alone provide confident probabilities for all the claims and counterclaims made about environmental problems. ... *uncertainties are so endemic* in these complex problems that suffer from missing data, incomplete theory and nonlinear interactions.[57] [italics added]

This is an intriguing, even astonishing comment from one of the major players on the global warming alarmist side—apparently there is "a maddening degree of uncertainty" within the climate discipline! And, in contrast to, say, Andrew Weaver's claim that there is no uncertainty within climate science about human-caused global warming (see Chapter 1), Schneider believes uncertainties are "endemic." It isn't possible, Schneider says, to rule out "either mild or catastrophic outcomes." This certainly isn't what the alarmists have been telling the public—as we saw in Chapter 6, for Gore there is total consensus and for Weaver no uncertainty at all on the science or the impending catastrophe that global warming will, they believe, produce.

However, while on the one hand Schneider propounds to the public an arch-alarmist view of this apparently undecided and uncertain question of how serious climate change will be, on the other he vehemently attacks Lomborg for daring to take the non-alarmist option. Moreover, if there ever was certainty, for Schneider it should not come from an uppity outsider. Lomborg's reply?

> Pointing out that it seems questionable that I should know better than the people who've devoted their lives to particular areas, though clearly circumstantial, nevertheless looks like a powerful point. Yet, any person who has devoted his or her life to a single issue will naturally come to consider this area one of the most crucial issues, and any problem inside the area will likely be seen as necessary to solve. And this is exactly my point – we should take the science of these

people seriously, but we should not uncritically adopt their *evaluation* of the problems.[58] [italics Lomborg's]

In other words, it's possible for experts to become tunnel-visioned, victims of tribalistic groupthink, a point we will examine at greater length in Chapter 12. The experts become so trapped in their paradigm (in this case, the alarmist paradigm) that they are literally incapable of seeing the validity of any other point of view.

It's also worth noting that in *The Skeptical Environmentalist*, Lomborg never says the world is perfect: "Note carefully what I am saying here: that by far the majority of indicators show that mankind's lot has *vastly improved*. This does not, however, mean that everything is *good enough*."[59] That the world isn't perfect but may be improving is a fact that environmental and climate alarmists prefer to deny, downplay or ignore because it harms their case. Since the *facts* don't support their *theory*, they are forced to fall back on bluster, name calling, *ad hominem* attacks, and the phony claim of "consensus."

To repeat: the big problem the alarmists have with Lomborg's book is that he uses the same data they do (official statistics), but draws different, less alarming conclusions from the data. Because those conclusions don't support the reigning alarmist paradigm, Lomborg must be discredited and expelled from the scientific club. Meanwhile, by only recognizing, listening to and allowing debate between "recognized" experts (i.e., those within the paradigm), the paradigm controls (or tries to control) what will be heard by the public, media and politicians in order to perpetuate the paradigm's existence and funding.

The Lomborg case has been described in some detail because it raises the question of how "open" alarmist climate scientists are to hearing reasonable arguments that question their "consensus" paradigm—i.e., how open they are to the lifeblood of good science, which is open inquiry and testing of the evidence. The record in the Lomborg case and others I've described shows that many climate scientists, like Schneider, Hansen and Weaver, are more concerned to stifle the global warming debate that engage in it.

In short, the alarmist climate paradigm meets all the criteria for what Popper called a "closed" society, or, in this case, a closed, tribalized scientific paradigm: it is actively hostile to criticism because it believes it has

The Truth (the Doctrine of Certainty). And, when criticized, it fights back viciously, rather than respecting reasonable opposing points of view.

MIXING SCIENCE AND POLITICS

Further, as we saw above in the *Scientific American* attack on Lomborg, the alarmist paradigm's attacks are based not on science but on politics and ideology—a perilous practice for scientists. As climatologist Roger A. Pielke, Jr. of the University of Colorado notes in his analysis of the attack on Lomborg's work: "If scientists evaluate the research findings of their peers on the basis of political perspectives, then 'scientific' debate among academics risks morphing into political debates."[60] Pilke, Jr., suggests the subtitle for the *Scientific American* articles should have been not "Science defends itself against *The Skeptical Environmentalist*" but "Our *political perspective* defends itself against *The Skeptical Environmentalist*."[61] [italics added] And he warns that this mixing of science and politics is a dangerous strategy for scientists:

> That some scientists engage in political activities is neither new nor problematic; they are after all citizens. A problem exists when, in the case of their opposition to [*The Skeptical Environmentalist*], scientists *implicitly or explicitly equate scientific arguments with political arguments,* and in the process reinforce a simplistic and misleading view of how science supports policy. In the process they damage the potential positive contributions of their own special expertise to effective decision making. *Scientists seeking political victories through science* may find this strategy expedient in the short term, but over the long run it may diminish the constructive role that scientific expertise can play in the policy process. [62] [italics added]

"At best," Pilke adds, this mixing "will further confuse already contentious and complex public debates. At worst, it is *an unethical misrepresentation of personal values as if they were science information.*"[63] [italics added] Or, as the saying goes, a scientist's opinion is not necessarily a scientific opinion. This mixing of personal politics and science is precisely what Lomborg's *Scientific American* critics have done, and precisely what many if not most of the attacks on non-alarmist, skeptical climate science have done. The alarmists can't find fault with (falsify) the skeptics' *science* (because it is often *their* science), so they condemn their opponents' politics, source of funding or personal character. This, too, is *stasis*, a corruption of science.

Yet, the alarmist paradigm's lack of civility and respect for differing, reasonably held points of view—to the point where those who disagree feel they cannot speak up without threat to their careers and reputations—is baffling. Lack of respect for opposing views is the life-blood of *politics*, which is a zero-sum game in which one party wins or loses depending on how well it portrays its opponents as unworthy. But respect for reasonable differing scientific views is an essential part of an open *scientific* process, even if the competition to achieve dominance for one's scientific paradigm remains fierce.

Furthermore, the alarmists protest too much when confronted by skeptics like Lomborg. Obviously, he's hit a nerve—he's challenging not just their science but their belief system. And so, like the Medieval Catholic Church, the paradigm strikes back, declaring people like Lomborg as heretics. Indeed, in an attempt to ensure that the public would not hear the skeptical side of the argument, some in the paradigm even asked Cambridge University Press to cease publishing Lomborg's *Skeptical Environmentalist* because, although Lomborg uses the same science as the IPCC and other official organizations, the book draws the "wrong" conclusions. Fortunately, Cambridge Press refused. Meanwhile, alarmists like Al Gore, whose films and books are filled with errors, are given a free pass because at least they are endorsing the paradigmic True Faith. All of this points to, in Popper's terms, primitivistic, tribalistic, "closed" thinking, rather than the scientific, rational, and "open" thinking we should expect from science.

Nature has her language, and she is not unveracious; but we don't know all the intricacies of her syntax just yet, and in a hasty reading we may happen to extract the very opposite of her real meaning.

—George Eliot, Adam Bede, Chapter 15

The essence of science is, without a doubt, the ability to predict phenomena.

—Baron Georges Cuvier

Predicting the climate to come is a gambling game that tempts many scientists. Some predictors speak with more prudence and restraint than others, and most admit honestly that they are making an educated guess. Everyone wants to be right, or preferably famous. And to be famous while one is still alive (although he be known as a false prophet later) is seemingly a much-desired satisfaction.

—Gwen Schultz[1]

It's almost naive, scientifically speaking, to think we can give relatively accurate predictions for future climate. There are so many unknowns that it's wrong to do it.

—James Lovelock[2]

Thus, further cooling is likely. Only a fool would gamble it will not continue, for what is being wagered is our future on this planet.

—Lowell Ponte, The Cooling, 1976[3]

Chapter 8

CRYING WOLF: A BRIEF HISTORY OF ALARMISM

Chapter 7 discussed how the alarmist view of global warming has crystallized into a "closed" scientific paradigm—a way of seeing the world that, due to the Doctrine of Certainty, tries to exclude different, less alarmist ways of evaluating the same climate data. For example, alarmist climate scientists tried to suppress Bjorn Lomborg's book *The Skeptical Environmentalist* not because Lomborg was using faulty data—Lomborg used the same IPCC and UN science as the alarmists—but because Lomborg's interpretation of the data was not alarmist. Unlike the alarmists, Lomborg was not crying wolf.

Not that there is anything wrong with crying wolf, scientifically speaking, if there is a real wolf to be alarmed about: alarmists can sometimes be right in their gloomy predictions. Historian Arnold Toynbee's *A Study of History* catalogues more than two dozen civilizations that have collapsed, and Toynbee warns that Western civilization, too, could be next. But looking at the historical record, we see that in predicting the doom of Western civilization in the last few centuries the alarmists have been far more often wrong than right. For example, a recent study examined 26 alarms similar to the global warming alarm, such as scares over DDT, overpopulation and famine, fluoride in drinking water and the like. The authors' conclusions?

> None of the 26 alarming forecasts that we examined was accurate. Based on analyses to date, 19 of the forecasts were categorically wrong (the direction of the effect was opposite to the alarming forecast), and the remaining 7 of the forecast effects were wrong in degree (no effect or only minor effects actually occurred).[4]

In other words, alarmism doesn't have an open line to the truths of history or science, as its practitioners believe—far from it. And, as we shall see in detail in Chapter 12, even intelligent, honest, principled people—as most climate scientists surely are—can and do hold views based not on "science" (the evidence) but on ideological or personality biases toward pessimism or optimism regardless of the facts.

It's useful to look at the alarmists' predictions with Karl Popper's principle of falsification in mind. As we saw in Chapters 6 and 7, well-crafted scientific hypotheses can be falsified—they offer predictions that can be tested. Repeated failures to accurately predict results falsify or at least call into serious question the hypothesis that led to those faulty predictions. With this approach in mind, let's take a brief look at the history of alarmism in the Western world. Have the doomsayers' predictions over the past few centuries and, more recently, decades, been confirmed or falsified? Does the history of alarmism in the West show the doomsayers to be prophetic, or mostly wrong? And if so, are our modern climate alarmists following in a long and not very fruitful tradition?

THOMAS MALTHUS: CRYING WOLF ABOUT OVERPOPULATION

The best-known Western alarmist is probably Thomas Malthus (1766-1834), author of *An Essay on the Principle of Population*, first published in 1798.

Malthus's theory was that as population grows, more and more land must be brought into cultivation to feed the growing numbers. However, since the best land is already being cultivated, increasingly marginal land is brought into production. That means scantier harvests, which means less food, which means that the growing population can't be adequately fed, which leads to famine, illness, or some other disaster that reduces the population to numbers that can be supported. The problem, Malthus believed, was that "population, when unchecked, increased in a geometric ratio, and subsistence for man in an arithmetical ratio."[5]

His theory was simple and eminently logical based on what was known about population and agriculture in 1798: in Europe, rising prosperity plus growing numbers continually overwhelmed the capacity of agriculture to feed those numbers and the result was famine, again and again. Malthus's problem was that he published his book at the end of the 18th century when the world was still in the grip of the Little Ice Age, a tough time for humanity. Also, he didn't foresee, and couldn't have foreseen, the immense productive power of the Industrial Revolution. For the first time in the history of humanity the masses had an alternative to soldiering, agriculture, craftwork or begging—they could work in the factories and mills. Industrial work was grim, but it gave millions work to do and food to eat. Meanwhile, farming also became more productive thanks to the innovation and invention that were both the spark for, and the outcome of, industrialization. In other words, there was an agricultural revolution as well as an industrial revolution.

Today's neo-Malthusian alarmists, such as biologist Paul Ehrlich (*The Population Bomb*) and The Club of Rome (*The Limits to Growth*), make the same mistake as Malthus: they extrapolate the past into the present without allowing for human ingenuity and inventiveness—an ingenuity that, in the Europe of the 19th century, was slowly evolving into a system of technological innovation and scientific discovery.[6] Indeed, if he were alive today, Malthus would, likely to his great astonishment, see population *falling* in all the advanced countries thanks to economic and social factors that his generation would have considered hopelessly utopian: greater prosperity, universal medical care, effective birth control and excellent sanitary conditions, all of which encourage and allow families to have fewer children because parents know more children will survive.

KARL MARX:
CRYING WOLF OVER CAPITALIST EXPLOITATION

Western civilization's next great alarmist was Karl Marx (1818-1883). Although Marx had the greatest respect for capitalism's *productive* powers, Marx severely underestimated its *adaptive* powers. Unlike Malthus, Marx was well aware of the Industrial Revolution. But, while he acknowledged and even appreciated its powers of material production, he hated capitalism because it made some people (the workers) subservient to others (the capitalists) who, he believed, exploited the workers unmercifully in what Marx saw as a new version of slavery based on wages. For Marx, the development of material abundance through capitalism meant there was no excuse for anyone to be hungry or live in poverty.

Marx's hostility to capitalism was personal as well as ideological: his family's poverty indirectly caused the death of several of his children. And, it is undoubtedly true that the early stages of the Industrial Revolution were very hard on workers. Therefore, Marx perhaps can't be blamed for thinking capitalism could only make life increasingly worse for laborers, whom he called the proletariat after the lower classes of ancient Rome. "Whilst the division of labor raises the productive power of labor and increases the wealth and refinement of society," Marx wrote, "it impoverishes the worker and reduces him to a machine. ... Society in a state of maximum wealth ... means for the workers *static misery*."[7] Elsewhere, he and collaborator Friedrich Engels describe the proletariat as a class "which has to bear all the burdens of society without enjoying its advantages."[8]

Nor did Marx believe capitalism could be reformed: he and Engels wrote that "in order ... to assert themselves as individuals, [the proletariat] must overthrow the State,"[9] which would occur when, inevitably, capitalism had become "intolerable" to the point of producing "in all nations simultaneously the phenomenon of the 'propertyless' mass."[10] Moreover, the revolution would bring an end to psychological alienation because, in communist society, nobody would have to work at anything he or she disliked.[11]

So how has a century and half of experience dealt with Marx's predictions, based on Marxist theory? Robert Heilbroner was a Marxist economist who realized Marxism did not work. The result was an essay in *The New Yorker* magazine entitled "The Triumph of Capitalism." It began:

> Less than seventy-five years after it officially began, the contest between capitalism and socialism is over: capitalism has won. The Soviet Union, China, and Eastern Europe have given us the clearest possible proof that capitalism organizes the material affairs of humankind more satisfactorily than socialism: that however inequitably or irresponsibly the marketplace may distribute goods, it does so better than the queues of a planned economy; however mindless the culture of commercialism, it is more attractive than state moralism; and however deceptive the ideology of a business civilization, it is more believable than that of a socialist one.[12]

It's no coincidence that many in the today's environmental movement hold Marxist or Marxist-inspired views: as we shall see in Chapter 13, it is the immensely productive capitalist system that many of the environmental movements hope to overthrow under the smokescreen of attacking global warming. There is much irony here: as journalist Peter Foster notes, "Whereas the knock against capitalism was once that it impoverished people, now the claim is that it is making them so fecklessly rich that they are destroying the planet."[13] Unfortunately, mass delusions like belief in Marxism are common throughout history and Chapters 11 and 12 will suggest that fears of global warming may fall into the same delusive category: beliefs firmly held in the face of massive evidence to the contrary.

PAUL R. EHRLICH: CRYING WOLF OVER THE POPULATION 'BOMB'

The most prominent of the modern neo-Malthusian alarmists is biologist Paul R. Ehrlich, and one of the bibles of modern eco-alarmism is his 1968 book *The Population Bomb.* Ehrlich's concern wasn't, in 1968, climate *per se* but overpopulation, although he and others believed then that global *cooling*, not warming, by reducing harvests would be one of the factors leading to mass starvation. Therefore, Ehrlich echoes Malthus when he writes: "It is *impossible* to increase food production enough to cope with continued population growth"[14] [italics added]. And, in *Population Bomb*, Ehrlich makes some rather spectacular predictions of doom due to overpopulation. His most famous prediction begins his Prologue:

> The battle to feed humanity is over. In the 1970s the world will un-

> dergo famines—*hundreds of millions of people are going to starve to death* in spite of any crash programs embarked upon now. At this late date *nothing* can prevent a substantial increase in the world death rate."[15] [italics added]

Many people on our planet are, of course, dying of starvation and starvation-related diseases, but in nowhere near the absolute numbers or proportion of population that Ehrlich predicted. On the contrary: in complete refutation of Ehrlich's prediction that "nothing can prevent a substantial increase in the death rate," most of the planet's more than six billion people *are* being adequately fed, more or less, according to the United Nations figures. Not perfectly fed, for sure, but they are not starving in the hundreds of millions that Ehrlich expected, nor are they likely to—unless the planet decides to go into a cooling phase. As history as shown, warm times are much better than cold times for feeding humanity.

So, the 1970s came and went without Ehrlich's predicted hundreds of millions of deaths. Ehrlich stuck to his guns in a 1978 revision of *Population Bomb*, writing: "A minimum of ten million people, most of them children, will starve to death during each year of the 1970s. But this is a mere handful compared to the number that will be starving before the end of the century."[16] He returned to the topic in the 1990 book *The Population Explosion,* writing: "Since 1968, at least 200 million people—mostly children—have perished needlessly of hunger and hunger-related diseases."[17] Doing the math for the 22 years since the first *Population Bomb*, we end up with fewer than 10 million deaths a year at a time when the population is rising exponentially (i.e., the death *rate* due to starvation was falling, not rising), and most of these deaths were due not to starvation as such but "hunger-related diseases" aided by corrupt and incompetent governments. This is a massive tragedy, of course, but these deaths remain, to use Ehrlich's phrase, "a mere handful" compared to the many hundreds of millions he predicted would be starving "before the end of the century." Of course, as we will see below, if Ehrlich had had his way and imposed coerced family planning, as suggested in *The Population Bomb*, billions of children who *didn't* starve or face hunger-related diseases during this time would never have been born.

In the 1990 *Population Explosion,* Ehrlich and his wife Anne ruefully acknowledge that their gloomy predictions didn't come true as they expected, noting: "Grain production in Asia continues to increase faster than the population, partly because population growth rates are lower than in other developing regions and partly because of greater success with Green Revolution technologies."[18] The heading for this section of the 1990 book is a grudging "Asia's Food Production: So Far So Good." Yet, in 1968, Ehrlich foresaw tens of millions of starvation deaths in India alone, and argued that preventing these deaths would be "impossible." And even when Ehrlich dabbles in "optimism," he gets it wrong. "On the bright side," he says in *Population Bomb*, "it is clear that fewer and fewer people in the future will be obese."[19] On the contrary, developed and developing nations around the world (including China and India) are today experiencing an epidemic of obesity because more food is available to more people than ever before in history.[20]

In *Population Bomb*, Ehrlich proposes totalitarian measures to curb the population explosion, just as today's exponents of global warming fears would take away citizens' freedoms to achieve green ends. For example, Andrew Weaver ominously warns that if we don't do what he wants—bring fossil-fuel emissions to zero—"I can't see how our western democracies will survive unless we move toward a War Measures Act or despot style of governance with centralized power in the hands of a few."[21] Similarly, Ehrlich wanted to see population control in the United States "by compulsion if voluntary methods fail,"[22] while forcing other countries to take similar drastic measures in return for foreign aid. Ehrlich even raised, and then rejected, the idea of putting sterility drugs into the U.S. water supply, not because that would be totalitarian but because "the option isn't even open to us thanks to the criminal inadequacy of biomedical research in this area."[23] "Criminal inadequacy" to refuse to do the research that would involuntarily sterilize millions of people? The non-alarmist mind boggles.

Instead, Ehrlich proposed a U.S. government agency called the Department of Population and Environment (DPE) to handle the messy details of forcing Americans to have fewer children and consume less. He suggested that the DPE would have "the power to take *whatever steps are necessary* to establish a reasonable population size in the United States and put an end to the steady deterioration of our environment."[24] [italics added] In common with the global warming alarmists, and in accord with Marxist principles of non-debate and a centralized state, Ehrlich considers his issue so important that human rights and democratic discussion must go by the wayside—the time for talk, as Al Gore and many other climate alarmists say today, was over.

Population *is* a critical issue in the climate debate. As environmental journalist Eugene Linden notes in his 2006 book *The Winds of Change*:

> Whenever climate conditions have permitted human numbers to expand, our species has expanded to and beyond the limits imposed by a hunting-and-gathering way of life, or early agriculture, or irrigation, right up to the present. The proclivity to procreate has periodically left cultures prone to crashes when climate has changed.[25]

However, it's worth remembering that the problem here is not climate change—climate is always changing and there's nothing human beings can do about that. The problem is humanity's ability (or inability) to *respond* to changes in climate, as we have had to do many times in our species' history. That said, it is easier for the human population to respond to warming than cooling; if the planet goes into another serious cooling phase like that of 1940-75 or the Little Ice Age, which could be happening given that the global temperature has not warmed for more than a decade, then we could be in trouble. However, unlike Ehrlich in 1968, the current crop of alarmists are all warning against warming, not cooling, even though, as noted before, warming has *always* been better for human survival than cooling.

Are we heading toward a population catastrophe of the sort Ehrlich envisions whether the earth warms or cools? The United Nations estimates that world population will level off at about nine billion by 2050 and then begin to decline.[26] This decline will occur not because of totalitarian measures of the sort Ehrlich advocates *but because the world is growing more prosperous*—exactly the remedy (affluence) that Ehrlich and many other eco-alarmists oppose.[27] As we've seen, population growth is decreasing because as people become more prosperous they can afford better medical care for their children. Therefore, they don't need a large number of children so a few will survive to look after the parents in their old age. Another reason is that in more affluent societies, women are better educated and have more political power and therefore more control over their own reproduction, aided by technological advances like effective birth control. As it is, the rate of population growth in *all* of the developed nations is slowing to below replacement levels, and the developing countries are beginning to follow suit as they, too, achieve a higher standard of living.

In the 1990 *Population Explosion,* the title of Chapter 1 is "Why Isn't Everyone As Scared As We Are?" Perhaps one reason is that, in the 22 years between *The Population Bomb* and *The Population Explosion, not one* of Paul and/or Anne Ehrlichs' major predictions of disaster came to pass; as hypotheses, *all* have been falsified. As of 2010, the Ehrlichs have been crying wolf for more than 40 years and so far no wolf has appeared. And yet, strangely, the Ehrlichs continue to enjoy an acclaim that, more properly, should go to prophets whose predictions actually pan out.

The weakness of Paul and Anne Ehrlich's "science" is on full display in a book written, ironically, to put the optimists' errors on full display: *Betrayal of Science and Reason: How Anti-Environmental Rhetoric Threatens Our Future*. In an appendix, the Ehrlichs attack what they call the "brownlash," the optimist literature predicting that maybe the future won't be so bad after all. But, like the alarmists' criticism of Bjorn Lomborg's *Skeptical Environmentalist* noted in Chapter 7, the Ehrlichs focus on trivialities because they can't find much that is substantive to attack in the optimist point of view. For example, the Ehrlichs take Gregg Easterbrook's optimistic (Easterbrook calls himself an "ecorealist") and brilliant *A Moment on the Earth* to task for stating that Lepidopterans are "insects of the moth family." In fact, the Ehrlichs gleefully point out, moths are insects of the Lepidoptra family. They criticize Easterbrook for calling the United States "the most carefully studied biosphere in the world."[28] There is only one biosphere, the Ehrlichs primly tell us—the earth itself. True as these criticism may be, aren't they rather picky? In a book like Easterbrook's, with more than 700 pages and thousands of facts, there will be a few errors. Critics of Bjorn Lomborg's *The Skeptical Environmentalist* have found seven errors, all acknowledged by Lomborg, in a book of more than 500 pages. The question is: do these errors destroy Lomborg's or Easterbrook's overall thesis? Yes, errors do damage a thesis—if they are *major* errors like those made by the Ehrlichs or Al Gore. The thesis usually stands if they are picky factual errors as were attributed to Lomborg and as the Ehrlichs direct at Easterbrook.

One of the Ehrlichs' few substantive criticisms is that Easterbrook misquotes Paul Ehrlich in saying Ehrlich predicted, in the 1968 *Population*

Bomb, "that general crop failures would 'certainly' result in mass starvation in the United States by the 1980s."[29] The Ehrlichs are right—Paul Ehrlich predicted mass starvation in the *underdeveloped* countries, not the United States, writing: "If the pessimists are correct, *massive famines* will occur soon, possibly in the early 1970s, certainly by the early 1980s." [italics added] However, in Easterbrook's defence, the clear implication of *Population Bomb* is that the United States would be massively affected by these mass starvations, as indeed it would be if Ehrlich's scary scenario had come true. It didn't. But—and this may have been what Easterbrook was thinking of—Ehrlich does say in *Population Bomb* that "if our [the United States'] current rape of the watersheds, our population growth, and our water use trends continue, *in 1984 the United States will quite literally be dying of thirst.*"[30] [italics added] Presumably, if the United States is "dying" of thirst, so are millions of its citizens, so perhaps Easterbrook hasn't misrepresented Ehrlich so badly after all.

What the Ehrlichs do in *Betrayal of Science and Reason* is attack Easterbrook for a, possibly, misuse of words, while completely ignoring that they had made a much bigger error—a prediction of "massive famines" and U.S. water shortages that haven't occurred, at least within the time frame the Ehrlichs set. In *Betrayal* the Ehrlichs grudgingly admit: "The 'optimists' proved more correct: 'only' some 250 million people have died from hunger-related causes since Paul made that statement [about mass starvation in *Population Bomb*]."[31] Meanwhile, more than six billion people have been fed, something that Paul Ehrlich considered "impossible" in 1968. Moreover, those who have starved didn't die from lack of food or overpopulation but the failure of their countries' political and economic systems to ensure that they got the food available—mass starvation is unknown in democratic nations.

So, whose error is the greater, Easterbrook's or the Ehrlichs'? Whose book would a rational person go to if he or she was looking for substantially accurate *evaluations* of the population or environmental data? Not, surely, the Ehrlichs'. And yet, despite a 40-year record of faulty predictions, "Ehrlich has earned respect and adulation in some circles" and professors continue "to assign the book [*Population Bomb*] to students long after its doomsday scenarios have been proven wrong."[32] And, lest we forget, in

1980 optimist economist Julian Simon bet Ehrlich that the prices of any five key commodities—Ehrlich got to choose the five—would become cheaper, not more expensive (i.e., scarce, which Ehrlich had predicted), by any date that Ehrlich chose over one year from the day of the bet. The prices went down and Ehrlich lost the bet, adding one more to a long string of failed predictions.[33]

Paul and Anne Ehrlich produce apparently highly researched books bristling with footnotes that are very convincing. They make predictions (evaluations) of apocalypse that don't come true, and then have the nerve to complain about optimist writers like Easterbrook who've gotten the outcomes right. The eminent physicist Richard Feynman has said: "If it [the hypothesis] disagrees with experiment it is wrong. In that simple statement is the key to science."[34] The real-life results have, at least so far, falsified the Ehrlichs' hypotheses. It is they, not Easterbrook, who are wrong. It is the alarmist Ehrlichs who are betraying science and reason.[35] This doesn't mean that, eventually, their fears won't come true, just as a stopped clock tells the right time twice a day. But an unbroken record of failed predictions should, for the rational person, cast doubt on the alarmists' basic premises.

JEREMY RIFKIN: CRYING WOLF ON WORK AND CLIMATE

Another perennial alarmist is economist Jeremy Rifkin. In 1995 he brought out a book entitled *The End of Work* that predicted "a near-workerless world" as machines replace human labor, leading to mass unemployment.[36] In 2007, when I began writing this book, almost every shop window in Victoria, B.C.'s, downtown was advertising for staff, in part because so many workers were being snapped up by the higher-paying resource industries in Alberta and B.C.'s hinterland—it wasn't just McJobs on offer.[37] And one of Western society's future dilemmas will be finding workers to replace the host of retiring Baby Boomers. But, then, for decades Rifkin has made a good living predicting the end of Western civilization, so it's not surprising that he jumped very early onto the global warming bandwagon with his 1980 book *Entropy: Entering the Greenhouse World.*

Rifkin begins *Entropy* with a scenario of New York in 2035 in which "palm trees line the Hudson River from 125th Street to the midtown exit. Recently, massive dikes have been built around the entire island of Manhattan in an concerted effort to hold back the rising sea water."[38] Phoenix, Arizona, in Rifkin's 2035 AD, is so hot the city is building an air-conditioned dome. The American mid-west is now a parched desert. And so on. This greenhouse world, Rifkin says, "threatens our survival"[39] and "our high-energy social and economic system is so fragile, so absolutely dependent upon continued inputs of nonrenewable resources, that *monumental collapse could come any time.*" [italics added] Therefore, what we should be heading for, Rifkin says, is a "low-energy society."[40] In other words, like Ehrlich, he not only envisions but encourages a society in which everyone is much poorer but, presumably, much happier.

Meanwhile, here's a reality check: even the alarmist Intergovernmental Panel on Climate Change, which still hadn't been created when Rifkin was writing the first edition of his book, predicts *at most* 59 centimetres (23 inches) of sea-level rise, and by 2100, not 2035. And that's the worst-case scenario; so far, as we've seen, the sea-level rise has been minimal for the last hundred years—just over two mm per year for a total of about 25 centimetres (10 inches) a century—and will likely continue to be so. But, then, writing that the oceans may invade a few inches of Coney Island beachfront isn't as scary as the prospect of New York under water. And, as climatologist Patrick Michaels notes, Rifkin's prospect of New York City becoming Miami North is unlikely given New York's geographic location: "For New York City to enjoy a climate like Miami's, the entire circulation system of the atmosphere and ocean must completely break down and be reorganized."[41] But then, Rifkin is an alarmist economist, not an alarmist geographer.

STEPHEN SCHNEIDER: 'SCARY SCENARIOS, SIMPLIFIED, DRAMATIC STATEMENTS'

Stanford University's Stephen Schneider is notorious for his admission that climate scientists will exaggerate a problem to get the public's attention:

> On the one hand, as scientists we are ethically bound to the scientific method, in effect promising to tell the truth, the whole truth, and nothing but — which means that we must include all the doubts, the caveats, the ifs, ands, and buts. On the other hand, we are not just scientists but human beings as well. And like most people we'd like to

> see the world a better place, which in this context translates into our working to reduce the risk of potentially disastrous climatic change.
>
> To do that we need to get some broad based support, to capture the public's imagination. That, of course, entails getting loads of media coverage. *So we have to offer up scary scenarios, make simplified, dramatic statements, and make little mention of any doubts we might have.* This "double ethical bind" we frequently find ourselves in cannot be solved by any formula. Each of us has to decide what *the right balance is between being effective and being honest.* I hope that means being both.[42] [italics added]

In other words, Schneider believes climate scientists are not bound to tell the whole, complicated and nuanced truth about climate change if they don't have proof but feel a compelling public interest is at stake. Therefore, it's acceptable to cry wolf—"offer up scary scenarios, make simplified, dramatic statements, and make little mention of any doubts we might have"—if scientists sincerely believe, even if they cannot prove, that there is a wolf. Indeed, in his 1976 *The Genesis Strategy*, Schneider writes that concerted political action does not require "knowledge of the exact location of each tree behind which a wolf may be hiding. Rather, knowledge of the *probability* that wolves do lurk in the forest should be sufficient information for deciding whether to take preventive action."[43] [italics added] Curiously, the preventive action that Schneider was urging in 1976 was against a future ice age, not warming, writing: "Today there are few people much concerned by the approach of the next ice age. And since ice ages take thousands of years to develop, why should we worry? There are several reasons to worry."[44] He had the wrong wolf then (cooling), but he's sure he's got the right wolf this time (warming).

AL GORE: 'OVER-REPRESENTATION OF FACTUAL PRESENTATIONS'

Lest readers think Schneider is alone in exaggerating the warming threat, let's look at remarks by two of the biggest players in promoting global warming alarmism, Al Gore and the Goddard Space Institute director James E. Hansen. Gore has said:

> In the United States of America, unfortunately we still live in a bub-

> ble of unreality [about global warming]. And the Category 5 denial is an enormous obstacle to any discussion of solutions. Nobody is interested in solutions if they don't think there's a problem. Given that starting point, *I believe it is appropriate to have an over-representation of factual presentations* on how dangerous it is, as a predicate for opening up the audience to listen to what the solutions are, and how hopeful it is that we are going to solve this crisis. Over time that mix will change. As the country comes to more accept the reality of the crisis, there's going to be much more receptivity to a full-blown discussion of the solutions.[45] [italics added]

Another, more common term for "over-representation of factual presentations" is exaggeration or outright lying, both qualities found in abundance in Gore's film and books. In other words, he is crying wolf. If Gore had evidence of a real wolf, there would be no need to exaggerate the threat, as he does so skilfully in *Inconvenient Truth.*

JAMES E. HANSEN: 'EMPHASIS ON EXTREME SCENARIOS'

James E. Hansen is one of the leading proponents of the catastrophist view of global warming, to the point of claiming that the oceans could rise as much 25 metres (82 feet) in the next century,[46] which is an ultra-extreme view even among global warming promoters. Should we believe him? Here's what he's said about some of his assertions in the past:

> *Emphasis on extreme scenarios* may have been appropriate at one time, when the public and decision-makers were relatively unaware of the global warming issue, and energy sources such as "synfuels," shale oil and tar sands were receiving strong consideration. Now, however, the need is for demonstrably objective climate forcing scenarios consistent with what is realistic under current conditions.[47] [italics added]

Needless to say, "emphasis on extreme scenarios" is another Gore-like euphemism for exaggeration or outright falsehood, albeit, as always with alarmists, in what Hansen sees as a good cause. Lacking scientific proof, Hansen, Gore, Schneider, et al., clearly feel they are in a bind: if they don't exaggerate/lie, they fear the public might not listen attentively. And so the alarmists resort to bluff and bluster, *ad hominem* attacks, and attempts to suppress (rather than answer) arguments against their position. Why? Because they don't have the empirical data, the smoking climate gun, to *prove* their point. If they did, they wouldn't need to mislead, exaggerate and lie. What the alarmists have instead of evidence is a firmly held *belief* that the data, mainly derived from computer models, points to disaster.

How do we know the alarmists don't have proof? Because they *say* they don't have evidence and thus fall back on the Precautionary Principle. As we've seen, climate alarmists like Gore state, quite plainly, that the situation is too urgent to require proof, and one alarmist economist even scorned "the fallacy of requiring proof."[48] Yet, the "fallacy" of requiring proof is the cornerstone of the scientific method—without it, there is no science. And so, while many alarmists cite their scientific credentials as a reason why the rest of us should fear global warming—in logical terms this is known as the Fallacy of Authority—they undermine science by denying the need for convincing empirical scientific proof. Instead, they urge governments to spend billions of dollars to take preventive action against a threat that may never occur, just as Paul Ehrlich wanted the U.S. government to spend billions, and destroy democratic freedoms, to defend against an overpopulation disaster in the United States that never occurred.

For the skeptics among us, comments like these by high-profile alarmists such as Gore, Hansen and Schneider raise the question of how credible these three are in anything they say, or indeed the credibility of any of the alarmist global warming "science." At the very least, an attitude of healthy skepticism seems appropriate toward all alarmist claims, especially when the alarmists are so insistent on shutting down discussion that might cast doubt on their theories. If they have the facts, why not just present them in open debate and make the skeptics look foolish? And, if the facts are not available—as they clearly aren't on the global warming issue or the proponents would not rely so heavily on exaggeration and the Appeal to Authority, i.e., "consensus"—then they could at least openly and honestly acknowledge that these "scientific" fears are *beliefs*, not science.

BELIEF, INTUITION, FAITH AND CERTAINTY ARE NOT EVIDENCE

In their better moments, consensus climate scientists *do* acknowledge that their fears are based on belief or intuition, not fact. Schneider said in 1989:

> We need 10 to 20 years to get an absolutely clear signal [that human-caused global warming is occurring]. I'll be surprised if it doesn't happen, but how do you assign a probability to something *when you have no objective means of doing so?* You base it on physical *intuition* and then state your assumptions. By my *intuitive reasoning*, the greenhouse signal has been detected at an 80% probability. My *faith* is based on the principle of heat trapping by greenhouse gases and the billions of observations that support it. All that objective stuff is based on *assumptions*. The future is not based on statistics, it's based on physics. Objectivity is overplayed.[49] [italics added]

It's worth keeping in mind, again, that Schneider's "intuition" and "faith" told him in 1976, when he brought out *The Genesis Strategy*, that the planet was cooling, not warming. James Lovelock, author of the Gaia hypothesis (more on Gaia in Chapter 11), writes: "We *sense* that our adding carbon dioxide to the air and soon doubling its abundance is seriously destabilizing an Earth system already struggling to maintain the desired temperature."[50] [italics added] Note: Lovelock offers not evidence but a "sense" of impending disaster. Professional academic alarmist Thomas Homer-Dixon tells us that "every shred of my intuition" convinces him we are heading toward disaster, including climate disaster.[51] Andrew Weaver notes, "There's *almost a sense* that there's a forest fire nearby, and you wait until the fire gets to your house before you leap, and that's too late."[52] [italics added] The forest fire is, of course, the global warming catastrophe that Weaver preaches to the public and the fact that he has "almost a sense" that catastrophe *might* occur is, for him, enough of a reason to take very expensive, indeed, economically ruinous, counter-action. Lester Brown, who publishes the yearly alarmist *State of the World*, is reported as saying: "Studying these annual temperature data, one gets the unmistakable *feeling* that the temperature is rising and that the rising is gaining momentum."[53] [italics added] The introduction to a Canadian book on global warming, *Hard Choices*, is written by a professor of philosophy who is also a poet, Jan Zwicky. Zwicky tells us that the weather on her Ontario farm, and other clues, have given her strong "intuitive alarm bells" that global warming is occurring, although she adds that these alarm bells "are not grounds for thinking the climate is changing. They are, after all, merely intuitions. They are not science."[54] Finally, IPCC contributor Mike Hulme has noted, in trying to contain the fallout from the Climategate emails:

> Climate scientists frequently have to reach their conclusions on the basis of the *partial, and sometimes poorly tested, evidence* and models available to them. And when their paymasters—elected (or non-elected) politicians—ask them for advice, as in the case of the IPCC, *opinion and belief become essential* for interpreting facts and evidence. Or rather, *incomplete evidence and models* have to be worked on *using opinions and beliefs* to reach considered judgments about what may be true.[55] [italics added]

However, intuitions, feelings, opinions and beliefs are not empirical science, as climatologist Patrick Michaels emphasizes again and again in his book *Meltdown*, which is skeptical of global warming fears. Michaels cites many cases of media reports on weather and climate in which reporters also "intuit" or "sense" or just assume that the story they are covering relates to global warming, such as changes in the population of butterflies in a particular area. However, as Michaels shows, a bit of research into the actual temperature data reveals warming cannot be the cause of the butterfly declines since the temperature in that region hasn't, by actual measurement, changed.[56] As Michaels concludes: "'Feelings' aren't a good metric for science or [news] reporting."[57]

This doesn't mean that intuition and feeling are to be scorned, especially when they are the intuitions of experts; intuition is an important part of the scientific process. As Carl Sagan notes, "Major scientific insights are characteristically intuitive" and a hunch is often what gets a fruitful line of scientific inquiry going. Sagan adds, however: "In *new areas* ... intuitive reasoning must be *diffident in its claims* and willing to accommodate the insights that rational thinking wrests from Nature."[58] [italics added] Climate science, which is only about 40 years old, surely qualifies as a "new area" and, before we take action, especially extreme action of the sort the alarmists demand, the intuitions of climatologists, alarmist and skeptic, must be backed up by real empirical data. As well, before we accept the intuitions of the "experts," we should feel confident that they are being honest with us. As we saw above, many, perhaps most, leading climate alarmists are *not* being honest, and are relying on exaggeration and, in Gore's case, outright fabrication to appeal to the public. Above all, since they lack evidence, alarmist climate believers ask the public to accept climate change as a "crisis" when the existence of a "crisis" is precisely what consensus climate science needs to *prove* with empirical evidence (in logic, this is known as the fallacy of begging the question). So far, this evidence has not been forthcoming; as Schneider admits, the only "proof" exists in climate models.

So, "sensing" something or "having an intuition" about something or "almost having a sense" of something or having a "feeling" can be valuable—there may well be a wolf out there—but they aren't the same as providing the hard scientific data *proving* there's a wolf out there. For their part, most climate skeptics, including many climatologists, "sense" or "feel" or "have an intuition" that warming won't be that much of a problem and perhaps even a blessing. Are skeptical climate beliefs, too, to be automatically accepted because skeptics really, really believe in what they're saying, or are only the beliefs and intuitions of alarmists valid?

In his *Critique of Pure Reason,* German philosopher Immanuel Kant (1724-1804) had some valuable advice about how to approach firmly held but unproved and perhaps unprovable beliefs. As a product of the European Enlightenment of the 17th century, Kant believed in the power of reason. But, as a Christian, Kant had a serious problem. He was fully aware that reason and evidence could not, and could never, *prove* the existence of God. Yet Kant also wanted to keep a belief in God. Kant's solution was similar to that of the global warming alarmists: in the absence of empirical proof, fall back on strongly held intuition. So, Kant writes:

> My conviction is not logical, but moral certainty; and since it rests on subjective grounds (of the moral sentiment), I must not even say: *It is* morally certain that there is a God, etc., but: *I am* morally certain.[59] [italics the author's]

Similarly, there is no doubt that many climate alarmists have a strong *moral certainty* that what they believe is objectively true and that, if their warnings are not heeded, the world is heading into "oblivion." But, like Kant, they should be intellectually honest and say "*I am*" certain, not, as they do so often to alarm the public, "*it is*" certain. And given the alarmists' moral certainty and sense of urgency, it is almost inevitable that they will be victims of confirmation bias, selecting data that supports this certainty while ignoring data that doesn't, just as Kant found non-rational support for his belief in God and discarded evidence that God doesn't exist. The result for climatology is computer models that, not surprisingly, support the modelers' moral certainty and that are, therefore, likely to be biased rather than objective. But, then, all prophets are working from moral certainty, all believe they have evidence to support their certainties, all believe that the world faces disaster if their warnings are ignored, and almost all of these alarmist prophets have been wrong. Why? Because they lack *empirical evidence* to support their moral certainties. Yet, as philosopher Walter Kaufmann writes: "What distinguishes knowledge is not certainty but evidence."[60]

In a healthy scientific discourse, an honest scientist who lacks empirical evidence will always make it clear to both fellow scientists *and to the public* that his or her views on climate are based on subjective beliefs or intuitions—as Kant put it, not "it is" certain but "I am" certain. Gore, Schneider, Hansen and others do at times admit that what they believe is just that, belief or intuition, as shown in the examples quoted above. Yet these alarmist believers also tell the public most of the time that what they are warning about is scientific fact rather than the product of a moral certainty (i.e., an unproven hypothesis).

LEARNING FROM THE PAST: ALARMISTS ARE ALMOST ALWAYS WRONG

Are the fears of global warming as justified as the alarmists preach, or do the optimists make a stronger case? Since we cannot know the future, we must get what illumination we can from the past. The alarmists have been preaching the economic, political, social and climatic doom of Western civilization for not just decades but centuries. What is their track record? How accurate have they been? If the alarmists have been successful in predicting doom in the past, then perhaps they should be trusted now. On the other hand, if the *optimists* have been consistently wrong, then perhaps their words deserve to be ignored, as the alarmists wish. What does the record show?

The record shows, unequivocally, that the alarmists have almost always been wrong in their predictions. Spectacularly wrong. Incredibly wrong. Almost unrelievedly wrong. In other words, their warnings have been, almost without exception, false alarms. And yet, for some reason, the doomsayers are still taken seriously. Economist Julian Simon has commented that people seem to be inoculated against the facts. The *facts*, as we will see in Chapter 9, show that in both the developed and most of the developing world the lot of humanity has been improving, at least in material ways, for the past century and a half. And, in the developed world, the environment has also been steadily improving for the past few decades. Yet most of the public don't believe this good news because they accept the alarmist propaganda that the planet's environment is deteriorating, aided by the media, which "give much attention to alarmists, but little to those who are skeptical of their claims."[61] Indeed, it's as if human beings would always prefer to believe bad news rather than good news, and this is true of even highly educated people who should know better. For example, Seymour Garte, a professor of environmental studies at the University of Pittsburgh, describes a conference at which a presenter lectured on pollution levels. Garte writes:

> On every slide showing pollution over time, the graph went down. In other words, every year, for every chemical, at every site, and for every method of measurement, the amount of pollution was decreasing. I raised my hand and asked [the presenter] if this was some sort of error or if it reflected reality. He looked at me with the weary patience that an expert in any field feels when asked a stupid question by a non-expert. He explained that of course it was real, "and everyone knows that air pollution levels are constantly decreasing everywhere." …
>
> I looked around the room. I was not the only non-expert there. Most of my other colleagues were also not atmospheric or air pollution scientists. Later I asked one of them, a close friend, if he had known that air pollution levels were constantly decreasing throughout Europe and the United States on a yearly basis. "I had no idea," he said. It was certainly news to me. Even though I was a professor of environmental health and had been actively involved in many aspects of pollution research for many years, *that simple fact had somehow*

> *escaped me.... I had certainly never seen it published in the media.*[62] [italics added]

The alarmist view of the state of the world was understandable at a time when industrialization was in its infancy and horrors, including pollution horrors, abounded. Today, however, when the evidence of our senses shows otherwise—most readers of this book have *good lives* that are, by any material and environmental standard, the best in human history—this alarmism is hard to understand. It's as if alarmism is genetically hard-wired into us, which may well be the case as we will discuss in Chapter 12. If this is so, however, if alarmism is genetically hard-wired into us, then we need to take a step back from ourselves, as it were, and look closely at our genetically hard-wired attitudes, just as we do with our other "instincts" when they get in the way of reason and civilization.

In the meantime, instead of giving doomsayers the benefit of the doubt, as so often happens, readers should be taking a critical look at alarmist claims. If someone tries to sell us something, we are normally a bit suspicious. Is the seller honest? Are we getting the best deal for our money? Is the product well-made or shoddy? What is this company's track record? We have a natural caution when it comes to spending our money in the marketplace. Yet when it comes to the environmental or scientific doomsayers, we throw caution away—they're environmentalists or scientists, after all; how could they be dishonest, mistaken or even deluded? And yet, no country has unlimited money, and, if met, the demands of the climate science alarmists *will* cost us, the public, the taxpayer, the average person, literally billions of dollars that we could be putting to other, far better uses. With so much at stake, surely it is our civic duty to look carefully at these alarmist claims, which includes asking probing questions and doing some comparison shopping. What are some of the other "sellers"—in this case, say, the climate optimists like Bjorn Lomborg—offering? What are their claims? How do their claims compare with the alarmist claims, based on logic and on our experience? Are the optimists offering more sensible ways of spending our money?

Above all, how likely is it that the West's technological civilization is on the verge of collapse, as the alarmists would like us to believe? Is this plausible? Or are the alarmists falling into the black-and-white logical fallacy—presenting us with an either-or picture of all-out disaster if we don't do exactly as *they* wish? Writing at a time, the late 1970s, when there were similar cultish fears of doom, *Time* magazine said of the late science-fiction and science writer Isaac Asimov:

> In a forthcoming book, *A Choice of Catastrophes*, the polymath popularizer [Asimov] seeks to soothe anxieties about global disaster. Says Asimov: "All the scenarios are either very low in probability, or very distant in the future."[63]

It's a pity that the *Time* of the 21st century, like so many other modern media alarmists, hasn't learned from past false alarms that while disaster scenarios may sell magazines, they are, in fact, "very low in probability."

To repeat: *not one* of the environmental or climate alarmist scenarios of the past 200 years has come to pass in the developed capitalist world, although the same can't be said for the communist world, which has been brutal on both its people and its environment. The IPCC warns us that the climate will be warming at an alarming rate over the next 50 to 100 years. Yet, at the moment (2010), the planet is not warming at all, and may not warm again for a decade or more. And even if it does begin to warm again, that is not empirical proof that humans are at fault or that warming will be disastrous. Nor is 0.6° Celsius, the increase over the 20th century, that alarming considering that the planet is coming out of an unusually cold period (the Little Ice Age) and warmed only .06° C per decade. This small increase is presented by the climate alarmists as somehow so threatening that we should bring Western industrial civilization to a halt.

In the next chapter we'll see in more detail that the end of the world not only isn't nigh but that, overall, and thanks to technology and the higher standard of living that many alarmists would like to bring down, the world is becoming a better place for both humanity and the biosphere.

Western civilization has given mankind the only economic system that works, a rationalist tradition that alone allows us material and technological progress, the sole political structure that ensures the freedom of the individual, a system of ethics and a religion that brings out the best in humankind.

—Historian Victor Davis Hanson[1]

Capitalism creates most of our jobs and we would all be desperately poor without the entrepreneurs who keep the economy alive and the financiers who invest in our corporations. But that's no reason, as Canadians see it, to look upon business with anything but suspicion.

—Columnist Robert Fulford[2]

The argument is often made that there is a fundamental contradiction between economic growth and the quality of life, so that to have one we must forsake the other. The answer is not to abandon growth, but to redirect it. For example, we should turn toward ending congestion and eliminating smog the same reservoir of inventive genius that created them in the first place.

—U.S. President Richard Nixon[3]

On what principle is it that when we see nothing but improvement behind us, we are to expect nothing but deterioration before us?

—Historian Thomas B. Macaulay[4]

Throughout the world the poor have very understandably always been more concerned about making money than about protecting the environment. In the past, any protectionist action was invariably taken by a class of society that … [was] well educated and well circumstanced … [and that] could afford such amenities and fine feelings.

—Novelist John Fowles[5]

Chapter 9

A BRIEF HISTORY OF PROGRESS, AND WHY WE DON'T BELIEVE IN IT

Cassandra, daughter of the King of Troy, suffered from a terrible curse: she could accurately prophesy disaster but was not believed. As a result, the term "a Cassandra" has come into the English language to describe someone who predicts disaster, i.e., usually a pessimist or alarmist. Today, optimists have exactly the opposite problem: they predict better times, but few believe them even though the data bears the optimists out. And so, comments continue appear like this one from a letter to the Victoria *Times Colonist*:

> Capitalism doesn't husband natural resources, but has polluted, contaminated, degraded and altered the environment for the worse. You could probably go anywhere in the world where capitalism has visited and find extreme examples of polluted water and air, contaminated soil and habitat destruction.[6]

The headline for this letter is "Capitalism brings global warming," so it's clear that for this writer, anti-capitalist alarmism and global warming alarmism are found together.

Unfortunately for the letter writer's thesis, the most polluted industrial environments on earth belong to the former Soviet communist empire as well as communist China. Non-industrialized peoples can also destroy their environments as we see on the fringes of the Sahara desert, which have been stripped of vegetation by livestock, adding to desertification; in the deforestation of the foothills of the Himalayas, a deforestation that is partially responsible for the sometimes devastating floods in Bangladesh; and in the destruction of endangered species such as gorillas, rhinos and elephants, largely caused by poaching in poorer countries (i.e., countries that are non-capitalist or are still developing capitalism).[7] A slightly more literate version of the letter writer's lament comes from a retired biology professor who complains that the complexity of "industrial and postindustrial society" has resulted "in massive overpopulation, in severe environmental degradation, in the exhaustion of natural resources, in the forced extinction of biological species, to name but a few."[8]

These two writers, and many others, present a grim but widely believed picture. The problem with this picture is that while what it may have been true 150 years ago, or even 50 years ago, the writers are largely wrong in describing the developed countries today. Yes, at some times and places, industry pollutes, contaminates, degrades and alters. But this now occurs primarily in nations where free markets are still a new development, like China and India, or where free markets have never developed at all. And, if capitalism has been the problem when it comes to environmental degradation, it is also the solution. In most of the developed world, as we will see in more detail below, capitalism has meant greater environmental protection for one simple reason: wealthy capitalist societies can afford to clean up their mess. Not only can they afford to clean up their mess but, as we will see, the citizens in wealthy countries *want* to clean up their mess—they are willing to pay more in taxes for a cleaner environment. On the other hand, environmental cleanup is, on the whole, a luxury that the poorer, developing nations cannot afford. That is, as novelist John Fowles has noted:

> Throughout the world the poor have very understandably always been more concerned about making money than about protecting the environment. In the past, any protectionist action was invariably taken by a class of society that ... [was] well educated and well circumstanced ... [and that] could afford such amenities and fine feelings.[9]

MACAULAY VS. SOUTHEY: OPTIMIST VS. PESSIMIST

The environmental and anti-capitalist Cassandras are, of course, nothing new in history. As we saw in Chapter 8, they've been dead wrong in the past, and they're almost certainly wrong now. One of the prominent (for his day) pessimists about the possibilities of capitalism was Robert Southey (1774-1843), the poet laureate of England from 1813 to 1843. Southey was writing at a time when industrial-strength capitalism was just beginning to develop in Britain, producing overcrowded, unsanitary cities, child and female labor, brutally long hours and unsafe conditions for many workers. In response, in 1824, Southey published a two-volume work entitled *Sir Thomas More: or, Colloquies on the Progress and Prospects of Society* that strongly attacked capitalism and looked back to a gentler time before the inhuman factories. In *Colloquies,* Southey (who refers to himself as Montesinos) is visited by the ghost of Sir Thomas More. More (1478-1535) was England's Lord Chancellor under Henry VIII, the author of one of the very first utopian novels, entitled *Utopia*, and the "man for all seasons" in the 1966 movie of the same name. More was beheaded for refusing to accept Henry as head of the Church of England.

Southey's work is, in effect, a dialogue between two religious and political conservatives asserting that More's 16th-century Renaissance times were much better for England than the early 19th century. So, for example, Southey has More ask: "Consider your fellow-countrymen, both in their physical and intellectual relations, and tell me whether a large portion of the community are in a happier or more hopeful condition at this time, than their forefathers were when Caesar set foot upon the island [in 55 BC]?"[10] More then answers his own question:

> I admit that improvements of the utmost value have been made, in the most important concerns: but I deny that the melioration has been general; and insist, on the contrary, that a considerable portion of the people are in a state, which, as relates to their physical condition, is greatly worsened, and, as touching their intellectual nature, is assuredly not improved. ... Are their bodily wants better, or more easily supplied? Are they subject to fewer calamities? Are they happier in childhood, youth, and manhood, and more comfortably or carefully provided for in old age, than when the land was unenclosed, and half covered with woods?

In other words, Southey (through More) is arguing that the condition of the average person in England was worse in the 1820s than it was not only in the 1500s but even in the first century BC when Julius Caesar invaded Britain. And Montesinos [Southey] dutifully replies: "Your position is undeniable. Were society to be stationary at its present point, the bulk of the people would, on the whole, have lost rather than gained by the alterations which have taken place during the last thousand years." More and Montesinos then present a long list of calamities that have befallen the people of England, including poorer health, poorer food, poorer clothing, and poorer shelter than when they were "hunters, fishers, and herdsmen." The pair look back with fondness on feudal times, and even slavery, since at least under feudalism and slavery the "labouring classes" had some measure of security.[11] Even the growing trade with other parts of the world is decried since it brings, they believe, not prosperity but foreign diseases.

More drives home his argument, triumphantly, as follows:

> Well, Montesinos, have you recollected or found any solid arguments for maintaining that the labouring classes, who form the great bulk of the population, are in a happier condition, physical, moral, or intellectual, in these times, than they were in mine?

Montesinos responds:

> Perhaps, Sir Thomas, their condition was better precisely *during your*

> *age than it ever has been either before or since.* The feudal system had well-nigh lost all its inhuman parts, and the worse inhumanity of the commercial system had not yet shown itself." [italics added]

A few decades later Karl Marx would take up the same anti-capitalist cudgel under the term "wage slavery," and Southey finds many echoes today in the environmentalist literature with its nostalgia for a "purer," more "natural" world before industrialization and technology.

One of the first optimists about the benefits of capitalism once it got past its growing pains was British essayist, poet, historian and Whig (liberal) politician Thomas B. Macaulay (1800-1859). Macaulay tackled the pessimistic view of capitalism and the Industrial Revolution in one of his most famous essays, a long review of Southey's *Colloquies* that appeared in the *Edinburgh Review* in 1830. Macaulay's review included the observation:

> Perhaps we could not select a better instance of the spirit which pervades the whole book [the *Colloquies*] than the passages in which Mr. Southey gives his opinion of the manufacturing system. There is nothing which he hates so bitterly. It is, according to him, a system more tyrannical than that of the feudal ages, a system of actual servitude, a system which destroys the bodies and degrades the minds of those who are engaged in it.[12]

Yet, Macaulay notes:

> Mr. Southey does not bring forward a single fact in support of these views. ... It is not from bills of mortality and statistical tables that Mr. Southey has learned his political creed. He cannot stoop to study the history of the system which he abuses, to strike the balance between the good and evil which it has produced, to compare district with district, or generation with generation.[13]

In other words, Southey is just offering belief and opinion, using Sir Thomas More to give his beliefs authority; he hasn't done any actual research to back up his assertions. In his review, Macaulay supplies the facts, based on research. And one of these indisputable facts, in the 19th century and today, is the death rate. Macaulay writes:

> The term of human life is decidedly longer in England than in any former age, respecting which we possess any information on which we can rely. ... *No test of the physical well-being of society can be*

> *named so decisive as that which is furnished by bills of mortality. …* That the lives of men should become longer and longer, while their bodily condition during life is becoming worse and worse, is utterly incredible.[14] [italics added]

While not disputing that "the lower orders in England … suffer severe hardships,"[15] Macaulay brings forward *facts and figures*—not beliefs and opinions—on the death rates in England since the development of capitalism showing that, on average, the English people were living longer, receiving better medical care and eating better than ever before in their history, even in the midst of (actually, ironically, because of) the horrors of the Industrial Revolution. As for the average person's material possessions:

> It is indeed a matter about which scarcely any doubt can exist in the most perverse mind that the improvements of machinery have lowered the price of manufactured articles, and have brought within the reach of the poorest some conveniences which Sir Thomas More or his master [Henry VIII] could not have obtained at any price.[16]

As a result, Macaulay notes:

> We might with some plausibility maintain that the people live longer because they are better fed, better lodged, better clothed, and better attended in sickness, and that these improvements are owing to that increase of national wealth which the manufacturing system has produced.[17]

Macaulay also makes a curious and perceptive observation that we will examine below in a modern context: that the better off people are, the more loudly they complain of their remaining ills:

> The labouring classes of this island, though they have their grievances and distresses … are on the whole better off as to physical comforts than the inhabitants of any equally extensive district of the old world. *For this very reason, suffering is more acutely felt and more loudly bewailed here than elsewhere.* We must take into the account the liberty of discussion, and the strong interest which the opponents of a ministry always have, to exaggerate the extent of the public disasters.[18] [italics added]

Rather than looking back, like Southey, Macaulay looks ahead: "If we were to prophesy that in the year 1930 a population of fifty millions, better fed, clad, and lodged than the English of our time, will cover these islands . . . many people would think us insane."[19] (Macaulay was close: in 1931 the population of Great Britain, including Northern Ireland, was about 46 million.) And Macaulay asks, in response to Southey's pessimism: "On what principle is it that when we see nothing but improvement behind us, we are to expect nothing but deterioration before us?"[20] Macaulay also comments, somewhat wryly: "The general view which Mr. Southey takes of the prospects of society is very gloomy; but we comfort ourselves with the consideration that Mr. Southey is no prophet."[21]

With the benefit of almost 200 years of hindsight we can see that Macaulay was right and Southey wrong, but what is more important is Macaulay's method. While Southey relied on belief, opinion and intuition, Macaulay got the *facts* to back up his case, and the facts showed that even in the early stages of the Industrial Revolution material life was being improved, not diminished, for the average person. At the same time, as Macaulay notes, and as we will see below, this doesn't mean the average person is *happier* then or now—if anything, he or she complains *more*. But the average person in the developed world is without a shadow of a doubt better off both physically and materially than any other group of human beings at any time in history.

I have quoted at length from an almost two-centuries-old debate to show how today's attacks on capitalism are nothing new and that, at least so far, history has shown that the optimists like Macaulay, rather than the pessimists and alarmists like Southey, have been the more accurate prophets. The debate also suggests how much the opinions of the pessimists and alarmists, like Southey in the 1800s and the *Times Colonist* letter writer today, are based not on facts but on opinion, ideology, belief, and values. When we look at the *facts* today, is the industrial, technological world the horror that the letter-writer says it is? Are we polluting and destroying the planet to the point of creating a "climate catastrophe," as one Canadian alarmist has put it?[22] While many, perhaps even a majority of Canadians, believe this environmental destruction is the case, the facts tell a different story.

THE IMPROVING STATE OF THE WORLD

As we saw in Chapter 7, the research that produced Bjorn Lomborg's *The Skeptical Environmentalist* began as an open-minded examination of what seemed an open-and-shut case to both Lomborg and his students: they firmly believed that the environment was being destroyed and the planet's people impoverished by capitalist depredation and pollution. That's certainly what we're being told, over and over by alarmists on climate and economic globalization. However, when Lomborg's team looked at the actual statistics, all supplied by the United Nations and other reputable institutions, what they found "conflicted with what we believed ourselves to know. The air in the developed world is becoming less, not more, polluted; people in the developing countries are not starving more, but less, and so on."[23]

Yet, a majority of the population in Canada and other developed countries continue to believe otherwise. For example, The Fraser Institute reports:

> Through surveys of college students, The Fraser Institute has found a strong disconnect between Canadian student perceptions of environmental trends (mostly negative) and the reality of environmental trends (mostly positive): 65% of the students attending Fraser Institute seminars believe that air quality is deteriorating. Fifty-eight percent of students are convinced that annual forest harvests exceed re-growth. Seventy-three percent of students believe we need to expand recycling programs and further control waste to avoid a "trash crisis."[24]

None of these student perceptions is factually correct. Yet how can this be? How can so many people, including educated people, be so badly misinformed? Before we look into why so many people hold views that are not true, let's look at what the facts, rather than the alarmists' opinions and beliefs, tell us about the improving state of the world.

The Improving State of the World is, in fact, the title of a book by Indur Goklany, an environmental policy analyst who has been a U.S. delegate to the Intergovernmental Panel on Climate Change. The book's subtitle is: *Why We're Living Longer, Healthier, More Comfortable Lives on a Cleaner Planet.* In it, and like Lomborg, Goklany marshals a wealth of data from sources like the World Bank, the United Nations, the World Health Organization, peer-reviewed scientific literature, and much more to show that the doomsday scenario painted by the globalization and environmental pessimists is, at least so far, completely wrong. Here is how Goklany states the alarmists' case:

> Neo-Malthusians contend that the human enterprise as currently constituted is unsustainable in the long run, unless the population shrinks; we diminish, if not reverse, economic development; and apply the precautionary principle to new technologies, which, in their view, essentially embodies a presumption against further technological change unless the technology involved is proven absolutely safe and clean.[25]

In support of the optimist, capitalist side, Goklany echoes Macaulay 150 years earlier in noting that "life expectancy is undoubtedly the single-most important indicator of human-well-being"[26] (Macaulay wrote: "No test of the physical well-being of society can be named so decisive as that which is furnished by bills of mortality"). What Goklany finds, not surprisingly, is that as wealth increases, so does life expectancy. And globally, and even in the poorest countries, life expectancy is going up, sometimes dramatically. So, Goklany notes, in the United States the life expectancy in 1820 was 40 years; by 2003 it was 78. Life expectancy in India in 1820 was 20; today it is 58. The only regions where life expectancy is currently falling are some parts of the former Soviet Union[27] and sub-Saharan Africa, although even there life expectancy has gone from age 20 in 1820 to 35 today.[28]

This increase in lifespan is occurring because, globally, people are becoming better off, even in most of the poorer countries. Citing the UN's *Human Development Report 2004,* Goklany notes that between 1990 and 2002, the Human Development Index (HDI), which combines three indicators—life expectancy, levels of education, and levels of economic development—improved in 108 of 138 countries for which data was available. Of the 20 countries with a decline in HDI, most were, again, in sub-Saharan Africa and the former Soviet Union.[29] This improvement occurred, it is worth noting, during a decade for which alarmists like Paul Ehrlich predicted massive famines and misery, and in which the global population went from 5.2 billion to over six billion—the largest population increase in human history.

But what about the toll on the environment? Surely this individual increase in well-being, combined with almost a billion more people in just over a decade, has meant environmental disaster for the planet in terms of both pollution and global warming (it's worth remembering, again, that carbon dioxide is not "pollution" but a necessary building block of life). The answer depends on where on the earth you look. In the developing countries, such as China and India, environmental problems are acute as these nations try to reach Western standards of industrial and technological development. News stories about China before the 2008 Beijing summer Olympics revealed a capital city sometimes as choked with air pollution as, say, London, England, was in the 19th century, when citizens on the street sometimes literally had to grope from door to door because the fog combined with coal smoke was so thick.[30] Today, London still has its traffic congestion problems but it is a very livable city in terms of air quality.

MEETING 'POST-MATERIAL' NEEDS

The story is much more positive in the developed nations. Logic suggests that more economic development would mean more environmental problems, and so, as environmentalists Ted Nordhaus and Michael Shellenberg note in their book *Break Through*: *From the Death of Environmentalism to the Politics of Possibility*:

> Environmentalists … have tended to view economic growth as the *cause* but not the *solution* to ecological crisis. Environmentalists like to emphasize the ways in which the economy depends on ecology, but they often miss the ways in which thinking ecologically depends on prospering economically.[31]

Environmentalists believe, Nordhaus and Shellenberg write, that any improvements in the environment are due to their efforts, which therefore must be unrelenting—even if the situation improves, the environment remains in constant danger. In fact, the authors say, environmentalism has triumphed because *values* have changed in the developed nations, and those values were shaped not by environmentalist disaster narratives but by what Nordhaus and Shellenberg call "post-materialism":

> Environmentalism and other progressive social movements of the 1960s were born of the prosperity of the postwar era and the widespread emergence of higher-order post-materialist needs. As Americans became increasingly wealthy, secure, and optimistic, they started to care more about problems such as air and water pollution and the protection of the wilderness and open space. This powerful correlation between increasing affluence and the emergence of qual-

> ity-of-life and fulfillment values has been documented in developed and undeveloped countries around the world.[32]

In other words, and in contrast to what many or most in the environmental movement believe, "The satisfaction of the material needs of food and water and shelter is not an *obstacle to* but rather the *precondition for* the modern appreciation of the nonhuman world."[33]

One of the benchmarks of the relationship between affluence and environmental quality is forests, in that forests, in many people's minds, equal "nature." It is true that the destruction of tropical rainforests continues, causing 25 per cent of human-caused greenhouse gases—although, as Nordhaus and Shellenberger note, ironically, "some of the deforestation in Indonesia and Brazil is driven by the rising demand for land to grow biofuels,"[34] which is a Green initiative. Lomborg cites figures on deforestation indicating that the world has lost about 20 per cent of its forests since humans began clearing land; the IPCC 2001 report estimates a loss of 20 per cent of forest land from 1850 to 1990.[35] However, as Lomborg points out, this loss of forest is nowhere near the devastation claimed by many environmental groups.

Now the good news: In both North America and many parts of Europe, the acreage devoted to forests is increasing—exactly the opposite of what the critics of capitalism fear. For example, Europe's forests have expanded by 10 per cent since 1990, although, for some reason, this news has not been heralded as the environmental triumph it is.[36] Forests in the United States are also expanding. As environmental writer Gregg Easterbrook notes in his aptly named *The Progress Paradox*, Connecticut now has more forest than it did in the 1800s—59 per cent forested compared to 35 per cent a century ago. Meanwhile the population of the state has tripled and its agricultural output increased fourfold.[37] Canada's forests, too, are holding or even expanding. According to a Fraser Institute study: "Canada's forests are being replenished at a rate equivalent to, or greater than, the rate they are being harvested."[38]

Why are forests increasing, not decreasing, in the developed world? Because more efficient farming techniques mean less land is needed for agriculture. Since about 98 per cent of people in the developed world now live in cities and towns, some of this excess land can be returned to nature. Up to 56 million acres in the United States alone was returned to forest between

1980 and 1990, or about 2.4 per cent of the U.S. land area.[39] In short, with more affluence comes more nature, not less. Furthermore, we are beginning to see the same positive trend for forests in parts of the developing world as more people leave agriculture and migrate to the cities. For example, the *New York Times* reports, in a story headlined "New jungles prompt a debate on rain forests":

> [In Panama], and in other tropical countries around the world, small holdings ... —and much larger swaths of farmland—are reverting to nature, as people abandon their land and move to the cities in search of better livings. These new "secondary" forests are emerging in Latin America, Asia and other tropical regions at such a fast pace that the trend has set off a serious debate about whether saving primeval rain forest — an iconic environmental cause—may be less urgent than once thought. By one estimate, *for every acre of rain forest cut down each year, more than 50 acres of new forest are growing in the tropics on land that was once farmed, logged or ravaged by natural disaster.*[40] [italics added]

A peer-reviewed 2003 study of satellite data showed that Net Primary Production, a measure of the planet's plant growth, had increased 6 per cent between 1982 and 1999; the biggest increase, 42 per cent of the total, occurred in the Amazon rainforest. A primary reason for this increase in biomass, according to the article's authors? Global warming—hence, a longer growing season—along with higher levels of carbon dioxide which, as we've noted, acts as a fertilizer. The authors write: "Most of the observed climatic changes [i.e., warming] *have been in the direction of reducing climatic constraints to plant growth.*[41] [italics added] In short, a warmer, more carbon-dioxide-rich planet is a greener planet, the exact opposite of what the public is told by climate alarmists.

Overall, notes Easterbrook, "Despite the hysterical tone of environmental reporting in the media, in the United States and Europe all environmental trends except for greenhouse-gas accumulation are positive and have been positive for decades."[42] Among the other good news stories: smog is down one-third since 1970, despite a doubling of cars; water quality is up, including even Lake Erie, which was described by Rachel Carson in 1962 in *Silent Spring* as one of a host of "irrecoverable" and "irreversible" natural disasters caused by chemical pollution[43]; and toxic emissions by industry were down 51 per cent from 1988 to 2002. And the same good news comes from most parts of Western Europe[44] and Canada. For example, the Fraser Institute reports that of 37 environmental indicators in Canada, 31 are stable or improving.[45]

Among that good news is greater life expectancy, which is the main reason why population is increasing. Why is greater life expectancy the reason for population increase? Because fertility in the developed nations is down, way down, and often below replacement in countries like Italy, Canada and Japan. In other words, affluence reduces the birth rate, which reduces overpopulation, which reduces resource pressure on the planet even though each individual is enjoying more material comforts. Incidentally, this decline in population is exactly the opposite of what alarmists like Paul Ehrlich predicted (see Chapter 8)—Ehrlich's view was (and probably still is, since he doesn't seem to learn from experience) that only draconian controls could hold population levels steady or decrease them.

Once again, the same almost uniformly good news cannot be said of the developing world, although even there the situation is in most respects improving thanks to globalization. Easterbrook notes:

> Hoping to rationalize away the fact that most of the developing world is better off with globalization than without it, opponents protest that globalization causes inequity; that "inequality is accelerating everyone on Earth," according to the anti-globalization International Forum on Globalization.[46]

But, once again, the statistics show the alarmists are mistaken—the developing world's share of global income is rising at a faster rate than its population. Living standards are rising particularly quickly in some of the Asian economies, notably China, India, and parts of Southeast Asia such as Thailand and Malaysia.[47] In other words, even for the developing countries, the world is becoming a better, or at least materially more comfortable, place. The nations that are lagging behind, according to Easterbrook, are those that don't have market economies (a.k.a., capitalism): "Developing nations with market systems have three times the economic growth rate of developing nations with closed systems, and higher living standards for typical people."[48]

There isn't space here to go into all the details of why and how the planet is becoming a better, richer, and more environmentally responsible place, at least in the developed nations, and how this improvement will eventually occur in the developing nations as well. For those interested in more information, the books by Goklany, Lomborg, and Easterbrook, cited above, make a very strong and even, I would say, irrefutable case. And perhaps there's no point in elaborating further: those who want to believe the world is getting worse will do so no matter how much evidence is presented to the contrary. Indeed, the more interesting question is why so many *want* to believe the worst, in the face of so much evidence that things are getting better.

This puzzle of why people want to believe the worst preoccupied the late Julian Simon, the optimist economist, to the point where he devoted a book to this issue entitled *Hoodwinking the Nation*. He suggested the reasons for pessimism while life is getting better can be found in the following factors:

- There is a tremendous financial and psychological incentive for scholars to present bad, rather than good, news. After all, as a scholar, you likely won't get funding for research if what you are researching isn't "bad news" in some way.
- There is a tremendous financial and psychological incentive for the media to report almost exclusively bad news; "good news" doesn't sell newspapers or bring viewership or listenership. This is why the media "give much attention to alarmists, but little to those who are skeptical of their claims."[49]
- Human beings have an apparently inborn tendency to view the past as "better" than today ("In my day we didn't have as much pollution/crime, etc.....").
- Activist groups of all stripes, not just environmental, believe the public needs a steady diet of bad news, even if it's not true, lest it become complacent.
- We are hard-wired by our evolution to be wary—better safe than sorry.
- Some people, such as activist groups, simply out-and-out lie because they think the end justifies the means, or because exaggerating a problem brings them funding (i.e., without a "problem," and preferably a "crisis" or even a "planetary emergency," they won't get any funding from government or the public).[50] Al Gore falls

> into this category with his self-admitted "over-representation of factual presentations" (i.e., misrepresentations, exaggerations, or outright untruths) about the dangers of global warming lest we not get the right message.

In short, we get caught in abstractions; our views don't come from the direct perceptions of our own reality. Thomas Macaulay identified this phenomenon during a speech in 1849: "Ever since I began to make observations on the state of my country, I have been seeing nothing but growth, and I have been hearing nothing but decay."[51] This isn't to deny, of course, that many people's environments and lives *are* horrible. However, in the developed nations, the lives of the great majority would be considered sumptuous by the standards of previous centuries. For this reliance on abstractions the media must take some responsibility, so that *National Post* columnist Robert Fulford writes: "Journalists create an atmosphere of melancholy. This consistency suggests that *our style of thinking overwhelms our powers of observation.*"[52] [italics added] When we do objectively assess our own lives, most of us in the developed world can rightly conclude that things are, over all, pretty good for us and those in our immediate circles. Therefore, why wouldn't this be true for others in the developed world as well, even if we can't be sure by direct observation of how they're doing?

Easterbrook identifies another psychological tendency leading to pessimism that he calls "collapse anxiety":

> Deep-seated in the minds of Americans and Europeans—perhaps in the minds of most—is a fear that the West cannot sustain its current elevated living standards and liberal personal freedom. We fear that the economy will collapse; not just sputter, but cease functioning. We fear that resources will run out. … Some amount of never-ending anxiety may be rational—keeping us on our guard—or even hard-wired into our nature by evolution, as perhaps the most fretful of our ancestors, the ones always warily scanning the horizon, were the ones most likely to survive. … But if a collapse were coming, its signs ought to be somewhere. That is not what trends show.[53]

So, yes, we do want to be aware of bad news—if there's an erupting volcano or an escaped tiger in the neighborhood, we may need that information for our very survival. But, as Easterbrook notes: "It's one thing …

to highlight when the bad happens—as the media should—and another to pretend that the good does not happen, as the media also do."[54] The result—perhaps evolution-based—is to tend to look at the dark side rather than the bright side, to be alarmist rather than optimistic, even if there's no cause for alarm.

HOW GREENS TRY TO DESTROY TECHNOLOGICAL CIVILIZATION

Ironically, the force most likely to lead to the collapse of Western civilization is not the failings of Western civilization itself but environmentalism's attempts to undermine Western civilization. As geologist Ian Plimer notes:

> Many Western cities are now having water shortages and irregular electricity supply because no new dams and coal- or nuclear-fired power stations have been built as a result of decades of green political pressure.[55]

New Orleans has already experienced the results of putting the environmentalist agenda ahead of the needs of people. In order to preserve wetlands, environmentalists fought improvements to the city's levee system.[56] The result? Eighty per cent of a city of almost half a million people was flooded, resulting in more than 1,800 deaths and property damage estimated at more than $100-billion.[57] The disaster might have happened anyway—most of New Orleans is below sea level—but the green legal action made it certain disaster would happen.

We in British Columbia have first-hand experience of environmentalist attempts to block water-reservoir and hydroelectric projects. While B.C. is justly proud of its large supply of inexpensive, non-polluting hydroelectric power, no new hydroelectric plant has been built since the 1980s, and B.C., which had traditionally been a net energy exporter, now imports 15 per cent of its energy.[58] The one new hydroelectric plant planned, Site C on the Peace River, has been blocked for three decades by green opposition.[59] Yet, even though the demand for electricity in B.C. is expected to rise by 45 per cent in the next 20 years, a Green opponent to the Site C dam writes:

> After hundreds of hours of consultations around the province, the BC Utilities Commission ultimately couldn't justify Site C *in part because of environmental impacts*. Since then, given the loss of spe-

> cies and habitats around the province, the excessive habitat disruption that already exists in the northeast, the threat to fresh water, and especially the uncertainty of climate change, what on Earth could make this a good idea today? We have the technologies to develop truly "green" energy. Let's quit living in the past and get on with it.[60] [italics added]

The problem, which this environmentalist writer doesn't seem to understand, is that humanity cannot exist without making *some* impact on the environment. And the previous network of hydroelectric dams has not damaged the province's environment—B.C. vehicle licence plates still have the word "Beautiful" on them and most British Columbians would agree. On the other hand, the province won't be very beautiful for its more than four million people if the power goes out. The anti-Site C writer wants close to zero impact for human activity. In other words, she puts the environment (or, more accurately, her alarmist view of possible damage to the environment) ahead of the needs of humanity.

Nor do we have "green" energy technologies that can come even close to meeting B.C.'s demand for power over the next few decades. A 2004 study of energy use in B.C. found that only 33 per cent of B.C.'s energy needs were supplied by "renewable" sources, and that 94 per cent of these renewable sources was hydro. A mere six per cent of B.C.'s power comes from "alternative" energies like wind and solar. The rest of B.C.'s power—at least two-thirds—comes from fossil fuels like coal and natural gas.[61] Meanwhile, again thanks to Green pressure, Liberal B.C. Premier Gordon Campbell has vetoed any move to nuclear power or drilling for off-shore oil. Yet, if the future demand for electricity isn't met, the alternative is brownouts and blackouts, leading to loss of industrial capacity and tens of thousands of unemployed—an economic collapse that, like that of New Orleans, will be almost entirely caused by environmentalist pressure.

PROGRESS DOESN'T MAKE US HAPPY, JUST MORE COMFORTABLE

Another reason why we are sure things are getting worse, Easterbrook suggests, is that material progress doesn't necessarily make us happy—a problem that Macaulay identified as well (because the English people were much better off than other nations, Macaulay wrote, "suffering is more acutely felt and more loudly bewailed here than elsewhere"). That is what Easterbrook calls the "progress paradox": "Americans and Europeans have more of everything except happiness."[62] So, yes, the philosophers and religious teachers were right when they warned us that material wealth doesn't bring happiness, because as Easterbrook notes, the things that make us truly happy—love of family and friends, the esteem of co-workers and society in general, etc.—can't be bought with money. At the same time, Easterbrook points out, people in the developed nations are told that more material possessions *will* make them happier—that's the point of life, we're told in advertisements: to accumulate more and more means more and more happiness. Yet, there rarely comes a point when we say, "Enough. I've got everything I want and need." As Easterbrook (and countless others) have observed: While our "needs"—food, shelter, health, companionship—may be finite, our wants are infinite, and so we can never, if we don't deliberately restrain our desires, find the permanent satisfaction we all seek, regardless of what the ads tell us. If we fall into this materialist trap, we create for ourselves a never-ending cycle of dissatisfaction and frustration.

Does this mean we should abandon capitalism for a lifestyle that is supposedly richer, in non-material terms, and more rewarding? For example, British novelist Geoffrey Household has written: "Is the secret of happiness a mixture of passionate fornication and somewhat chancy hunting? If so, we human beings have been continually frustrated by urban life ever since we were fools enough to invent it."[63] Sigmund Freud wrote a book on civilization and its discontents. Today, everywhere we look, we are told that people are "alienated," a term popularized by Karl Marx in his critiques of capitalism. In other words, even if capitalism-based civilization meets most or all of our material needs—hence, "post-materialism"—we are not living in a psychological utopia.

But is there such a thing as a psychological utopia? Was there a "pure," natural state in the past when we were all happier and from which we have "fallen"? Some, like the poet Robert Southey, believe we were happier in past ages, and many environmentalists today are convinced that if we could only return to nature, our souls would be filled and fulfilled. If you examine this idea closely, it becomes obvious that it is a myth. Human beings have probably always been unhappy with their lot—and this discontent is the impetus for progress, for trying to make the world a better place. Perhaps a hunter-gatherer society, with its "mixture of passionate fornication and somewhat chancy hunting" was more psychologically satisfying than modern civilization. Materially and physically, however, the conditions for our hunter-gatherer ancestors were often wretched, with a life expectancy in the mid-30s for most of humanity for much of human history.[64] Despite whatever discontent we feel with technological civilization, would any of us seriously want to go back to an earlier time? Easterbrook asks this question in *The Progress Paradox:*

> If the means existed, would you exchange places with a typical person living in any year before your birth? Exchange places permanently—not, say, observe the Battle of Hastings and then rematerialize in the present. ... In this deal you'd be transported back to the year and society of your choosing to live out the rest of your life as an ordinary person. ... A good guess is that hardly anyone in the United States or European Union today would accept a one-way ticket to the everyday life of the past.[65]

That seems like a good guess. No matter how psychologically unhappy we in the developed world may be, few of us would willingly give up our modern material comforts—hot showers, painless dentistry, warm, dry houses, a varied, nutritious diet from all around the world, a 7½-hour work day—for the relatively impoverished existence of almost anyone in the past, including the upper classes and royalty—as Easterbrook points out, "Average Americans and Europeans ... live better than most of the royalty of history, if only owing to antibiotics."[66] Or, as comedian Bill Maher puts it: "If you think you've got it bad now, you haven't read history." If nothing else, as citizens of the modern world, we live much longer than our ancestors and, as Julian Simon has pointed out, "The most important and amazing demographic fact—the greatest human achievement in history, in my view—is the decrease in the world's death rate."[67] So: would we be happier if we lived closer to "nature" (i.e., were poorer), as many environmentalists believe? It's doubtful—we'd just be unhappy about other things, most prominently not having enough creature comforts and dying decades earlier than we would in the technological world.

Another question: How much do we want to blame our *civilization* because we are not happy? Putting this another way: Do we really want a society that controls our lives in order to make us happier, like the citizens of Aldous Huxley's *Brave New World* or George Orwell's *Nineteen Eighty-Four*? Another name for the *state* deciding what makes us happy or not is totalitarianism, and, indeed, the last lines of Orwell's *Nineteen Eighty-four* are: "But it was all right, everything was all right, the struggle was finished. He had won the victory over himself. He loved Big Brother."[68] But the "happiness" of Orwell's protagonist, Winston Smith, came after intensive brainwashing and torture. Big Brother has been tried many times in human history and has never made anyone except the ruling elites happier, assuming even they are happy given the need to constantly and forcefully maintain their power against the resistance of the masses.

Another reason for the current malaise of Western civilization can be found in the 20th century. The 19th century in Europe and North America was a time when, despite the supposed "evils" of capitalism, there was real optimism that social problems could be solved through technology, combined with anxiety that the spiritual riches of the past might be lost—fear of this spiritual loss is the message of Matthew Arnold's famous poem "Dover Beach." In short, the 19th-century European and American believed in progress, but with an eye to the past. This belief in progress was crushed by the First World War (1914-1918), the bloodiest and most extensive war in history up to that time although, incredibly, the Second World War (1939-1945) was even worse. In short, battered by the most dynamic century of change in human history, the 20th, Western civilization became pessimistic. Western culture is still pessimistic, and we are approaching the prospect of climate change with a similar spirit of defeatism and despair, to the point of not recognizing the astounding advances we have made in almost all aspects of life, including environmental improvement. We are also ignoring Western civilization's almost unlimited ability to respond and adapt to not just global warming but almost any challenge—for example, it took less than a year for the United States to recover economically from the shattering Islamist attack of September 11, 2001—although the ability of northern nations like Canada to cope with a return of the glaciers is another matter. Former U.S. president

Richard Nixon was quite correct when he said, paradoxically, in his 1970 State of the Union address:

> The argument is often made that there is a fundamental contradiction between economic growth and the quality of life, so that to have one we must forsake the other. The answer is not to abandon growth, but to redirect it. For example, we should turn toward ending congestion and eliminating smog the same reservoir of inventive genius that created them in the first place.

And, with all its faults—and Western civilization has many—it is still, for the average man and woman in the developed world and increasingly in the developing world, infinitely better than anything in material terms that the average person has enjoyed—or, more accurately, not enjoyed—at any time in the past. As historian Victor Davis Hanson has written:

> Western civilization has given mankind *the only economic system that works*, a rationalist tradition that alone allows us material and technological progress, the sole political structure that ensures the freedom of the individual, a system of ethics and a religion that brings out the best in humankind....[69] [italics added]

But, if life is materially better than at any time in human history for most people in the world, and certainly within the developed world, can we be sure this improvement in living standards will continue? In 2008-2009, the world economy was battered by, among other things, an apparent shortage of food (caused, in part, by the Green-led demand for biofuels, which reduced the agricultural acreage available for food crops), and near-record oil prices. These signs could, of course, presage the end of Western civilization as we know it. Or, they could simply be a challenge to be overcome, as we've overcome similar challenges in the past.

INTELLECTUAL CAPITAL: THE 'ULTIMATE RESOURCE'

Above all, we can expect the future to be better because, assuming we avoid a third world war, each generation adds to the physical and intellectual capital of the next generation. For example, Canadians today don't have to pay for the building of thousands of miles of roads and railroads—their ancestors paid the bills, and did without many material comforts we take for granted in order to make our transportation network happen. And the same is true of intellectual capital—we don't have to rediscover the laws of nuclear physics, biology and chemistry, with all their benefits; they have been given to us, and we will add to them for our children's and grandchildren's benefit. As a result, we are making progress in spite of ourselves, while, paradoxically, demanding less and less from the environment.

Economist Julian Simon, in particular, stressed the importance of intellectual capital. For Simon, what constrains any people or nation is rarely the lack of physical resources. What hampers a people or nation is the lack of intellectual resources combined with a lack of hard work to make the best use of the physical and mental resources available. Japan, Hong Kong and Singapore are well-known examples of resource-poor nations that have become wealthy based on the intellectual and physical capital of their people. And, as Simon noted in many of his writings, intellectual capital—brainpower—is *unlimited*, to the point where Simon believed the intellect could stretch material resources to near infinity. A while ago, my teenage daughter showed me her latest gadget, an iPod that held more than a thousand songs. It fit into her palm. It even plays movies. To achieve the same capabilities a few decades ago would have required a roomful of computers. What the iPod contains in abundance is "intellectual capital," not material capital. For Simon, intellectual capital is the "ultimate resource," and he is right.

And so, each generation, in employing and expanding its resources, generates material and intellectual capital for the next generation. Indeed, that is what each generation strives for—to make its children better off. The fact that past generations have succeeded in making their children better off should not, therefore surprise us. What should surprise us is that so many people apparently don't understand this basic economic fact. However, based on this fact, Simon argued (again, echoing Macaulay):

> With reasonable surety one can expect that the material conditions of life will continue to get better for most people, in most countries, most of the time, indefinitely. Within a century or two, all nations and most of humanity will be at or above today's Western living standards.[70]

Even the IPCC estimates that people around the world will be "generally more affluent than today" by 2100, even if (actually, especially if) we do nothing about climate change.[71] If this is true—and history so far indicates it is true—then our children and children's children will be richer than we are, with more resources than at present to tackle the environmental problems that global warming (or cooling) presents to them, just as current generations in the developed world have tackled and generally overcome the environmental problems we once faced, such as pollution of our air and water, even though alarmists like Rachel Carson said the damage was "irreparable."

On the other hand, if we continue the folly of crippling industrial and technological civilization in the name of curbing carbon emissions, an effort that even its supporters acknowledge is symbolic rather than substantive—Andrew Weaver notes that the Kyoto Accord would have had "zero effect" on warming[72]—or just out of a Marxist-inspired desire to destroy market capitalism, the developed nations will become poorer—and so will the developing nations. Poorer citizens will be less willing to spend vast sums on environmental protection; instead, they'll be scrimping to make ends meet. Under this anti-carbon scenario, the environment will, paradoxically, suffer more from our current misguided efforts to save it than if we carried on so that everyone on the planet, including those in the developing nations, becomes wealthier. Why? Because wealthier people are more willing, and financially more capable, of protecting the environment and, as a bonus, more capable of coping with whatever climate change this progress might have caused.

If nothing else, this must be the first time in history that democratically elected politicians have deliberately set out to make their people poorer, rather than more prosperous. And herein lies a danger: Very few people *want* to be poorer. So the only way this poverty-creating scheme can work is by creating a kind of environmental/political dictatorship of the sort proposed by Paul Ehrlich to control population (Chapter 8). It may be a fault in human nature, but the average person is willing to go only so far in cutting his or her lifestyle to deal with a threat like global warming that is, at this point, nothing but a scientific abstraction. To make these cuts, the enviro-political elite will have to increasingly use force, and democracy will suffer. As philosopher Jeffrey Foss notes in explaining his opposition to Kyoto-type interventions into the economies of the developed world:

> Increasing the central control of economies [which Kyoto would have done] has generally depressed those economies. ... The rise of

> democracy and human freedom has gone hand-in-hand with prosperity. … Humanity must secure its own necessities before it can care properly about nature as a whole. Our struggle with nature must be won before we can sympathize with our former competitors [other species]. We are on the verge of securing natural freedom for nearly all of humankind. Freedom from death, suffering, and excessive work is a necessity if our concern for nature is to be effective.[73]

Instead, many environmentalists want us to go backwards, in some cases as far back as a hunter-gatherer economy.[74] Yet, as Foss writes:

> The romantic notion that we must repent, reject the path of knowledge, and return to the path of innocence is out of touch with reality. To reject our intelligence, technology and science would be to reject the natural processes that gave birth to human nature, an act of self-mutilation that would be every bit as much an insult to nature as to human nature. We did not steal fire from the gods. That is just a romantic myth. In reality, nature gave us the spiritual fire of intelligence that enabled us to employ physical fire. … Our *sense* of alienation is real and understandable. It does not follow that we should undo ourselves.[75]

Worse: not only do many environmentalists want us to retreat back to the caves, but like Ehrlich they are willing to propose totalitarian methods to achieve this result. But once democracy is gone, a sacrifice the alarmists seem willing and even happy to make as part of their crusade against global warming, it will be difficult to get freedom back. And, as noted above, once freedom and prosperity are gone, the good of the environment will cease to be a concern of the average citizen or cash-strapped governments; humanity will, once again, be too busy making a living to want to spend precious resources on environmental protection. Thus, more prosperity for all of humanity, not less, is the way forward for both us and the environment. Progress is good, not bad, for the environment.

And, progress is good for humanity's spiritual and moral environment as well. For example, philosopher of science Bryson Brown notes that economic progress and moral progress are intimately related:

> Happily, we don't need utopian delusions to recognize the advances that practical, secular and, yes, scientific developments have produced, many of them of great moral significance. Safe water, vacci-

> nations, basic nutritional standards, decent sewers and hygiene have ensured, for the first time in history, that the vast majority of children in our fortunate country [Canada] live to be adults. The development of the rule of law and democratic institutions in many countries over the past 200 years is also remarkable. It is stunning to see this progress so lightly dismissed.[76]

In other words, in making material progress we have *also* made more moral progress than at any time in human history; for example, never before in human history have human beings been as concerned and protective of both the environment *and* of their fellow human beings. Therefore, yes, it is shocking to see the developed world's progress in protecting the environment so easily dismissed by environmentalists and alarmist climate scientists. Surely it is time to recognize the facts rather than unsubstantiated beliefs and opinions when we assess what can only be described as incredible progress on the environmental front, a progress that is due almost entirely to what dogmatic environmentalists see as the main cause of environmental degradation—the affluence that has come from capitalism. And there is no reason whatsoever to believe that the world will not continue to get better for most of humanity, if we can prevent the damage that the pessimists and alarmists are determined to cause. A huge part of that damage is to the integrity and reputation of science, as we will see in the next chapter.

The evidence ... however properly reached, may always be more or less wrong, the best information being never complete, and the best reasoning being liable to fallacy.

—Thomas H. Huxley[1]

A theory that is not refutable by any conceivable event is not scientific.

Karl Popper[2]

Always ask whether the hypothesis can be, at least in principle, falsified. Propositions that are untestable, unfalsifiable are not worth much.

Carl Sagan[3]

Fudging the data in any way whatsoever is quite literally a sin against the holy ghost of science, I'm not religious, but I put it that way because I feel so strongly. It's the one thing you do not ever do. You've got to have standards.

—James Lovelock[4]

The idea that human beings have changed and are changing the basic climate system of the Earth through their industrial activities and burning of fossil fuels—the essence of the Greens' theory of global warming—has about as much basis in science as Marxism and Freudianism. Global warming, like Marxism, is a political theory of actions, demanding compliance with its rules.

—Historian Paul Johnson[5]

Chapter 10

IS ALARMIST CLIMATE SCIENCE A 'SCIENCE'?

Alarmist climate scientists know that, for their global warming alarms to be credible, they must present themselves to the public as "hard" physical scientists, working with hard physical data and making hard, valid predictions. But is climate science a "hard" physical science? Is it even a science at all in the strictest sense? To answer that question we need to be clear about what is meant by "science."

No list of what constitutes "science" will satisfy everyone, but most scientists would probably agree with the following[6]:

- Science does not ultimately rely on authority or consensus but on evidence.
- Science does not claim certainty.

- Science makes predictions that are testable and falsifiable.
- Science does not bring in *ad hoc* auxiliary hypotheses to fit the data.
- Science does not prejudge the results; it open to competing explanations.
- Scientific claims should be replicable by others; therefore, scientific data must be open to all researchers.
- Science demands strong evidence for extreme claims.
- Science regards truth as its highest value.

Let's go through the list and see how well alarmist climate science fares.

SCIENCE DOES NOT RELY ON AUTHORITY OR CONSENSUS

On authority in science, philosopher Bertrand Russell writes:

> The triumphs of science are due to the substitution of observation and inference for authority. *Every attempt to revive authority in intellectual matters is a retrograde step.*[7] [italics added]

While alarmist climate science has many observations and a lot of inference, it also relies heavily on authority (the "consensus") in presenting its findings to the public. Yet, as Russell puts it, "The whole attitude of accepting a belief unquestioningly on the basis of authority is contrary to the scientific spirit."[8] Astronomer Carl Sagan writes: "Arguments from authority carry little weight—'authorities' have made mistakes in the past. They will do so again in the future."[9] However, this is what the public, media and politicians are being asked to do: accept the IPCC reports without criticism, questions or, in the case of governments, public debate or due diligence, based on the IPCC's authority. This is not to say, of course, that scientists cannot and do not speak with authority. However, they lose credibility and undermine their authority when they try to persuade the public that authority alone (i.e., a "consensus" of experts) is enough reason to accept a theory's validity, as alarmist climate science does. And "experts" lose authority when they resort to exaggeration, misleading information and outright falsehoods to support their claims, as alarmist climate science does all too often (see Chapter 8).

SCIENCE DOES NOT CLAIM CERTAINTY

For almost 200 years after Isaac Newton (1643-1727), philosophers and scientists believed that "once a scientific fact or law was discovered, it was not open to change. This certainty was believed to be the distinguishing characteristic of science."[10] Scientific certainty vanished in 1905 with Einstein's Theory of Relativity, which showed that even the science of Isaac Newton was not fixed, immutable and certain. In other words, "the search for certainty that had been the central preoccupation of Western philosophy [and science] since Descartes was an error."[11] Today, therefore, writes Russell, "it is part of the scientific attitude that the pronouncements of science do not claim to be certain, but only to be the most probable on present evidence."[12]

Yet, as we've seen, many high-profile alarmist climate scientists do claim, at least to the public, to be not just probably certain but utterly certain that climate change is primarily the result of human activities and that warming will be disastrous. Some even say there is no debate—no debate at all—in climate-science circles on these propositions. To again quote Andrew Weaver:

> There is *no such debate* [over the human causes of global warming] in the atmospheric or climate scientific community, and … making the public believe that such a debate exists is precisely the goal of the denial industry. … Scientific debate over global warming *would therefore imply uncertainty*.[13] [italics added]

At least, certainty is what is presented to the public. In their academic journal publications, consensus climate scientists are more willing to admit their anthropogenic hypothesis might be tentative. For example, in an academic paper on what might happen to climate if CO_2 levels doubled, David Rind of the Goddard Institute for Space Studies writes:

> The experiments conducted here indicate that *many uncertainties still exist* ... Until model differences are resolved, *various basic questions about the climate of the doubled CO_2 world will remain uncertain. ...* The problems ... are *formidable*.[14] [italics added]

However, Rind later said during a TV interview: "When Pearl Harbor occurred, we didn't say, 'Well, we'll only take the sorts of action that won't affect our economy'." That is, he compared global warming to the Japanese attack on Pearl Harbor—hardly the nuanced position that appeared in his scientific paper.[15] In Chapter 8, Stanford University's Stephen Schneider is quoted as saying that "*uncertainties* so infuse the issue of climate change that it is still impossible to rule out either mild or catastrophic outcomes"[16] and that in exploring climate change "we end up with a maddening degree of *uncertainty*."[17] [italics added] So, alarmist climate science has one vocabulary for scientific insiders in which uncertainties with the science are admitted, and another vocabulary for the public, media and politicians in which certainty is claimed. In other words, the public, media and politicians are being deliberately misled on the certainty that consensus climate scientists actually feel about their findings and about the quality of the empirical evidence—as Rind put it, "various basic questions ... remain *uncertain*"—behind this certainty.

This is not to say that science cannot be fairly certain—reach a high probability—about some matters; of course it can or a definitive statement about anything would be impossible. However, at least in many if not most of the alarmist climate writings for the public, media and politicians, there is rarely any admission that the writer might, conceivably, be wrong—an astonishing position for scientists to take.

SCIENCE MAKES ACCURATE PREDICTIONS

Baron Georges Cuvier (1769-1832), one of the pioneers of biological classification, wrote in 1822: "The true mark of a theory is without doubt its ability to predict phenomena."[18] Philosopher of science Karl Popper also considered making predictions the fundamental task of both the natural and social sciences: "The task of the social sciences is fundamentally the same as that of the natural sciences—to make predictions."[19] For Popper, and for most scientists, a scientific theory that can't make accurate predictions is a dead end.

Science emphasizes the importance of predictions because successful predictions show that a theory has explanatory power. For example, quantum physics theory is bizarre and sometimes quite unbelievable. Yet quantum physics is accepted because it has produced "stunningly successful predictions."[20] On the other hand, if predictions consistently fail, then the hypothesis is falsified—it has not met the test of truth. However, a string of successful predictions does not ensure that a theory is established as more than provisionally true—*one* failed prediction can bring the whole theory down. And so, Popper notes, Albert Einstein "was looking for crucial experiments whose agreement with his predictions would by no means establish his theory; while a disagreement, as he was the first to stress, would show his theory to be untenable."[21] Einstein was willing to let his hypothesis of relativity stand or fall based on one crucial prediction: that light would bend in a gravitational field. Light did bend, and Einstein's hypothesis was well on its way to becoming a theory. If light had not bent, then Einstein's hypothesis would have failed—it would have, in Popper's terms, been falsified.

Of course, some scientists do not accept that prediction and falsification are the defining characteristics of science. For example, economist Eric D. Beinhocker writes:

> The hallmark of science is not its ability to forecast the future, but *its ability to explain things*—to increase our understanding of the workings of the universe. The role of predictions in science is to help us distinguish between competing explanations. ... Science is full of examples of fields where researchers can explain phenomena and test the validity of their explanations, without necessarily being able to make accurate forecasts. For example, biologists can explain but not forecast the folding of proteins, and physicists can explain but not forecast the exact motion of a turbulent fluid.[22] [italics added]

So, some sciences are primarily "explanatory," while others are primarily "predictive." The social sciences, like Beinhocker's economics, offer predictions but rarely "hard" predictions. For example, no rational person would bet his life on the election predictions of a political scientist in the way that astronauts bet their lives on, say, the predictions of the physicist who programmed a flight to Mars or the engineer who designed the rocket. Everyone understands that a political, economic or sociological forecast is, at best, a highly informed guess rather than a hard scientific prediction. Therefore, no respectable sociologist or political scientist or economist would assert to the public, media or politicians that her predictions were "certain," as alarmist climate science does. Social-science predictions are not "certain"—and social scientists do not claim their predictions to be certain—because human social life, like climate, is a chaotic, non-linear, stochastic (random) system, with too many factors to make precise prediction possible. Social science predictions may, of course, sometimes prove correct, but in that case they often fall into the category of a lucky guess. Therefore, not surprisingly, a study of predictions by 284 political scientists, economists, journalists and others found that, over a period of 20 years, their predictive accuracy was no better than random chance.[23]

Similarly, although climate science deals with "hard" physical data like solar radiation, ocean current temperatures and the like, the predictions of climate science—alarmist and skeptical—are like those of the social sciences, an informed guess about a system that even the IPCC admits is too complicated to forecast except in the most general terms. Like economics, climatology does best when it tries to *explain* the many factors making up climate rather than make firm predictions as do the "hard" physical sciences like physics and chemistry. This view of climate science as explanatory rather than predictive is supported by IPCC contributor Mike Hulme, who writes:

> The problem is that in areas of science which are seeking to understand the behaviour of large complex systems which can't be replicated in the lab, it is very hard if not impossible to apply the scientific litmus test of *falsification through experimentation. And climate change is one such area of science.* We have scientific theory, we have empirical observations. What we haven't got are lots of different Earths that can be experimented on in controlled conditions. *Virtual climates created inside computer models are the best we've got.*
>
> All of this means that climate scientists frequently have to reach their conclusions on the basis of the *partial, and sometimes poorly tested, evidence and models available to them.* And when their paymasters—elected (or non-elected) politicians—ask them for advice, as in the case of the IPCC, *opinion and belief become essential for interpreting facts and evidence.* Or rather, *incomplete evidence and models* have to be worked on *using opinions and beliefs to reach considered judgments about what may be true.*[24] [italics added]

Yet, despite the mainly explanatory rather than predictive nature of climate science, alarmist climate science, indeed, environmentalism as a mass movement, is all about predictions—predictions of doom if we don't mend our carbon-emitting ways—because alarmist climate science conceives itself as a hard physical science rather than a "soft" science like the social sciences. Alarmist climate science claims the ability to go beyond explanation

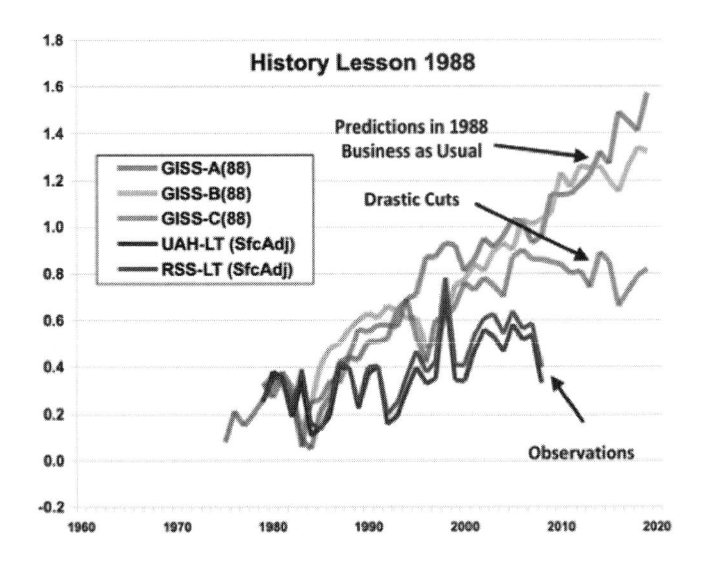

Figure 10.1: James Hansen's predictions compared to actual temperatures as of 2008.

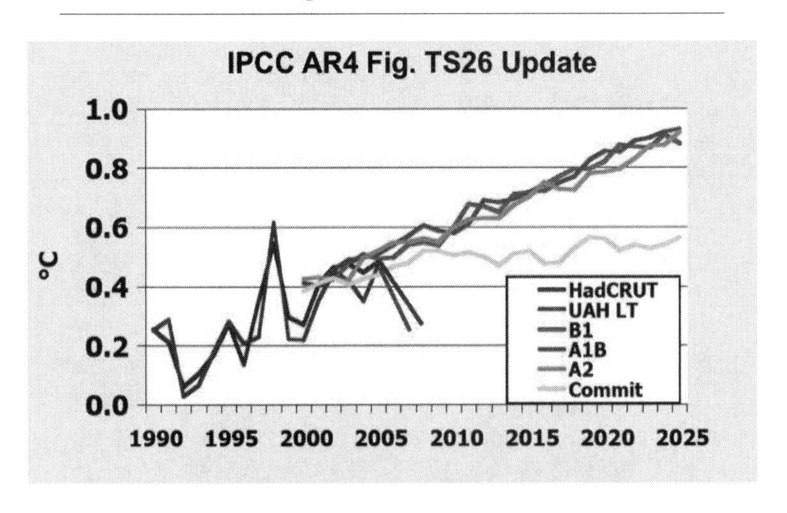

Figure 10.2: IPCC 2007 predictions (rising lines) and actual temperatures (bottom two, falling lines). The horizontal "Commit" line is the predicted temperature if the IPCC's recommendations were followed. The actual temperature is well below this line.

to robust predictions. At the same time, and unlike the hard sciences, alarmist climate science will not declare the anthropogenic hypothesis falsified if the results fail to match the predictions, e.g., if the planet doesn't warm as predicted by the models. In other words, unlike, say, Einstein, alarmist climate science wants to have its cake and eat it, too.

How accurate have alarmist climate science's "robust" apocalyptic predictions been, at least so far? Based on its hypothesis that warming is principally caused by anthropogenic carbon dioxide, consensus climate science makes predictions of future temperatures that prove to be not only too high, but often not even close to reality. So, for example, the Goddard Institute's James Hansen predicted in 1988 that temperatures would soar in the following 30 years. Instead, and even with a very large El Niño-caused peak in 1998, the global temperature has not warmed since the late 1990s and, in 2008, 20 years after Hansen's prediction, the planet had even cooled. Figure 10.1, which compares Hansen's predictions to the actual temperatures up to 2008, comes from climatologist John Christy, who is also a contributor to the IPCC. Note that the actual temperature is even lower than Hansen predicted after "drastic cuts" in carbon emissions.

Christy concludes:

> All [climate] model projections show high sensitivity to CO_2 *while the actual atmosphere does not.* It is noteworthy that the model projection for drastic CO_2 cuts still overshot the observations. This would be considered a *failed hypothesis test* for the models from 1988.[25] [italics added]

As we saw in Chapter 4, not one of the IPCC's "scenarios" has been even close to accurate in predicting the current temperature. Figure 10.2 shows a graph of temperatures predicted by the IPCC and the actual temperatures (the two falling lines) in the 21st century so far. The Climatic Research Unit's hacked emails ("Climategate") reveal clearly that senior IPCC climatologists were deeply concerned that the planet's current cooling was not reflected in their models—i.e., the models did not accurately predict what was occurring. As James Lovelock observes:

> The great climate science centres around the world are more than well aware how weak their science is. If you talk to them privately they're scared stiff of the fact that they don't really know what the

> clouds and the aerosols are doing. They [clouds and aerosols] could be absolutely running the show. We haven't got the physics worked out yet.[26]

Thus, Kevin Trenberth, a lead IPCC author, laments in one email: "The fact is that we cannot account for the lack of warming at the moment and it's a travesty that we can't."[27] And Tim Flannery, an arch-alarmist, notes:

> In the last few years, where there hasn't been a continuation of that warming trend, we don't understand all of the factors that create Earth's climate, so there are some things we don't understand, that's what the scientists were emailing about. ... These people [the scientists] work with models, computer modeling. When the computer modeling and the real world data disagree you have a problem.[28]

Yes, you do have a problem. But, rather than seriously acknowledging to the public the problem with their models, alarmist climatologists prefer to stick with the theory and ignore the reality, then expect the public, media and politicians to accept major political and economic changes based on these faulty models with their failed predictions. One climatologist, Chris Folland of the Hadley Climate Centre, even stated this preference directly: "The data don't matter... *We're not basing our recommendations on the data.* We're basing them on the climate models."[29] [italics added] In the Climategate emails, Folland complains that tree-ring data offered by fellow climatologist Keith Briffa shows that the Medieval Warm Period was, in fact, warmer than today, a fact, Folland correctly notes, "that dilutes the message rather significantly."[30] Scientists with integrity don't put the "message" ahead of the facts—they accept the facts. In the end, as we will see below in more detail, Briffa changed his data to conform to the "message" (the IPCC's anthropogenic global warming hypothesis).

This doesn't mean that temperatures couldn't suddenly zoom upward at some point in the future, even the near future; it just means that the IPCC climate models in the past have so far failed to predict what would be happening *now*, the first decade of the 21st century. And, if the climate did suddenly warm up, as the AGW hypothesis predicts, there is no evidence that this warming was due to the causes (*anthropogenic* greenhouse gases) claimed by the climate modellers—there are many possible reasons why the climate might warm apart from CO_2, much less human-created CO_2. In other words, if IPCC models did, finally, get the warming right, that could simply be a lucky guess. After all, if the models couldn't predict the climate a decade ahead, why would we expect a higher ratio of success for longer-range predictions?

So, if the climate did become warmer again, would that prove the models are right after all? Popper would say no—theories can't be conclusively *proved* by empirical tests, only *falsified.* Thus, the climate models may be like a stopped clock that has the correct time, but only twice a day, getting it right from time to time more by luck than design. Climatologist Jim Renwick, head of New Zealand's National Institute of Water and Atmospheric Research and a lead contributor to the IPCC, admits as much when he acknowledges: "Climate prediction is hard; *half of the variability in the climate system is not predictable* so we don't expect to do terrifically well."[31] [italics added] In other words, a climatologist is just as likely to accurately predict future climate by flipping a coin as by investing millions of dollars in a complicated computer model. Indeed, a study of the accuracy of computer climate models over the past 100 years found them consistently wanting. The study's Abstract states:

> The results show that models perform poorly, even at a climatic (30-year) scale. Thus local model projections *cannot be credible*, whereas a common argument that models can perform better at larger spatial scales is unsupported.[32] [italics added]

Even climate alarmist Lovelock agrees:

> We do need scepticism about the predictions about what will happen to the climate in 50 years, or whatever. It's almost naive, scientifically speaking, to think we can give relatively accurate predictions for future climate. There are so many unknowns that it's wrong to do it.[33]

In other words, the computer climate models have not shown themselves to be reliable predictors, yet these models are the *only* basis ("our only available tool," as Stephen Schneider puts it,[34] "virtual climates inside computer models are the best we've got," as Hulme writes[35]) for alarmist climate science's contention that humans are at fault for warming and that catastrophe looms.

And yet, we continue be told by climate modelers that their models are robust and valid. For example, Andrew Weaver states, during a debate with MIT climatologist Richard Lindzen:

> These climate models are not as uncertain as [Lindzen] makes out. There has been great success in recent years comparing the paleo-record of climate with modeling results of the same period. We've found that, yes, indeed, the models do a very fine job of capturing global climate changes in the past.[36]

Unfortunately, while they may be able to model the *past,* the models are not successful at capturing global climate changes in the *present* and *future*, perhaps because climate can't be accurately predicted, a fact that, as we've seen before, the IPCC itself acknowledges:

> In climate research and modelling, we should recognize that we are dealing with a coupled non-linear chaotic system, and therefore that *the long-term prediction of future climate states is not possible*. The most we can expect to achieve is the prediction of the probability distribution of the system's future possible states by the generation of ensembles of model solutions.[37] [italics added]

Of course, alarmist climatologists then ignore this statement in making their doomsday prophecies. Why? Perhaps because, as hydrologist Jay H. Lehr notes:

> Too often today we substitute mathematical models for observations. *Observed facts always take precedence over theory* no matter how beautiful the logic.[38] [italics added]

That is, although models may work in theory, the observed *fact* is that climatologists' predictions consistently fail to mirror reality.

However, alarmist climate scientists do not accept the "message" of actual temperature observations—that, as Christy puts it, the sensitivity of the atmosphere to carbon dioxide is much less than the models predict and that the anthropogenic hypothesis, at least in its extreme version, has been falsified. Some alarmists even continue to claim that warming is *accelerating* when, in fact, warming has stalled. For example, IPCC chairman Rajendra Pachauri told an audience in Australia in 2008: "We're at a stage where warming is taking place *at a much faster rate* [than before]."[39] [italics added] If the *head* of the IPCC can't get his facts right, what are we to think of the rest of the organization? Vicky Pope, of the Hadley Meteorological Centre, said in February 2010, "If anything, the world is warming *even more quickly* than we had thought."[40] For Pope, the world is warming "more quickly" at a time when the world isn't warming at all. This is Alice in Wonderland logic, not science. Meanwhile, for those willing to accept the evidence of their senses, i.e., the empirical data, the conclusions of the computer climate models have been falsified, or at least shown worthy of a healthy skepticism. In short, alarmist climate science fails the prediction test that is the essence of "hard" science.

SCIENCE OFFERS TESTABLE AND FALSIFIABLE HYPOTHESES

As we saw in Chapter 6, the worth of an hypothesis is tied to its ability to be tested and falsified, not its ability to be tested and confirmed. Popper states:

> I shall certainly admit a system as empirical or scientific only if it is capable of being tested by experience. These considerations suggest that not the *verifiability* but the *falsifiability* of a system is to be taken as a criterion of demarcation. ... It must be possible for an empirical scientific system to be refuted by experience.[41]

Physicist Victor J. Stenger offers another way of saying this:

> While failure to pass a required test is sufficient to falsify a model, the passing of a test is not sufficient to verify the model. This is because we have no way of knowing *a priori* that other, competing models might be found someday that lead to the same empirical consequences as the one tested.[42]

Thus, for example, a scientist may hypothesize that human-caused carbon emissions are causing the planet to warm. He may try to establish several predictive tests of his hypothesis with a computer model, and the hypothesis might pass those tests. The hypothesis that human carbon emissions are the prime cause of warming might become an accepted theory—"the science is settled," as the consensus likes to say.[43] But what if, while CO_2 emissions continue to climb, the planet stops warming and even begins to cool, as happened in the late 1990s? At that point, the theory is in danger of being falsified, if it hasn't already been falsified. If falsified, then clearly some other theory is needed to explain the warming that occurred from the mid-1970s to the late 1990s and the lack of warming and even cooling after the late

1990s, because the anthropogenic carbon-dioxide theory has failed a crucial test and, in science, the absence of failures, not a string of successes, is the mark of a strong theory. That said, and as we saw above, a major problem for alarmist climate science is that AGW is not an empirically *testable* hypothesis, in part because you can't put the earth in a test tube. Therefore, "Computer modeling is our *only available tool* to perform what-if experiments such as the human impact on the future."[44] [italics added] We've already seen how accurate these computer climate model predictions are—not very.

An important claim of the alarmist hypothesis is that our carbon emissions will eventually lead us to disaster. Since we can't get information from the future this hypothesis is not testable or falsifiable, although its proponents regard a disastrous future as almost certain based, again, solely on computer modeling ("our only available tool"). The reality is, we do not know if this hypothesis is true and won't know until disaster either happens or, far more likely, doesn't. As the Scripps Institute of Oceanography website notes:

> Thus [carbon emissions are] an experiment *whose course and outcome are uncertain.* ... We are passengers on a voyage into fog-shrouded uncharted waters. Some say, there are reefs ahead. Others say, there are none. *The truth is, no one knows.*[45] [italics added]

In other words, predictions of certain doom ("oblivion") are pure speculation, not the scientific fact that alarmist climate science claims. Therefore, it is reasonable to advise caution in predicting what the future might bring; even the highly respected Scripps Institute believes it is not reasonable to present this disaster to the public as *certain* ("the course and outcome are *uncertain";* "the truth is, *no one knows*"), as alarmist climate science does through the Doctrine of Certainty. Just because we don't know what the future might bring doesn't mean the future has to be disastrous, as the alarmists believe. Not all change is bad and the public should be hearing more than one opinion on the possible effects of climate change, not just the alarmists'.

Because the AGW hypothesis is not empirically testable and therefore not falsifiable except in computer models, this hypothesis does not belong to what we normally think of as "hard" physical science. But as we've seen, for those wedded to the anthropogenic global warming hypothesis, no amount of contrary evidence can persuade them that their overall hypothesis has been falsified. Or, as geologist Ian Plimer notes: "What would the IPCC and warmers accept as disconfirming evidence for their current popular paradigm? I have asked warmers many times. It is not possible to get an answer."[46]

SCIENCE DOES NOT BRING IN *AD HOC*, AUXILIARY HYPOTHESES

Another requirement of a strong hypothesis is that "the hypothesis being tested must be established clearly and explicitly before data taking begins, and not changed midway through the process or after looking at the data."[47] Fortunately, the anthropogenic global-warming hypothesis is admirably clear and explicit: the planet is warming *primarily* due to human fossil-fuel carbon emissions and deforestation; natural climate factors like the sun and oceans play minor roles. To falsify this hypothesis, it's only necessary to show that, despite increasing carbon dioxide levels, whether mainly from humans or not, the planet is not warming. If the planet isn't warming, then clearly carbon emissions cannot be the *principal* cause of warming, and human-caused carbon emissions, less than five per cent of the planet's natural annual carbon budget, *definitely* cannot be the "principal" cause of warming. And, in fact, this is exactly the case. Unfortunately for the alarmist paradigm, the planet has not warmed in the 21st century and IPCC lead author Phil Jones, a key figure in the Climategate emails, even suggests there has been no "statistically significant" warming since 1995.[48] However, rather than accept that the consensus hypothesis has been falsified, the hypothesis is "changed midway through the process." Thus, as environmental writer Mark Lynas, author of the hyper-alarmist book *Six Degrees: Our Future on a Hotter Planet*, writes:

> Although CO_2 levels in the atmosphere are increasing each year, *no one ever argued that temperatures would do likewise.* Why? Because the planet's atmosphere is a chaotic system, which expresses a great deal of inter-annual variability due to the interplay of many complex and interconnected variables. Some years are warmer and cooler than others.[49] [italics added]

No one ever argued that temperatures would not increase each year? That increased carbon emissions and increased warming go hand-in-hand is *pre-*

cisely what alarmist climate science argues; this *is* the alarmist hypothesis. *All* of the IPCC's computer climate models predict a steadily increasing temperature as CO_2 levels increase, as Figure 10.2 shows; they do not predict periods of non-warming and cooling like that of the first decade of the 21st century. For the alarmists, natural variation has taken a back seat to human intervention when it comes to climate. Until the planet stops warming, that is. When the planet stops warming, the consensus takes another tack: it adds an *ad hoc*, auxiliary hypothesis—when it's cooling, nature, not humanity, becomes the "principal" driver of climate.

Therefore, what initially appeared to be a clear and explicit hypothesis is no longer clear, and so the AGW hypothesis has "escaped refutation." As a final bolt hole, the alarmists can always say, and do, that even if the planet isn't warming *now*, it will in the future, which is a very safe bet. So is: the planet will cool in the future. We don't need expensive computer programs to predict that the planet will both warm and cool in the future and that, as Lynas rather fatuously puts it, "some years are warmer and cooler than others." What we need to know is *when* the planet will warm or cool and by *how much*, both questions that alarmist climate science cannot, accurately, tell us. In this respect, again, consensus climate science behaves more like a social science than the hard physical science it claims to be. On the reluctance to give up a failing hypothesis, urban theorist Jane Jacobs notes:

> Most people don't enjoy having their entire worldview discredited; it sets them uncomfortably adrift. Scientists are no exception. A paradigm tends to be so greatly cherished that, as new knowledge or evidence turns up that contradicts it or calls it into question, the paradigm is embroidered with qualifications and exceptions, along with labored pseudo-explanations [i.e., *ad hoc*, auxiliary explanations]—anything, no matter how intellectually disreputable or craven, to avoid losing the paradigm.[50]

SCIENCE DOES NOT PREJUDGE THE RESULTS

Another rule for scientific discovery is that those doing the studies must not be biased; instead, they must find and analyze the data "without any prejudice of how the results should come out."[51] As a corollary, scientists are not allowed to manipulate—"cherry-pick"—their data to get the result they expect or, in the case of alarmist climate science, want. But, of course, as we saw in Chapter 4, the IPCC and alarmist climate science in general *have* prejudged the results and any data that doesn't fit the desired results is discarded. The Michael Mann hockey stick, which eliminated the Medieval Warm Period and the Little Ice Age, is clearly an example of prejudgment, as is the IPCC's mission statement, which aims to uncover *only* "human-induced" climate change. And as we saw in Chapter 4, the body of the 1995 IPCC report found no discernible anthropogenic warming; however, the writers of the report's Summary for Policymakers, the part of IPCC reports that is subject to "review by governments," wanted anthropogenic warming. Therefore, key paragraphs of the main report that found no human-caused warming were removed to conform with the (politically controlled) summary, rather than, as one would expect, the summary conforming to the text of the full report.

Explicit evidence of prejudgment and cherry-picking by top IPCC climate scientists, including Mann, is clearly evident in the Climatic Research Unit (Climategate) emails.[52] For example, Phil Jones, then head of the CRU (he resigned after the scandal), comments (Nov. 16, 1999): "I've just completed Mike's [Michael Mann's] *Nature* [journal] trick of adding in the real temps to each series for the last 20 years (i.e., from 1981 onwards) and from 1961 for Keith's [climatologist Keith Briffa] to *hide the decline* [since the tree-ring data showed cooling, not warming, after 1961]." [italics added] Mann says in an email to Phil Jones and others (June 4, 2003): "It would be nice to try to 'contain' the putative 'MWP' [Medieval Warm Period], even if we don't yet have a hemispheric mean reconstruction available that far back." Climatologist Tom Wigley asks Jones (Oct. 5, 2009): "How does Keith [Briffa] explain the [Steve] McIntyre plot that compares Yamal-12 with Yamal-all? And how does he explain the apparent 'selection' of the less well-replicated chronology rather that the later (better replicated) chronology?" (The issue here is that the version of the hockey stick graph created by Briffa produced the flat "shaft" of the stick—hence, no Medieval Warm Period or Little Ice Age—by using a *partial* selection of tree core data for climate dating—Yamal-12—rather than the full selection of Yamal readings. However, as McIntyre demonstrated, the full Yamal data did *not* produce the flat hockey-stick shaft and, therefore, did not show 20th century warming as significantly higher than previous centuries.[53])

Similarly, Briffa wrote to Mann and Jones (Sept. 22, 1999): "I know there is pressure to present a nice tidy story as regards 'apparent unprecedented warming in a thousand years or more'." However, Briffa added: "In reality the situation is not quite so simple—I believe that *the recent warmth was probably matched about 1,000 years ago*." [italics added] In other words, the Medieval Warm Period was as warm or warmer than today; therefore, today's temperatures are not unprecedented or even unusual and the "hockey stick" is wrong. In the end, though, "Briffa changed the way he computed his data and submitted a revised version. This brought his work into line for earlier centuries, and 'cooled' them significantly."[54] On April 29, 2007, Briffa wrote to Michael Mann: "I tried hard to balance the needs of the science and the IPCC, *which were not always the same*. I worried that you might think I gave the impression *of not supporting you well enough* while trying to report on the issues and uncertainties." [italics added] Needless to say, for a true scientist, the needs of science *always* come ahead of an institutional ideology like that of the IPCC.

In summary, and as with so much of alarmist climate science, the CRU scientists cherry-picked data to reach a predetermined conclusion, the conclusion the IPCC wanted, a clear violation of this rule for scientific research.

SCIENCE ENCOURAGES OTHER RESEARCHERS TO REPLICATE FINDINGS

In science, "reported results must be of such a nature that they can be independently replicated. Not until they are repeated under similar conditions by different (preferably) skeptical investigators will they be finally accepted into the ranks of scientific knowledge."[55] Here, again, we find at least two leading figures in alarmist climate science reluctant to reveal the data that led to their results: Michael Mann, inventor of the "hockey stick," and then-CRU head Phil Jones, who has collaborated with Mann and whose institute supplies temperature data to Britain's Hadley Meteorological Centre.

As we saw in Chapter 4, the so-called "peer reviewers" of Mann et al.'s 1998 hockey stick article in *Nature* made no attempt to check the article's figures, even though those figures wiped out two prominent climate features of the past thousand years, the Medieval Warm Period and the Little Ice Age. Given this, it is astonishing that red flags weren't raised for the Mann paper reviewers.

Canadian statistician Steve McIntyre *did* wonder how the MWP and LIA could have been so cavalierly dismissed and asked Mann for his data so it could be checked. To make an eight-year story short, Mann released some of his data but not all of it, and McIntyre, now working with Canadian economist Ross McKitrick, was unable to reproduce Mann's results.[56] Eventually, the U.S. Congress's House Energy and Commerce committee stepped in to demand, as required by U.S. law for publicly funded research, that Mann reveal all the data that led to his 1998 *Nature* paper's astonishing hockey-stick conclusions. As for Jones: when asked in 2004 to reveal the raw data for the temperatures given out by the Climatic Research Unit, he replied: "We have 25 or so years invested in the work. Why should I make the data available to you, when your aim is to find something wrong with it?"[57] Eventually, in 2007, Jones was forced by the UK Information Commissioner's Office to release the data.[58] More recently, the Climategate emails revealed several efforts by Jones to evade Freedom of Information requests for his data, as required by British law. For example, Jones writes to Tom Wigley (Jan. 21, 2005): "If FOIA [the Freedom of Information Act] does ever get used by anyone, there is also IPR [Intellectual Property Rights] to consider as well. Data is covered by all the agreements we sign with people, *so I will be hiding behind them*." [italics added]

Mann didn't want to release his data because his 1998 paper said what the climate alarmists wanted to hear—that temperatures today are warmer than any time in the past millennium—and he didn't want the paradigm challenged by an outsider. Within the paradigm, he was protected because peer reviewers have work of their own and limited time to assess others' work. Therefore, climate science peer review is often just a matter of ensuring that the paper concurs with "the popular paradigm."[59] Indeed, the Climategate emails show very clearly that peer review in alarmist climate science consists of a clique of well-acquainted scientists who agree on the AGW hypothesis and are eager to promote papers that support it, while blocking papers that don't support it. And scientists who might have doubts are reluctant to go against the majority, as exemplified by one of Jones's emails (July 5, 2005) in which he states: "*The scientific community would come down on me in no uncertain terms* if I said the world had cooled from 1998. OK it has but it is only seven years of data and it isn't statistically significant." [italics added] And, as we saw above, Briffa tailored his Yamal results to avoid offending fellow alarmist believers. As a *Daily Telegraph* columnist notes: "What the CRU's hacked emails convincingly demonstrate is that climate scientists in the AGW camp have corrupted the peer-review process."[60]

One would hope Jones and Mann are the exceptions and that most alarmist climate scientists are willing to release their data even to those whose "aim is to find something wrong with it," because finding something wrong with the data, i.e., trying to falsify it, is exactly what scientists are supposed to do. Yet Mann and Jones are two *leading* IPCC climatologists who, apparently, don't understand basic scientific principles and didn't trust their data enough to have it brought under critical scrutiny outside their peer-review clique. As Ian Plimer notes of the hockey stick episode:

> After reading the history of the "hockey stick," no one could ever again trust the IPCC or the scientists and environmental extremists who author the climate assessments. The IPCC has encouraged a collapse of rigor, objectivity and honesty that were once the hallmarks of the scientific community.[61]

SCIENCE DEMANDS STRONG EVIDENCE FOR EXTREME CLAIMS

Alarmist climate scientists often make extreme claims that the public, media and politicians are supposed to accept because, well, the claims are coming from scientists and if you can't trust a scientist, who can you trust? On extreme claims, the pioneering 19th-century scientist Thomas H. Huxley (1825-1895) wrote: "It is a canon of common sense, to say nothing of science, that the more improbable a supposed occurrence, the more cogent ought to be the evidence in its favor."[62] In the 20th century, astronomer Carl Sagan observed, in the same vein: "Apocalyptic predictions require, to be taken seriously, higher standards of evidence than do assertions on other matters where the stakes are not as great."[63] Sagan's comment often appears as "extreme claims require extreme proof," but Huxley beat him to the punch.

If this is so, how much credence we should place in *extreme* climate claims of the type that Gore, Suzuki, Hansen, Weaver, Schneider and other prominent climate alarmists present as scientific fact, or at least as scientifically plausible? Among these extreme claims is Weaver's ominous prediction of a "sixth extinction" that will wipe out "between 40 per cent and 70 per cent of the world's species" should the global temperature rise above 3.3 degrees Celsius (a rise that would be, for Weaver, entirely humanity's fault).[64] Weaver has also called for a complete ban on fossil-fuel use,[65] which would be catastrophic for developed economies and their people. Similarly, the Goddard Institute's James Hansen warns of possible sea level rises of five metres (16 feet) in the next 85 years.[66] And, of course, there are Gore's apocalyptic predictions, such as New York under six metres of water in the near future, no summer Arctic ice by 2014,[67] etc., if we fail to follow his draconian political and economic program. By contrast, even the alarmist IPCC's 2007 report predicts a minimum sea-level rise of 18 cm (6 inches) and a maximum of 59 cm (23 inches) by 2100, depending on the scenario.[68] Meanwhile, thanks to a cooling planet, the Arctic ice is returning to what is considered "normal."[69] In fact, as we saw in Chapter 8, so far, none—not one—of the environmentalists' extreme and apocalyptic predictions, from Thomas Malthus to Paul Ehrlich (mass starvation in the 1970s) to Suzuki, Weaver and Gore, has come to pass. Or, as columnist and broadcaster Rex Murphy notes:

> So much of what the alarmists promised was supposed to be happening now isn't happening. So many events are running counter to their near-term projections, they've decided to go all Armageddon with their long-term ones, projections for a future that none of us will be around to check.[70]

By any standard, the claims of Gore, Weaver, Hansen, et al., are extreme. Yet we are expected to accept these extreme claims with almost no public debate, scrutiny, or criticism—after all, the science is "settled," there is a "consensus," the alarmist climate scientists are the experts and the situation is in "crisis"—based on no empirical evidence, unless mathematical models are considered the equivalent of firm empirical evidence.

However, when it comes to climate science, as we've seen, computer models are *not* the equivalent of empirical evidence—so far they have not accurately predicted the *actual* behavior of the climate and perhaps never will. As one skeptical journalist has noted:

> One could fill pages either of global warming's manifest absence from our lives or its failure to show up on schedule or as expected. Where are the hurricanes, the sea level increases, the floods in Europe, the steady signs of warming?[71]

Answer: these predicted disasters aren't there. Therefore, lacking empirical support, climate alarmists abandon scientific principles of evidence to fall back on the Precautionary Principle—if the worst *could* happen we must act as if it *will* happen. This attitude is nicely summed up by environmental writer Jonathan Schell, who warns: "Now, in a widening sphere of decisions, the costs of error are so exorbitant that *we need to act on theory alone*. It follows that the reputation of scientific prediction needs to be enhanced."[72] [italics the author's] One way to enhance the reputation of scientific prediction would be, of course, to make more accurate predictions based on better theories. Since that's not going to happen in alarmist climate science, anyone who dares to challenge the alarmists' claims, no matter how extreme, and asks for proof more convincing than the flawed predictions of computer models, is labeled a "denier" and, if possible, silenced.

SCIENCE REGARDS THE TRUTH AS SACRED

The final and by far the most important feature of science is a commitment to truth. This is, or should be, science's highest value. In other words, just as life is sacred for the good medical professional and honor is sacred for the good soldier, truth is sacred for the good scientist. A commitment to truth *is* science; science is based on its integrity, honesty and objectivity. The public's trust in science, when many other institutions such as religion and politics have lost this trust, is based on the belief that science is aiming for the non-biased truth as much as humanly possible. This doesn't mean science is value-free. As so often, Friedrich Nietzsche saw deep into the soul of humanity and of science when he wrote:

> It is still a *metaphysical faith* upon which our faith in science rests—that even we seekers after knowledge today, we godless ones and anti-metaphysicians still take our fire, too, from the flame lit by a faith that is thousands of years old, that Christian faith which was

> also the faith of Plato, that God is the truth, that truth is divine.[73] [italics Nietzsche's]

In short, science is based on an underlying value that, like all underlying values, can't be proved but must be taken on faith. And that underlying value—rooted, for Nietzsche, in Christianity and the ancient Greeks and held by all ethical scientists regardless of their religious beliefs or lack of them—is that *the truth is sacred* (Nietzsche uses the term "divine"). Lovelock makes the sacredness of scientific truth quite clear when he observes:

> Fudging the data in any way whatsoever is quite literally *a sin against the holy ghost of science*. I'm not religious, but I put it that way because I feel so strongly. It's the one thing you do not ever do. You've got to have standards.[74] [italics added]

And yet, as we saw in Chapter 8, leading climate alarmists, including Stephen Schneider ("scary scenarios"), James Hansen ("emphasis on extreme scenarios"), Phil Jones ("hide the decline") and Al Gore ("over-representation of factual presentations") are willing to compromise the truth if the unvarnished evidence doesn't produce the terrified reaction they want from the public, media and politicians. Further: alarmists have been bludgeoning the public about global warming for the entire first decade of the 21st century—a time *when no warming has occurred*. Just to repeat: During the entire decade when we've been repeatedly lectured about global warming—the science is "settled," the evidence is "overwhelming," the warming is "unequivocal," etc.—*there has been no warming*. And, when the lack of warming became obvious in about 2007, and as the Climategate emails clearly show, alarmist climatologists deliberately tried to conceal this news from the public and even other researchers ("hide the decline"). In other words, many alarmist climatologists have been either deliberately dishonest with the public, or incompetent at prediction, or both.

This willingness to subvert the truth isn't just confined to leading alarmist climate scientists. For example, a 2008 poll of 367 climate scientists asked respondents the following question: "Some scientists present extreme accounts of catastrophic impacts related to climate change in a popular format with the claim that their task is to alert the public. How much do you agree with this tactic?" *More than 14 per cent* of respondents, or 51 climate scientists, considered this "extreme" tactic acceptable; an additional 12.5

per cent were neutral, for a total of 27 per cent, or 99 climate scientists.[75] In other words, at least *one in four* climate scientists who answered this poll would be willing to mislead the public themselves, or would not consider it wrong for other scientists to mislead the public (i.e., they were neutral), by presenting extreme scenarios. One in four. This is an astonishing percentage in a profession—science—that relies so much on the public perception of its integrity. It's hard to imagine so many scientists in any other discipline admitting that they would be willing to mislead the public if they thought it necessary to compel belief in their scientific hypotheses—why not just present the evidence? If it's strong enough, the public will believe. As for the 75 per cent of climate scientists who still care about honesty, there has never, to my knowledge, been a discipline-wide condemnation of alarmist climate promoters who make extreme and misleading claims, such as Hansen, Schneider and particularly Gore. Indeed, physicist Richard A. Muller defends Gore's exaggerations, as well as Gore's inability to live the simple, low-carbon lifestyle he preaches for the rest of us. Miller states:

> If [Al Gore] reaches more people and convinces the world that global warming is real, *even if he does it through exaggeration and distortion— which he does, but he's very effective at it*—then let him fly any plane he wants.[76] [italics added]

And yet, the public and governments are expected to accept what alarmist global warming believers claim based not on evidence, which these alarmists openly admit they have distorted and exaggerated[77] or simply don't have (hence, reliance on "intuition," see Chapter 7, and on the Precautionary Principle), but on their authority. We wouldn't accept the "trust me" of a used-car salesman who used deceptive tactics to make a sale; why would we accept the word of Gore, Hansen, Schneider, et al., when we know they are willing to exaggerate or lie? As philosopher Walter Kaufmann has observed:

> It is by no means immoral for a scientist, historian, or philosopher to hope that some proposition may be proved true, or to feel strongly about it. But it is considered immoral for him to be partial to the point of suppressing relevant evidence, and it is a sign of incompetence, *if not a violation of professional ethics,* if he fails to undertake a relevant investigation for fear that its results might be fatal for a belief he cherishes.[78] [italics added]

Indeed, the practice of bending the truth when presenting science to the public violates the proposed scientific code of ethics. The code, similar to the Hippocratic Oath that binds physicians, includes as one of its seven tenets the following:

> *Do not knowingly mislead*, or *allow others to be misled*, about scientific matters. Present and review scientific evidence, theory or interpretation *honestly and accurately*.[79] [italics added].

When the public, media and politicians are misled by scientific claims that are not honest or accurate, as *leading* alarmist climate scientists do far too often and as the environmental movement does routinely, and when other climate scientists allow the public to be misled because they fear losing research grants (where are the cries of protest within the "consensus" camp against climate exaggeration from scientists?), science loses something very precious: its soul. At that point it is no longer science but political and social advocacy. It loses something else very precious: its credibility with the public. At that point science has ceased to be, as Carl Sagan put it, a "candle in the darkness." Instead, it becomes part of the darkness.

Climate science, at least in its alarmist form, has lost its soul. It still has some credibility with the public, but only because—in the belief that the normal rules of scientific proof don't apply because it is warning of a "crisis"—alarmist climate science does everything it can to keep the public, media and politicians from learning about the flaws in its arguments. However, as the truth comes out through avenues like Climategate, public support for climate alarmism has plummeted. For example, a March 2010 Gallup poll shows worry about global warming among U.S. respondents is at the bottom of a list of eight environmental issues,[80] and rightly so, while British ultra-alarmist George Monbiot, recognizing that the public isn't buying his "science," has declared: "There goes my life's work."[81]

But *why* has alarmist climate science lost its commitment to truth at all costs? One reason is that alarmist climate science has made a huge *philosophical* error.

SOCRATES HAD IT WRONG: THE GOOD IS NOT NECESSARILY THE TRUE

One of the most famous poems in the English language, "Ode on a Grecian

Urn" by John Keats (1795-1821), ends with two of the most famous lines in English poetry:

> "Beauty is truth, truth beauty,"—that is all
>
> Ye know on earth, and all ye need to know.

The idea that truth and beauty are synonymous comes from the Greeks, through Socrates/Plato, who added a third term: the Good. So, for Plato, the good, the true and the beautiful are synonymous: "The divine is beauty, wisdom, goodness, and the like."[82] As an idea, this trinity of the beautiful, the good and the true has a powerful attraction and most of us would agree with Keats's version of it: of course what is beautiful is true and what is true is beautiful! Unfortunately, in science, this fusion of the true, the good and the beautiful is pernicious because what we consider beautiful and/or good is a *value*, whereas truth is a *fact*, and science is, ideally, value-neutral. For science, the ultimate value is Truth, not the Good or Beautiful, although, of course, there are many times when what we regard as good and/or beautiful is also true. However, when it comes to science, as philosopher of science Michael Ruse notes:

> An important mark of professional science is that *it tries to divorce itself from the immediate concerns and values of the society that is producing it.* ... Successful professional science stands alone, its truth separate from the interests or cultural background of the person who produces it or who accepts it.[83] [italics added]

Bertrand Russell goes so far as to chastise Plato for his "identification of the good with the truly real," thereby giving "a legislative function to the good," that is, allowing values (the perceived Good) to be involved in determining what is regarded as factual truth.[84] As philosopher of science Jeffrey Foss writes:

> Science has traditionally assumed a position of value neutrality. The value neutrality of science, moreover, is not a mere add-on for science, not an optional feature. Value neutrality has been *essential* to science from the start, and still is. It has enabled science both to make astounding progress in the pursuit of knowledge about nature and to gain high respect as an authority about the natural world. ...
>
> The only value that motivates scientific activity as such is that of the knowledge of nature. *No matter how worthy or noble in them-*

> *selves, all other values are a hazard to scientific objectivity.* Religious, political, aesthetic, or environmental ideals may distort scientific judgment, leading to scientific claims being made because they seem worthy of acceptance given these ideals, even where unjustified by the evidence.[85] [italics added]

A science that fails to make truth its main concern—that fails to regard truth as sacred—is no longer science. Instead, it has become an advocate for a particular value system regarded as transcendentally Good.

The Good that the UN, IPCC and alarmist climate science in general are promoting over Truth is, as Foss terms it, "the health of the environment," i.e., environmentalism. In effect, as we will see in Chapter 11, environmentalism has become a religion and for Believers it is the health of the environment, not factual truth, that is sacred. And, many of those who hold environmentalism as a religion are also scientists, including climate scientists, whether they acknowledge their faith publicly or not.

Now, of course, we all care about the health of the environment, or should. And in a free society individuals and environmental groups have the absolute right to make the health of the environment their primary value, although it would be better if they didn't distort the facts to support their case, as, say, Al Gore does so skillfully in *An Inconvenient Truth.* However, a *scientist*, even one whose strongest belief is the religion of environmentalism, must not make the health of the environment his or her *primary* Good; Truth must still come first. Instead, many leading alarmist climatologists and climate spokespeople like Gore are centred on environmental values, not environmental facts. Yet, if Copernicus had put the Good of his time—the religious "truth" that the sun revolved around the earth—ahead of the empirical facts, the development of science might have been set back decades or centuries.

This book documents many cases in which alarmist climate science has put environmental ideology ahead of scientific truthfulness. For example, as we saw in Chapter 6, environmentalist Norman Myers declared that we could expect 40,000 species extinctions a year, a figure he later admitted he had more or less made up "to get the issue of extinction onto scientific and political agendas."[86] Similarly, Rajendra Pachauri, the president of the IPCC as of early 2010,[87] told an Australian audience in 2008 that global warming was *accelerating*, at a time when global warming had not occurred since the 1990s.[88] Global warming may well accelerate in the future, but that's not what Pachauri said. In other words, in the name of the greater perceived environmental Good, even the *president* of the IPCC is willing to *quite openly* distort the climate facts. As we saw in Chapter 4, a claim that the Himalayan glaciers would almost disappear by 2035 was known to be false but was included in the 2007 IPCC report to put pressure on politicians.[89]

Similarly, the IPCC continues to claim that global warming will cause more and more violent hurricanes, even though this is not true and the IPCC's scientists *know* it isn't true. Why make this claim, then? Because the IPCC's ultimate value, environmental health, is "better served by the general acceptance of the hypothesis that global warming will bring dire results, such as more violent hurricanes—whether or not this is true."[90] In late 2009 and early 2010, a raft of other IPCC errors were revealed—errors the IPCC should have caught but didn't because the false or exaggerated claims agreed with its environmental "message."

All human institutions, including science, are based in values, some of which are considered to be "sacred." Nor is it possible for science to be completely divorced from society. As literary critic Northrop Frye writes:

> Science can never exist outside a society, and that society, whether deliberately or unconsciously, directs its course. Still, the importance of keeping science "free," i.e., unconsciously rather than deliberately directed, is immense.[91]

To deliberately direct science based on a moral or social agenda, as the IPCC does and as much of alarmist climate science does, is to give the Good "legislative power." At that point, science is no longer "free." Instead, it has become "deliberately directed" based on someone's agenda, in this case the agenda of the IPCC and of the politicians who oversee the IPCC's conclusions. If science has any other value but one, empirical truth, then it cannot be, and cannot claim to be, unbiased and objective. In practice, from the values point of view, alarmist climate science acts more like a religion than a science, as we will see in more detail in Chapter 11.

WHAT WE KNOW

So, is consensus climate science a "science"? Of course, it is a science in the sense that economics and sociology and biology are sciences. However, as we've seen, these "soft" sciences are more suited to collecting and explaining data than making hard predictions. So the question, more specifically, is: is consensus climate science a "hard" physical science that can make robust, testable, falsifiable predictions, as its practitioners would like the public, media and governments to believe and as politicians would like the public to believe? Here the answer can only be no, both theoretically (you can't design falsifiable tests at the planetary level) and practically (alarmist climate predictions have consistently been wrong). Nor, as we've seen, does climate science, at least in its alarmist form, meet the basic requirements for good science expected of even the "soft" sciences.

POSTSCRIPT: THE CLIMATEGATE SCANDAL

After the CRU Climategate emails, alarmist climate science had a chance to show that it was concerned for its integrity as a discipline and the integrity and reputation of science as a whole. In most scientific disciplines, a scandal like the CRU emails would prompt a thorough soul-searching and house-cleaning, including the expulsion and public disgrace of scientists who had been revealed as manipulating data (i.e., scientific fraud), suppressing (as opposed to openly engaging) opposing viewpoints, refusing to release data for independent verification, and rigging the peer-review system in their favor. In this respect, it's worth keeping in mind that Jones, Mann, Trenberth and others named in the emails are not minor figures, but *leading* IPCC authors and supporters. However, rather than take direct action against clear scientific abuse and bias by some of its key contributors, the IPCC decided instead to create an "independent" panel to investigate criticism of its reports.[92] The United Nations will also be examining the 2007 IPCC report's errors.[93] In short, alarmist climate science is clearly a discipline that cannot regulate itself, in part because it fully supports the "message" of the Climategate scientists. It would, therefore, be unfair to punish them for doing what the IPCC asked them to do: create a fearful image of uncontrolled global warming.

Fortunately, post-Climategate, some scientists, including climatologists, have stepped up to defend the integrity of climate science and of science in general (although, one wonders where they were before Climategate, when the same corruption of science was just as evident, as I discovered when I began researching this book in 2007). So, Tulane University physicist Frank J. Tipler writes:

> The now non-secret data prove what many of us had only strongly suspected—that most of the evidence of global warming was simply made up. That is, *not only are the global warming computer models unreliable, the experimental data upon which these models are built are also unreliable.* .... This deliberate destruction of data [by the CRU[94]] and the making up of data out of whole cloth is the real crime — the real story of Climategate. *It is an act of treason against science.* It is also an act of treason against humanity, since it has been used to justify an attempt to destroy the world economy.[95] [italics added]

Climatologist and IPCC contributor Mike Hulme, who teaches at East Anglia University, also calls for a reappraisal of climate science and the highly politicized IPCC in the wake of the CRU emails:

> This event might signal a crack that allows for processes of re-structuring scientific knowledge about climate change. It is possible that some areas of climate science have become sclerotic. It is possible that climate science has become too partisan, too centralized. The tribalism that some of the leaked emails display is something more usually associated with social organization within primitive cultures; it is not attractive when we find it at work inside science.
>
> It is also possible that the institutional innovation that has been the IPCC has run its course. Yes, there will be an AR5 [a report in 2012 or so] but for what purpose? The IPCC itself, *through its structural tendency to politicize climate change science*, has perhaps helped to foster *a more authoritarian and exclusive form of knowledge production*—just at a time when a globalizing and wired cosmopolitan culture is demanding of science something much more open and inclusive.[96] [italics added]

Georgia Tech climatologist Judith Curry writes:

> What has been *noticeably absent* so far in the Climategate discussion is *a public reaffirmation by climate researchers of our basic research values*: the rigors of the scientific method (including reproducibility), research integrity and ethics, open minds, and critical thinking.

> Under no circumstances should we ever sacrifice any of these values; the CRU emails, however, *appear to violate them.*[97] [italics added]

The Institute of Physics, which represents 36,000 physicists worldwide, presented a brief to the British House of Commons committee investigating the Climategate scandal that appears to understand, unlike the IPCC, that there is a serious problem with alarmist climate science as currently practised. The Institute's statement reads, in part:

> The CRU e-mails as published on the internet provide *prima facie* evidence of determined and co-ordinated *refusals to comply with honourable scientific traditions* and freedom of information law. … There is also reason for concern at the *intolerance to challenge* displayed in the e-mails. This impedes the process of scientific "self correction", which is vital to the integrity of the scientific process as a whole, and not just to the research itself. In that context, those CRU e-mails relating to the peer-review process suggest *a need for a review of its adequacy and objectivity* as practised in this field and *its potential vulnerability to bias or manipulation.*[98] [italics added]

Remote Sensing Systems climatologist Petr Chylek has observed:

> For me, science is the search for truth, the never-ending path towards finding out how things are arranged in this world so that they can work as they do. That search is never finished. *It seems that the climate research community has betrayed that mighty goal in science. They have substituted the search for truth with an attempt at proving one point of view.* …
>
> Yes, there have been cases of misbehavior and direct fraud committed by scientists in other fields: physics, medicine, and biology to name a few. However, *it was misbehavior of individuals, not of a considerable part of the scientific community.* The damage has been done. The public trust in climate science has been eroded. At least a part of the IPCC 2007 report has been put in question. We cannot blame it on a few irresponsible individuals. *The entire esteemed climate research community has to take responsibility.* …
>
> Let us stop making *unjustified claims and exaggerated projections about the future* even if the editors of some eminent journals are just waiting to publish them. Let us admit that our understanding of the climate *is less perfect than we have tried to make the public be-*

> *lieve. Let us drastically modify or temporarily discontinue the IPCC.* Let us get back to work.[99] [italics added]

Even Andrew Weaver, a leading IPCC contributor and strong promoter of the apocalyptic warming hypothesis, acknowledged that the IPCC was guilty of "a dangerous crossing of that line" between science and political advocacy.[100] (The irony here is that Weaver himself also frequently crosses that line, to the point where a review of his book *Keeping Our Cool* was headlined "Climate author goes political."[101])

However, most of the alarmist climate science establishment have refused to recognize that is has a severe problem with integrity and instituted changes to deal with the problem. Instead, alarmist climate science has circled the wagons and preferred to focus on the "illegality" of making the "stolen" emails public, rather than reacting to the content of the emails themselves. For example, Working Group I of the IPCC issued this statement:

> IPCC WGI condemns the *illegal act* which led to private emails being posted on the Internet and firmly stands by the findings of the AR4 [Assessment Report 4, i.e., 2007 report] and by the community of researchers worldwide whose professional standards and careful scientific work over many years have provided the basis for these conclusions. … The internal consistency from multiple lines of evidence *strongly supports* the work of the scientific community, including those individuals singled out in these email exchanges.[102] [italics added]

As for the reaction from the pinnacle of alarmist climate science: IPCC chairman Rajendra Pachauri, too, has defended the email authors and condemned the release of the emails as an attempt to discredit climate science.[103] Yet what is obvious to anyone outside the alarmist paradigm is that what discredits climate science is not the *release* of the emails but their *content*.

The content of those emails is so damaging that, in an interview with the BBC, even disgraced former CRU head Phil Jones appears to have recanted at least some of his extreme views and directly or implicitly admits he and his Climategate colleagues distorted the climate data in presenting it to the public. For example, Jones agrees with the BBC interviewer that there had been "no statistically significant warming" since 1995, whereas in the emails he was at pains to hide this fact from the public and even fellow researchers.[104] He admits that the warming from 1975-1998, which the IPCC has claimed was *unusually rapid* due to human carbon emissions, was no more accelerated that previous warmings from 1860-1880 and 1910-1940 when, the IPCC says, natural factors, not humans, were the cause. Jones admits that from 2002-2009 the planet has been cooling slightly (-.12°C per decade), although he contends that "this trend is not statistically significant." Still, this is more honest than trying to make the public believe that warming is, as Pachauri and other alarmists continue to claim, accelerating. And he admits that the Medieval Warm Period might, maybe, perhaps have been warmer than today—a big step for someone who, previously, was a party to eliminating the MWP entirely through the hockey stick and was utterly committed to proving that the late 20th century's warming was "unprecedented."

However, Jones says he still believes humans were at fault for the late 20th century warming because "we can't explain the warming from the 1950s by solar and volcanic forcing." As one commentator on Jones's interview notes: "It seems the [anthropogenic] belief is based more on the absence of knowledge than the presence of proof."[105] In other words, as we've noted so often in this book and which Jones here implicitly acknowledges, there is *no* empirical proof that humans are to blame for the recent warming. The IPCC just can't find another culprit.

Jones's admissions, given only after his organization's betrayal of scientific truth was exposed in the emails, plus the refusal of alarmist climate science to deal seriously with what the Climategate emails reveal about key IPCC scientists and their manipulation of data, are sure indictors that climate science in its alarmist form has lost its commitment to truth and is more interested in defending an ideology—the alarmist "message"—than presenting the not-so-alarming facts. In short, alarmist climate science has ceased to be a science in the true sense. In the next two chapters, we'll discuss in more detail why this astonishing betrayal of scientific principles might have occurred.

There is so much of a blind and casual acceptance of global warming as the crisis of our time, or its high "moral essence," and such an overwhelming pressure to accept its tenets and claims as to amount to a stampede. There is, in other words, so much that is antithetical to the spirit of real science in the various campaigns being waged as to immensely discount the principal claim that it is science that is driving them all.

—Rex Murphy[1]

Some individuals are absolutely convinced from their very limited reading that we are heading for disaster via global warming. Strangely, no amount of evidence seems to shake this crowd. They appear to have a religious attachment to the issue: they read a few reports about the potential threats associated with the greenhouse effect, and they are sold. They have a set of policies that are strongly supported by the perception of a greenhouse catastrophe, and they are not going to accept anything but the threat of disaster. ... They care about the environment more than they care about science.

— Robert C. Balling Jr.[2]

Truth consists in some form of correspondence between belief and fact.

—Bertrand Russell[3]

When people cease to believe in God, it is commonly supposed they believe in nothing. Alas, the truth is worse. They believe in anything.

—Attributed to G.K. Chesterton[4]

Chapter 11

ENVIRONMENTALISM, RELIGION, AND GLOBAL WARMING

Many observers today argue that environmentalism has become a religion. Indeed, this fact is glaringly obvious to non-believers—that is, people who believe in environmental protection but not in environmentalism as the Highest Good. As science writer Gregg Easterbrook puts it, "Environmental debates have a much stronger religious aspect than most participants like to let on."[5] The theoretical physicist Freeman Dyson has elaborated on this theme in the *New York Times Review of Books*:

> There is a worldwide secular religion which we may call environmentalism, holding that we are stewards of the earth, that despoiling

> the planet with waste products of our luxurious living is a sin, and that the path of righteousness is to live as frugally as possible. The ethics of environmentalism are being taught to children in kindergartens, schools, and colleges all over the world. Environmentalism has replaced socialism as the leading secular religion.[6]

Dyson is actually quite positive about the idea of environmentalism as a religion: "Environmentalism, as a religion of hope and respect for nature, is here to stay. This is a religion that we can all share, whether or not we believe that global warming is harmful." However, he also notes that the environmentalist religion shares with many if not most other religions a tendency to adopt a "dogmatic tone" in its dealings with unbelievers (in a global warming context, skeptics or "deniers"). He writes:

> Unfortunately, some members of the environmental movement have also adopted as an article of faith the belief that global warming is the greatest threat to the ecology of our planet. That is one reason why the arguments about global warming have become bitter and passionate. Much of the public has come to believe that anyone who is skeptical about the dangers of global warming is an enemy of the environment. …
>
> Many of the skeptics are passionate environmentalists. They are horrified to see the obsession with global warming distracting public attention from what they see as more serious and more immediate dangers to the planet, including problems of nuclear weaponry, environmental degradation, and social injustice. Whether they [the skeptics] turn out to be right or wrong, *their arguments on these issues deserve to be heard.*[7] [italics added]

But, as this book has tried to show, other, skeptical arguments are not being heard, and even actively suppressed by alarmist climate science. For example, Dyson quotes a pamphlet entitled "Climate Change Controversies: A Simple Guide" by The Royal Society, Britain's national academy of science, that says:

> This [pamphlet] is not intended to provide exhaustive answers to every contentious argument that has been put forward by those who seek to *distort and undermine* the science of climate change and *deny the seriousness of the potential consequences of global warming.*[8] [italics added]

Compare this comment with what philosopher Karl Popper considered the true spirit of science:

> If you are interested in the problem which I tried to solve by my tentative assertion, you may help me *by criticizing it as severely as you can*; and if you can design some experimental test which you think might refute my assertion, I shall gladly, and to the best of my powers, *help you to refute it*.[9] [italics added]

For Popper, true science bends over backwards to consider alternative hypotheses. By contrast, the Royal Society pamphlet is closed to opposing ideas—as closed as any dogma-based theocracy.

However, those caught up in the extreme environmentalist ethos do not see their vision as "religion"—many would consider themselves non-religious in the traditional sense—but as scientific Truth or Reality. Indeed, many are insulted at the suggestion that they believe in a religion. For example, a letter from an environmentalist to the *Times Colonist* noted that to be called religious was to be lumped in with the "crazies, ignorant, fear-mongers, or anti-progress Luddites."[10] Instead, for this letter-writer, environmentalism is based on a *scientific* view that the planet is heading toward disaster and that our grandchildren "will all inherit a blighted earth." And just as no amount of evidence can persuade the deeply religious person that his or her views are not Truth, so is this often the case with many in the environmental movement. For example, economist Julian Simon once addressed a meeting of environmentalists and asked them the following questions:

> "How many people here believe that the earth is increasingly polluted and that our natural resources are being exhausted?" Everyone agreed. He said, "Is there any evidence that could dissuade you?" Nothing. Again: "Is there any evidence I could give you—anything at all—that would lead you to reconsider these assumptions?" Not a stir. Simon then said, "Well, excuse me, I'm not dressed for church."[11]

That is, Simon's environmentalist audience clearly fell into the "religious" category. If no amount of scientific evidence will persuade you that a belief is wrong, then your belief is not based in science, it is based in faith.

This refusal to change one's views is acceptable in religion because religion is not based on empirical evidence and doesn't claim to be. Therefore, for the true believer there is no possibility of falsification. When there is no possibility of falsification, we are in the realm of belief, including religious belief, not science. Therefore, environmentalism's views would not be an issue if it openly admitted to being a religion, but it claims to be based in science. However, if environmentalists' views are based in science, then environmentalists should be open to empirical testing, the possibility of falsification, and a potential change of views if the evidence does not support those views.

FACTS VERSUS ENVIRONMENTALISM

Here is one example (many more could be listed[12]) of environmentalist blindness to scientific fact. Environmentalists have waged a successful campaign to have the U.S. government declare polar bears an "endangered species" at a time when polar bear numbers are the highest on record. As *National Post* columnist Don Martin put it:

> The "threatened" status afforded the Canadian great white is intriguing. One might not associate that alarmist term with an animal whose Arctic population has doubled to 25,000 bears in the last 40 years, with only two of the 13 pockets of population experiencing any decline and the rest enjoying a boom. Yet somehow, despite that population surge, its long history of surviving even warmer climates and having lived off much reduced sea ice, the polar bear is now the world's photogenic canary in the global-warming coal mine.[13]

The disconnect between scientific fact and environmental rhetoric is so great that Martin even wonders if perhaps this human-caused global warming scare isn't all it's cracked up to be:

> Nothing is more confounding and contradictory than the global warming question, because the so-called "junk science" practised by the alleged "climate change deniers" is backed by evidence to suggest this is merely the latest incarnation of regular planetary warming periods.

For Martin at least, on the issue of polar bears the global-warming alarmist movement has drifted so far from reality that it raised doubts in his mind about other environmentalist claims, and rightfully so since most of these claims are either doubtful or outright false.

Of course, environmentalists are often right, including their demands in the past for non-polluted air and water, and we should be grateful to them when they are right. But all too often, as we've seen with the polar bears, environmental groups are not persuaded by reason, data, evidence, and facts. In Chapter 12 we'll discuss in greater detail why, apart from religion, this might be so. For now, let us ask, as does climate researcher Robert C. Balling Jr., why so many people are so eager to believe the alarmist position based on so little actual evidence. Balling writes:

> Some individuals are absolutely convinced from their very limited reading that we are heading for disaster via global warming. Strangely, no amount of evidence seems to shake this crowd. They appear to have a religious attachment to the issue: they read a few reports about the potential threats associated with the greenhouse effect, and they are sold. They have a set of policies that are strongly supported by the perception of a greenhouse catastrophe, and they are not going to accept anything but the threat of disaster. ... They care about the environment more than they care about science.[14]

Simon suggests that there may be "a human predisposition to be attracted by prophecies of doom—and in the case of some, a predisposition to make such prophecies."[15] In this respect, it's interesting that "the greatest crisis humanity has ever faced," for Al Gore, was first nuclear war (he was perhaps right on that), then the threat to the ozone layer[16] (not quite so dangerous), and now carbon emissions. In each case, Gore uses the same apocalyptic language, much like one of the old-time religious prophets of the Old Testament. It is no coincidence. As Simon notes: "Environmental prophets concur with religious prophets in accusing us of an excess of worldliness, and especially of enjoying the benefits of wealth."[17] However, as we saw in Chapter 6, while Gore wants us all to reduce our consumer-loving lifestyle, the crisis is not so severe that he leads by example in severely reducing his own lifestyle.

MAIN FEATURES OF RELIGIOUS BELIEF

What are the main features of the most common forms of religious belief, at least in the Western world? And does environmentalism have these features? Some, taken from various sources, are as follows:

- The belief that human beings are sinners, even if they, as individuals, have done nothing wrong (Original Sin). Even infants are, in this view, sinners.

- The belief in a Fall from a "perfect" state of spiritual being or, historically, a Fall from a previous Golden Age. Related to this is the belief that there is another, perfect state—either in this life if we could just control our sinful natures (Eden) or after death (heaven).
- The belief in some kind of supra-individual, conscious entity (God, Mother Earth, Nature, Gaia).
- The belief that this supra-individual entity rules the universe and should be worshipped and, in most religions, appeased through prayer, penance and sacrifice.
- The belief that, through religious belief and ritual, human beings can control their fate, including the ability to influence natural phenomena, e.g., praying for miracles or dancing for rain.
- Belief in an Apocalypse or some other cataclysmic event due to humanity's sinfulness.
- Prophets who announce this doom, which is always coming soon but never quite happens.
- A priesthood of religious experts.
- The belief that material enjoyments are wrong. In other words, Puritanism or, in Islamic terms, Talibanism, rather than full enjoyment of life.
- Belief in infallible or certain sources of knowledge, i.e., dogma.
- Belief that those who don't accept the prevailing dogma are heretics.
- In some religions, the possibility of buying "indulgences" that excuse sinful behaviour.
- And, last but not least, a burning desire to convert others to the Truth.

In comparing religion and science, we see immediately that true science has none of these attributes, except perhaps a priesthood of PhDs. Science does not believe in apocalypses caused by human sinfulness; that's a religious idea borrowed by alarmist climate science and propagated by the likes of Gore, Hansen and Schneider. Nor does science believe human beings are naturally sinful from birth, although as individuals and societies we are clearly far from perfect. Science believes human beings can control their environment (to the extent it's possible), but only through understanding physical systems, not by prayers or rituals. And, as we saw in Chapter 10, science is, or should be, value-neutral—it does not judge people's lifestyles as good or bad (although, scientists will, like all of us, make these judgments at a personal level).

However, what is astonishing, even shocking, is that *all* these religious archetypes are found both in modern-day environmentalism and in alarmist climate science. And so, as novelist Michael Crichton notes:

> Environmentalism is in fact a perfect 21st-century remapping of traditional Judeo-Christian beliefs and myths. There's an initial Eden, a paradise, a state of grace and unity with nature, there's a fall from grace into a state of pollution as a result of eating from the tree of knowledge, and as a result of our actions there is a judgment day coming for us all. We are all energy sinners, doomed to die, unless we seek salvation, which is now called sustainability. Sustainability is salvation in the church of the environment. Just as organic food is its communion, that pesticide-free wafer that the right people with the right beliefs imbibe.[18]

Crichton also warns that "the greatest challenge facing mankind is the challenge of distinguishing reality from fantasy, truth from propaganda."[19]

When it comes to religion, this disentangling in extremely difficult. Indeed, many thinkers have argued that religion is hard-wired into us.[20] And if religion is innate to human psychology—and there is a great deal of evidence that it is—then it follows that almost all human beings will end up believing in some form of religion, even if that "religion" is atheism or agnosticism or nationalism or communism, etc. As Catholic writer G.K. Chesterton is reported to have said: "When people cease to believe in God, it is commonly supposed they believe in nothing. Alas, the truth is worse. They believe in anything." The West is largely a secular society and many environmentalists, and scientists, have left traditional religion behind—they do not believe in the Judeo-Christian God, for example. But if religion is hard-wired into us then, as Chesterton suggests, inevitably some other set of beliefs will take the place of traditional religion. Novelist A.S. Byatt noted in a CBC radio interview that all human beings, including herself, need fantasy, like fairy tales, in their lives. And so, when Byatt was introduced to religion as a young girl, she says she recognized immediately that religion met that need for fantasy—except that religious tales, unlike fairy tales, were supposed to be accepted as factually True.[21] With the decline of the traditional, Judeo-Christian religions in much of the Western world, at least three new secular religions are vying to replace them: Marxism (albeit often in the watered-down form of socialism or "social justice"), environmentalism (the worship of nature and often a conscious or unconscious ally of Marxism), and, for some, the worship of science (and/or reason) itself.

THE ADAM AND EVE STORY

Below is an example of how religious myth-making deceives us and makes our lives more difficult than they need to be. Later we'll look at how this myth-making may be deceiving us, as well, in terms of climate change.

The Adam and Eve story from Genesis contains two basic ideas that have become embedded in Western thinking. One is that human beings are naturally sinful for eating the fruit of the "tree of knowledge"—that is, for developing self-conscious awareness and intelligence. In other words, the Jewish and Christian religions teach that we should feel guilty about having intelligence when it is our intelligence that makes us human. It makes sense to feel guilty about the evils that individual humans and societies *do*; it makes no sense to feel guilty about what makes us a species—what we *are*. Yet, this is what the Judeo-Christian religions ask of us.

A second idea is that there was some kind of perfect, paradisiacal state that we came from (Eden) and/or are going toward (Heaven). Many Utopian visions have arisen from this myth that all would be well if we could just return to the perfection of the past. A modern variation is the joy and relief we are all expected to feel if we can just get "back to nature." Almost everyone loves nature, but the vast majority of us would prefer to have the comforts of technological civilization at hand as well. Anyone who doubts this should watch an episode or two of the reality TV show *Survivor* where, after a couple of weeks in the jungle, the participants are ecstatic over the prospect of enjoying, say, a shower or a piece of pizza.

Where did these astonishing ideas that human beings are guilty for simply being human and that the past is perfect come from? Assuming they didn't come from a supernatural deity, these ideas must have evolved as part of our relationship with the natural world. In general, the less able a people are to face nature on equal terms, i.e., the less technologically developed they are, the more superstitious and/or religious they are. Or, as Dutch philosopher Baruch Spinoza (1632-1677) put it: "Superstition is engendered, preserved, and fostered by fear."[22] The force that our ancestors feared most was the power of nature, and behind nature, for most, was a God. Now, since nature often causes us pain, and because this God is both all-powerful and all-good, it makes sense that *we* must be responsible for this pain, i.e., God is punishing us, presumably for our own good, because we have sinned. As a corollary, we can reduce our pain by praying to, sacrificing to, and in other ways trying to appease this all-powerful, sometimes punishing divinity. Divine punishment for sin is the basis of virtually the entire Old Testament,[23] and the Old Testament is one of the cornerstones of Western civilization and thought (the others are the New Testament, ancient Greek philosophy, and science).

Since the belief in human guilt is so deeply ingrained, it's not surprising that our default, unconscious position when nature throws us into disaster, like the South Asian tsunami of 2004 or the Haiti earthquake of 2010, is that these calamities are "God's punishment."[24] This idea is, rationally, absurd and yet millions believe it.

Similarly, when it comes to climate change, the blame-us, Original Sin, Adam and Eve response is so deeply embedded that millions of otherwise rational people—including scientists!—are willing to believe that the planet is warming because we are sinful (that is, we emit carbon). That so many embrace as rational this irrational, even archetypal, belief in human blame speaks not of science but of religion. Although many fundamentalist Christians believe literally in the Adam and Eve story, rationalists understand that the Adam and Eve story is just that, a *story*, perhaps even a fairy tale. The Genesis story may reflect a deep psychological truth—perhaps there is something within us that needs to feel guilty about events we haven't caused—but it is not a scientific or empirical truth. This self-blame for natural events is a trait we need to get over not, as alarmist climate science does, encourage.

SEEING WHAT ISN'T THERE

Another piece of the religious puzzle is the Western approach to experience: if there is an experience, there must be a discrete *experiencer*; if there is a cause, there must be a discrete *causer*. As French philosopher Jacques Derrida writes, for the Western mind, "the notion of a structure lacking any centre represents the unthinkable itself."[25] It's difficult to imagine ourselves or the universe as not having a *permanent* central core (God) and ourselves as not having an equally *permanent* central core (the immortal soul). And so, whether these permanent, metaphysical centres exist in reality or not, we create them in our minds and cultures. As the American philosopher John Dewey writes in his essay "The influence of Darwinism on philosophy":

> For two thousand years, the conceptions that had become the familiar furniture of the mind rested on the assumption of the superiority of the fixed and final; they rested upon treating change and origin as signs of defect and unreality.[26]

This is one reason why Darwin's *Origin of Species* was so shocking when it came out in 1859—Darwin challenged the religious idea of a fixed universe in which an orderer (God) made the important decisions, like the creation of human beings and, indeed, all creatures. Suddenly it appeared that natural forces themselves had "created" us and that existence was a process of continuous change (evolution), an idea that was profoundly disturbing to Western philosophy and potentially deadly to Western religion.[27] Therefore, given the deep psychological need for permanence within Western culture (and perhaps all cultures), it's not surprising that many environmentalists want to believe the planet's climate today is, or should be, fixed and final—hence the desire to "stabilize the climate." That there might be other temperatures or other levels of carbon dioxide that are equally beneficial to human life isn't part of this environmentalist belief system, which sees any change of the present climate as "signs of defect and unreality."

ENVIRONMENTAL APOCALYPTICISM

The deeply held fear that we are heading into some sort of environmental Apocalypse speaks volumes in supporting the claim that environmentalism is largely, at this stage, a religion. The apocalyptic idea comes right out of the Old and New Testaments. Christianity rests almost entirely on the foundations of apocalypticism—the early Christians expected the Last Trump and Resurrection of the Elect literally at any moment. As religious scholar Bart Ehrman writes:

> For apocalypticists, cosmic forces of evil were loose in the world, and these evil forces were aligned against the righteous people of God ... making them suffer because they sided with God. But this state of affairs would not last much longer. God was soon to intervene in this world and overthrow the forces of evil; he would destroy the wicked kingdoms of this world and set up his own kingdom, the Kingdom of God, one in which God and his ways would rule supreme, where there would be no more pain, misery, or suffering.[28]

This religious Apocalyptic vision seems remarkably similar to the environmental vision that the world is evil because human beings pollute (which we do, in some cases, although carbon dioxide itself is not a toxic "pollutant" but essential for life), and that we will therefore be punished with "oblivion" or the "demise of the planet" or "thermageddon." Of course, alarmists have been predicting the West's demise for hundreds of years, as we saw in Chapter 8—Paul Ehrlich is a good example of the Old Testament prophet's thundering predictions. But, as Ehrman observes, the promised apocalypse not only didn't come for the early Christians, it has *never* come: "The fervent expectation that we must be living at the end of time has proved time after time—every time—to be wrong."[29] The biblical prophets may have been accurate in predicting *historical* human-caused disasters, such as the fall of the Kingdoms of Israel to, say, the Babylonians or Romans. But so far the supernatural God has been remarkably reluctant to intervene in our environmental affairs, for good or ill. As literary critic Greg Gerrard notes:

> Just like Christian millennialism, environmental apocalypticism has had to face the embarrassment of failed prophecy even as it has been unable to relinquish the trope altogether.[30]

Another religious idea taken over by environmentalism is the perfect Holy Kingdom, in environmental terms "pure nature" or the perfect environment that existed before humanity appeared. This, too, is a figment of our imaginations, not a reality that ever existed or ever will exist. All attempts in the past two centuries to create this perfect kingdom—communism, fascism, fanatical Islamism, religious fundamentalism—have led to totalitarianism and, as often as not, mass murder. This utopianism is a dangerous idea and yet it lies at the core of the environmentalist religious vision as a vision of "pure" Nature—purified of humans, that is.

SACRIFICE AND RITUAL

What about religious sacrifice and ritual—do these religious ideas have any counterparts in the environmental creed? Here is one example: the now-defunct Kyoto Accord. As Andrew Weaver notes:

> Suppose every country that actually signed Kyoto ratifies it and meets their target, including the U.S., but goes no further. That 2.08 degree [Celsius] warming becomes 2.0 degrees and the sea level rises 48.5 cm [by 2100]. Suppose we all reduce our emissions by another one per cent a year, after the Kyoto targets are reached. The Earth still warms by 1.8 degrees and sea levels rise 45.5 cm. So there is an element of 'We're hooped anyway' here.[31]

In other words, even alarmist climate scientists admit that Kyoto might, at best, have cut global temperatures by .08 degrees Celsius over the next 90 years, an amount so small as to be unnoticeable, at a cost to the Canadian economy of anywhere between $3 billion and $17 billion a year.[32] In his book *Keeping Our Cool*, Weaver writes: "On Kyoto, I agree wholeheartedly, *as would almost anyone in the scientific community*, that it will have *zero effect on global warming.*"[33] [italics added] Yet the environmental movement continues to insist on futile measures like Kyoto and its most recent incarnation, the Copenhagen Protocol, to try to mitigate (stop) global warming, rather than policies that would allow us to adapt to warming (assuming the planet is in fact warming, which was not the case at the time of writing in 2010).

From the skeptical side, journalist Christopher Monckton calculates that if Western society cut its carbon emissions by 30 per cent in the next decade, as suggested by the 2009 Copenhagen Accord, the actual reduction in temperature by 2020 would be .04°C.[34] To produce this negligible decrease in warming, environmentalists are asking the West to deliberately create an economic catastrophe, which is what a 30 per cent reduction in industrial output would mean. We will spend billions of dollars for no practical result, but we are still urged to carry on, even though we might as well sacrifice a bull for all the effect our sacrifices will have on the climate. What is this but *religious* ritualism and sacrifice? Presumably, if we sacrifice to the Kyoto or carbon gods, we'll at least feel we have some control, and, to quote psychologist William Glasser: "We feel good when we are in control and bad when we are not."[35] This, indeed, is one of the attractions of religion: it gives us the illusion of control over nature even if we don't have that control in reality.

But, surely, we should be beyond this primitive logic by now. Wouldn't we be better putting the nation's time, money and effort toward creating new, non-fossil-fuel technologies, rather than spending billions of dollars trying to meet a carbon curb deadline that will not do what Kyoto and Copenhagen are promoted as doing: reduce global warming? Kyoto may be dead, but most objective observers acknowledge that the currently mooted carbon taxes will also do little to stem the tide of carbon dioxide, especially if India and China don't reduce their emissions, which they won't because to do so would sacrifice their people's hopes for a materially better life. Cap and trade is another form of expensive sacrificial ritual, sanctioned by environmental thinking. Carbon trading will make some companies and individuals rich while making most products more expensive for the average consumer, but it will do almost nothing to reduce carbon emissions.[36]

And even if we went back to the Stone Age economically, the planet would almost certainly continue warming, as it did in previous interglacials that were 1-3°C warmer than today's interglacial, with sea levels at least six metres (20 feet) higher than today's levels, according to the IPCC itself. There is no reason whatsoever to believe that the Holocene interglacial won't reach the same temperature and sea level heights as previous interglacials whether humans were here or not. Nor is there any reason whatsoever to believe that, as with previous interglacials, ours will not eventually end in a slow death by ice. In other words, we would have made massive economic sacrifices for no good reason. And if we foolishly did put ourselves back to the Stone Age, millions of people would starve to death—the survival of more than six billion people depends almost entirely on science and technology and, until we come up more advanced forms of energy, that means carbon emissions.

GOD AND EDEN

Does the idea of God play any part in the environmentalist catechism? Not directly—many environmentalists have left traditional organized religion behind, or believe they have. But, in the process, some are creating a new religion based on a new god, or perhaps a very old god—Mother Nature, now called Gaia. We see this new environmental religion in the title of David Suzuki's *The Sacred Balance*, which tells us that if we can just get closer to nature, bring our reason and emotions into sync, we will find the "balance"—the Eden—we are all looking for.[37] According to Suzuki, "Acting differently from the rest of creation, separating ourselves from the divine will, we broke the harmony [of creation]."[38] In other words, Suzuki is offering the modern, scientized version of the Biblical Fall and the modern, scientized version of a psychological Eden.

Our hunter-gatherer ancestors were very close to nature; did they find that balance and harmony Suzuki promises? There is no evidence that they did—as neuropsychologist Steven Pinker and many others have noted, primitive society was, and still is, just as rife with war and suffering as our modern world, and perhaps more so, albeit on a smaller scale. For example, anthropologist Carol Ember estimates that "90 per cent of hunter-gatherer societies are known to engage in warfare, and 64 per cent wage war at least every two years."[39] Although the issue is still highly contested, it is possible that the first native peoples in North and South America wiped out all the large mammals 10,000 years ago,[40] and perhaps the mammoths and wooly rhinos in Europe and Asia as well.[41] Although climate change undoubtedly also played a role in these extinctions—the world was warming rapidly at that time, more rapidly than now—the "sacred balance" that prehistoric and primitive peoples are supposed to maintain that modern humans do not is yet another environmental myth, as is the "sacred balance" that nature is supposed to maintain. Environmentalists have simply dressed Jean-Jacques Rousseau's "noble savage"—an utterly discredited concept—in new, green terminology.

And if our ancestors did have this sacred balance, why did they lose it? Wouldn't they hold onto it for dear life? The idea that nature is "sacred" is powerful, but it is based on an idealized view of nature as somehow mostly beneficent to human beings. It is not. Human beings have had to fight nature in every step of our evolution. Nature is the source of our physical existence and of much beauty and joy—in small quantities and suitably tamed—but Nature is also the source of disease, disaster and death. The belief "that whatever happens in nature is good" is called the "naturalistic fallacy"[42] and this fallacy is strongly at work in environmentalist thinking. In truth, nature doesn't care about humans, or any other creature, as individuals or as species, no matter how much we worship her.

Nor is there a mystical natural "balance" that we are somehow disrupting, except in the environmental mythos; there is no *scientific* justification for this belief whatsoever. As noted in Chapter 6, the UN estimates that 95 per cent of all species that have ever appeared on earth are extinct. Where's the balance in a 95 per cent extinction rate? And yet the environmentalists blame—some even hate—humanity for what they consider an unacceptable level of current species extinctions (in reality, one or two species a year, although environmentalists believe the extinction rate is thousands per year). As for the climate—one look at the ups (interglacials) and downs (glacials) of Earth's ice age climate over the past two million years should dispense with this idea. Nature is a rollercoaster, not a balance board. This idea that Nature is in a "sacred balance" may be a nice image for attracting recruits to the environmental religion, but it is scientifically false—in effect, it is a fairy tale.

And what is Al Gore up to when, as we saw in Chapter 6, he suggests in *Earth in the Balance*:

> His [Francis Bacon's] moral confusion—the confusion at the heart of much of modern science—came from the assumption, echoing Plato, that human intellect could safely analyze and understand the natural world without reference to any moral principles defining our relationship and duties to both God and God's creation.[43]

What Gore is proposing is a reunion of science and religion, so that what we investigate scientifically would have to pass some kind of religious (or political?) scrutiny before going ahead. He is quite explicit about his religious agenda in *Earth in the Balance*, which has the *religious* subtitle "Ecology and the Human Spirit." And so he writes: "My own faith is rooted in the unshakeable belief in God as creator and sustainer."[44] In short, Gore appears to believe in Intelligent Design, a concept that is as unscientific as it's possible to be, while presenting his extreme environmental and fundamentalist religious views as "science."

The mixing of science and politics has already happened with the IPCC, whose reports are subject to "review by governments." Yet, as we saw in Chapters 6 and 7, the early scientists like Bacon struggled, and were sometimes imprisoned, tortured and even executed precisely to create the split between church and state, science and politics, and, yes, mind and emotion that Gore and Suzuki decry, so that reason could restrain our religious impulses—the emotional, perhaps instinctive and often destructive impulses that lie at the heart of so many conflicts in the world today and in the past. In his essay on theism, John Stuart Mill describes his view of the proper balance of reason and religious belief, which he considers a product of "imagination" rather than reason:

> When the reason is strongly cultivated, the imagination may safely follow its own end and do its best to make life pleasant and lovely inside the castle, in reliance on the fortifications raised and maintained by reason round the outward bounds.[45]

For all reason's faults, we are far better off in a world governed by reason than by religious enthusiasm. For both reason (including science) and imagination (including religion) to flourish, there must be some separation between the two faculties.

However, *this* balance is not what is being urged by Gore and Suzuki, among many others who make their appeals to emotion rather than reason in their environmental writings. Suzuki, in *The Sacred Balance*, urges his readers to "be confident in our gut reactions and *insist that the experts prove their case*."[46] [italics added] However, it apparently isn't necessary for environmentalists to prove *their* case scientifically, as we saw with the polar bear example—a gut reaction is enough. At the same time, Suzuki says he bases his approach on "science"—he proudly declares himself a "scientist" in his speeches—but only the science that fits his preconceived ideas. Anyone who's been to a Suzuki talk, as I have, can tell you that it's more like attending a religious revival than a scientific seminar, to the point where he was called "Saint Suzuki" in a *Maclean's* magazine profile.[47] There's no give and take of differing points of view, no dissent allowed, in Suzuki's religion, just faith-based adoration based on the Doctrine of Certainty. It is the opposite of good science, which (at least ideally) actively seeks out facts that might contradict the hypothesis in order to arrive more closely at the truth.

GAIA: THE GREEN GOD

The "God" in the green vision is usually a She—Mother Nature, the Mother Goddess, etc. The most blatant attempt to reintroduce the idea of God as a scientific concept is James Lovelock's Gaia hypothesis. For Lovelock, a medical doctor who also designs scientific instruments, Gaia is the earth as a sentient being. While he admits this hypothesis is "like a religious belief ... scientifically untestable," and that he is not "thinking of the Earth as alive in a sentient way, or even alive like an animal or a bacterium,"[48] Gaia is nonetheless a "vast being who in her entirety has the power to maintain our planet as a fit and comfortable habitat for life" and she is "now through us awake and aware of herself."[49]

As an example of Gaia's power, Lovelock notes that although the sun has increased its intensity by about 30 per cent since the earth was formed 4.5 billion years ago, Gaia has cleverly put the planet in an ice age so it won't burn up.[50] Apparently, Gaia somehow deliberately positioned the Antarctic continent at the south pole tens of millions of years ago, deliberately caused continents to collide and create the vast mountain ranges of the Rockies and Himalayas, and made the North and South American continents join up three million years ago, all of these geological factors working together to make the planet colder. This is pretty good planning on Gaia's part. On the other hand, a glaciation is not such a great thing for biodiversity—global cooling reduces biodiversity, global warming increases it—much less for human beings. But, then, for Lovelock, what we have to keep foremost in mind "is the health of the Earth, not the health of people." Indeed, for Lovelock, people are "perceptibly disabling the planet like a disease."[51] As a result of this disease (us), he asserts: "Before this century is over billions of us will die and the few breeding pairs of people that survive will be in the Arctic where the climate remains tolerable."[52]

Lovelock believes that human carbon emissions (the "fever") are overcoming Gaia's attempts to keep the planet cool in the face of a warming sun. But here's where Lovelock leaves the scientific method far, far behind: It's also possible that carbon emissions are part of Gaia's plan to keep the planet from getting too cold—we are within about five degrees Celsius of the next glaciation. Either explanation works perfectly well; there is no testable, falsifiable hypothesis here, so Lovelock's Gaia idea does not qualify as "science" but as a religion.

On the plus side, Lovelock acknowledges: "Pollution is not, as we are so often told, a product of moral turpitude. It is an inevitable consequence of life at work."[53] Human beings are not rabbits who can live in holes in the ground. At our current numbers, we need industry and technology to survive, and industry and technology inevitably produce carbon emissions. In this, we are no different from beavers that need to produce dams or wasps that need to build nests to survive. In other words, industry and technology are as much a part of human beings as our brains and opposable thumbs. Although many environmentalists see human beings as separate from nature, and even "unnatural," this view is yet another remnant of Judeo-Christian archetypes that view humanity as a special creation of God and therefore not part of nature. But Darwin demonstrated that we are as much a product of nature as any of Gaia's other creatures. If what we do is build civilizations, along with the inevitable carbon emissions, then that, too, surely, is part of Gaia's plan. Or not. As with most religious pronouncements, you can read Gaia many different ways to fit your point of view.

As we will see in more detail in Chapter 13, Lovelock also has some good things to say in favour of nuclear power to reduce carbon emissions and against the green movement's irrational fear of nuclear power.[54] But, in the end, the Gaia hypothesis is stock Old Testament stuff: humans are evil, God is angry (one of Lovelock's books is entitled *The Revenge of Gaia*), and the Apocalypse is coming soon, just as the first-century Christians believed. It's not surprising that so many environmentalists take Gaia seriously, given that so much of environmentalism is based in the traditional Western religious archetypes. However, it is astonishing that many presumably more sober climate scientists have also given Gaia their blessing. For example, in *Global Warming: The Complete Briefing*, one of the IPCC's lead authors, John Houghton, explicitly endorses Lovelock's "Gaia" hypothesis:

> Gaia remains a scientific theory. But some have been quick to see it as a religious idea, supporting ancient religious beliefs. ... As we pursue an understanding of the Earth's environment, *it is essential*

> *that scientific studies and technological inventions are not divorced from their ethical and religious context.*[55] [italics added]

One of the earlier alarmist books on global warming was *The Heat Is On* by journalist Ross Gelbspan. In it, Gelbspan criticizes MIT climatologist Richard Lindzen for charging that Houghton "is motivated by a religious need to oppose materialism."[56] Is Lindzen wrong? Here's what Houghton wrote in a letter to the World Council of Churches in 1996: "Such policies like cutting energy use by more than 50 per cent can contribute *powerfully to the material salvation of the planet from mankind's greed and indifference*."[57] [italics added] Unlike most environmentalists, at least Houghton is up front about his religious agenda, but which religious agenda? Given that Houghton is chairman of an organization called the John Ray Initiative, which aims at "connecting Environment, Science and Christianity," one might, like Lindzen, suspect Houghton has been unable to separate his science from his values. Thus, like Gore, he might well be "motivated by a religious need to oppose materialism" to seek, through environmental regulation, to impose a Christian-type, world-denying morality on those who would not agree with him if approached on a directly religious basis. The Catholic Inquisition worked quite hard to ensure that early science (then called "natural philosophy") remained firmly under the control of the "religious context," and this seems to be what Houghton and Gore want as well. Of course, science has ethical obligations, but do we really want to tie those ethical obligations specifically to religion, whether Christian or environmentalist? Do we really want to end the separation of religion and science that has been so fruitful in improving human material life over the past 400 years? Do we want our scientists manipulating their data to fit a religious ideology like Christianity or Gaia, as the CRU scientists manipulated their data to fit the AGW ideology?

ENVIRONMENTALISM HAS ALL THE FEATURES OF A RELIGION

In sum, when we survey the characteristic features of religious belief, we see *all* of them prominently displayed in modern environmentalism. There is the belief that human beings are sinners (Original Sin); that we have Fallen from an originally pure state; that enjoyment of material comforts is wrong ("not sustainable"); that we are under the control of some metaphysical Being (Gaia, Mother Nature) that rules the planet; that we need to appease this Being—i.e., try to control our world—through sacrifice, prayer and ritual (e.g., the Kyoto and Copenhagen Accords); belief in an Apocalypse or End of Days brought on by humanity's sinfulness; prophets who announce this doom if we don't mend our ways (Gore, Ehrlich, Hansen, Schneider, Weaver); a priesthood of experts (the climate scientists) whose words are dogmatically infallible and certain (e.g., the IPCC reports and computer models); a refusal to change beliefs in the face of scientific facts to the contrary; the casting out as heretics of those who don't accept the dogma (climate skeptics, i.e., "deniers"); and a burning desire to convert others to the environmental Truth. The environmental movement even has the medieval Catholic institution of indulgences through which sinners can buy their way back to grace. Hence, Al Gore buys carbon credits so his carbon-emitting lifestyle can continue without guilt. The parallels between environmentalism and religious belief are, quite simply, astonishing, and so obvious that, although of course environmentalism contains some elements of science, it is absurd to claim that environmentalism is not *primarily* a religion. It is.

The religious nature of environmentalism is not, in itself, a problem, and the argument in this book is not against religion as such—religion is a personal choice. Nor is this an argument against nature-worship as a religion—paganism has a long and honorable tradition.

What this book warns against is the attempt to impose the environmental religion on non-believers under the cloak of "objective" science. Environmentalism contains much scientific truth, of course; as noted above, environmentalists have done good work in making the public aware of the toxic pollution of the air, water and earth (again, carbon dioxide is a fertilizer not a pollutant). But, as we've seen, those problems are largely behind us, at least in the developed world, and one would therefore expect the environmental movement in the developed countries to have faded into the background. The fact that the environmental movement hasn't faded away, even though virtually all of its goals have been achieved, strongly suggests that much of today's environmentalism is not based on scientific fact but on the emotional need for religious belief and practice in a secular age. And, among the believers in the environmental religion are many alarmist climate scientists—how else to explain their prophecies of environmental apocalypse, for example, when there is not a shred of *empirical* evidence to support this claim? Or, as climate scientist Roy W. Spencer writes: "When scientists become emotionally attached to a specific theory, you know that more than science is involved."[58] When religious belief is promoted as scientific truth, as has happened with alarmist climate science, then both science and religion are corrupted and we, the public, should be concerned about this corruption.

But, one might argue, aren't the environmentalists asking for changes that will be beneficial in the long run, such as less dependence on fossil fuels and more efficient energy sources? As we saw in Chapter 4, many politicians believe that, as former Canadian Environment Minister Christine Stewart put it, "No matter if the science is all phony, there are collateral environmental benefits."[59] Perhaps. But if these collateral environmental benefits are so reasonable and desirable, why the need for a *religious* movement, playing on emotion more than reason—playing ultimately on fear, which Gore himself identifies as "the enemy of reason"—to promote these benefits? Why can't these measures be sold to the public on their own merits? Similarly, if less dependence on fossil fuels and more efficient energy use are desirable in themselves, as they are, why bring in the issue of climate as a smokescreen, using religious fervor as one of its tools? Why say we should use fewer fossil fuels to "save the planet" or "stop global warming"—which are essentially religious arguments—rather than because we are running out of fossil fuels and need to switch to alternative, more sustainable sources?

I wrote this book with the hope, in part, that some readers will see more clearly, if they didn't before, the religious and political underpinnings of environmentalism, including its irrational fears of global warming. I hoped that readers would thus view climate alarmism with the same caution, if not skepticism, that one would bring to any religious or political proselytizing effort, even if that effort arrives at your door in the guise of science. Why? Because, as I hope I've shown, global warming fears are being promoted on primarily moral and emotional rather than scientific grounds, despite what the public is told ("the science is settled"), because the empirical evidence for anthropogenic global warming and the following apocalypse is not there.

And if environmentalism is primarily a religion, which it clearly is, we should expect from environmentalists tolerance of *other* belief systems. This tolerance is rarely found in the environmental movement because its followers believe they are acting on the basis of *science*, and are therefore the bearers of something approaching absolute truth. They are, in Kant's terms, "morally certain" of what they believe and are unwilling to admit, as Kant did, that reason and science cannot provide the empirical support for their moral claims. Instead, environmentalists feel justified in forcing their truth on others in a sometimes intolerant way ("the time for discussion is over"), an intolerance that traditional religions in democratic societies have largely given up.

As for the science on global warming being "settled" and therefore beyond discussion—again, this is dogma, not fact. There is much evidence that modern climate change (warming from 1975-1998, now cooling) is primarily caused by natural planetary cycles. There is also evidence to support the theory that humans are driving the climate. The *scientific* issue of the cause of global warming and cooling remains open; environmental religionists like Gore, Hansen, Schneider, etc., would like to declare the case closed in the interests of dogma, not science. Science is always open to new evidence—or should be. Therefore, it would be helpful if more global-warming crusaders understood that their cause *is* at least partly a religion, and that religion and science is a volatile mix that will be harmful for both institutions, just as the IPCC's mixture of science and politics ("review by governments") has damaged if not destroyed its value as a scientific institution.

The public policy question when it comes to climate is: do we want to base our responses to climate change on what appears to be an environmental *religion*, or do we want to base our responses to climate change on *science*, that is, on empirical data? If the latter, then most of the fears cultivated by the environmental movement, supported by alarmist climate science, are clearly unfounded—there is no empirical evidence that humans are dangerously warming the planet. If the former—if we wish to base our public policy decisions on what is essentially religious belief, then at least those who propagate these beliefs should be up-front about the religious nature of their proposals.

WHAT WE KNOW

What can we be sure of when it comes to the relationship of environmentalism, climate alarmism and religion?

- We know that much of the environmental movement today is based not in science but ideology/belief/moral certainty (e.g., the supposedly endangered status of polar bears; many more examples could be offered).
- We know that militant environmentalism has *all* of the features of a religious, rather than scientific, outlook, including dogmatic certainty, apocalypticism, a belief in the inherent sinfulness of humanity, a belief in an environmental Utopia (either in the past or the future), heretics ("deniers"), and so on.
- We know that science and religion must be kept separate lest values corrupt the search for truth, as has occurred with alarmist climate science. Proof? As we saw in Chapter 10, the CRU emails clearly show top IPCC scientists preferring the environmentally perceived Good (we must make the world appear to be warming) over the scientific Truth (the world is not warming at present).
- In a liberal, democratic society, no religion should be preferred over any other when it comes to making public policy. Militant, religiously based environmentalism wishes to subvert this policy of religious toleration by claiming its beliefs are rooted in science.

In the next chapter we will look at some of the other reasons why alarmist climate science has chosen to abandon empirical evidence in favor of the anthropogenic and apocalyptic warming ideology.

One of the striking sociological features of science is a well-known political and social activity called "jumping on the bandwagon"—latching on to a trendy idea when it is rolling to get a free ride. Politicians do it when a particular candidate, party, or ideology has irresistible momentum to put them in power. Scientists do it when a particular theory or idea gives one a better change of getting a grant, publishing a paper, or landing a book contract.

—Donald Prothero[1]

Even a succession of professional scientists—including famous astronomers who had made other discoveries that are confirmed and now justly celebrated—can make serious, even profound errors in pattern recognition.

—Carl Sagan[2]

The history of philosophy is to a great extent that of a certain clash of human temperaments.

—William James[3]

Chapter 12

GLOBAL WARMING FEARS ARE BASED ON PSYCHOLOGY, NOT SCIENCE

A while ago, I received an email from a friend who asked:

> How can many, many respected, competitive, independent science folks be so wrong about [global warming] (if your premise is correct). I don't think it could be a conspiracy, or incompetence... Has there ever been another case when so many 'leading' scientific minds got it so wrong?

The answer to the second part of my friend's question—"Has there ever been another case where so many 'leading' scientific minds got it so wrong?"—is easy. Yes, there are many such cases, both within and outside climate science. In fact, the graveyard of science is littered with the bones of theories that were once thought "certain." For example, as recently as the 1970s at least some climate scientists feared global cooling, and before the 1960s geologists vigorously opposed Alfred Wegener's hypothesis of continental drift. A hundred years ago virtually all scientists believed that Newton's Laws were carved in stone—until Einstein proved them wrong. In other words, the examples of "leading" scientific minds getting it wrong are legion; otherwise, science would be static. As Carl Sagan has written: "Even a succession of professional scientists—including famous astronomers who had made other discoveries that are confirmed and now justly celebrated—can make *serious, even profound errors* in pattern recognition."[4] [italics added] There is no reason to believe that climate scientists are exempt from this possibility, as I've tried to show in this book.

That leaves the first question, which is *how* so many "respected, competitive, independent science folks [could] be so wrong" about the causes and dangers of global warming, assuming they are wrong. And here, I confess that even after more than two years of research into climate fears, this question continues to baffle me. It is not baffling that so many scientists believe humanity *might* be to blame in global warming. What is baffling is that alarmist climate scientists are so *certain* of their hypothesis—100 per cent certain, according to Gore, without a shred of uncertainty, according to Andrew Weaver, James Hansen, et al. Yet, as columnist Mark Steyn has noted (in agreement with virtually all other scientists outside of alarmist climate science):

> Science is never "settled," and certainly not on the basis of predictive models. And any scientist who says it is is no longer a scientist. And the dismissal of "skeptics" throughout the [Climategate] correspondence is most revealing: a real scientist is always a skeptic.[5]

This dogmatism, the Doctrine of Certainty, makes it clear, at least to the observer outside the alarmist climate paradigm, that something is very wrong with the state of alarmist climate science.

Also very baffling is why politicians in so many developed nations (Czech president Vaclav Klaus is a notable exception) should have taken up the anti-global-warming cause while knowing that the campaign against carbon dioxide would wreck their economies. Of course, they do know the danger, which is why—fortunately—in practice the politicians have done very little but talk. Indeed, if global warming was the "crisis" that the UN claimed at the Copenhagen Protocol gathering of December 2009, surely the politicians would have taken concrete action. Instead, the politicians debated, dithered, and departed Denmark having accomplished nothing, which was their intention all along. Therefore, we can assume that most of the world's politicians actually don't believe their own propaganda. So why did they get on the AGW bandwagon in the first place if they're not going to do anything concrete about warming?

The best explanation for this two-faced and self-defeating stance is that politicians are scared to death of losing the environmentalist vote and so they pretend to go along with the AGW hypothesis, while knowing (surely they know, do they not?) that what the environmentalists want is both politically and practically impossible. In short, most politicians are unwilling to *lead* on this issue by telling voters the truth about climate change—that there's no evidence outside of computer models that we're facing a pending catastrophe—because they fear not being perceived as environment-friendly.

However, the damage has been done. Politicians and their subservient ("review by governments") alarmist climate scientists have stoked the fears of climate change and these fears now threaten to wreck the most productive economies in human history to no useful end. Cutting carbon emissions will not reduce warming, if warming is how the planet's natural forces decide to go. But carbon curbs will, by reducing our economic strength, make it harder for us to deal with warming (or cooling) if and when it does occur. So, in attempting to understand where the leading alarmist climate scientists are coming from—"How can many, many respected, competitive, independent science folks be so wrong?"—let's begin with the political origins of the climate change scare, then move on to the psychological and self-interested reasons why climate scientists might want to climb onto the global-warming bandwagon, including those who know (surely they know, do they not?) that the problem isn't as serious as they would like the public to believe.

POLITICAL ORIGINS OF THE AGW HYPOTHESIS: A VERY BRIEF HISTORY

Two *politicians*—not scientists—are primarily responsible for the dominance of the anthropogenic global-warming hypothesis: former British Prime Minister Margaret Thatcher and former U.S. Vice-President Al Gore. According to skeptic Richard Courtney, in the late 1980s Thatcher wanted to reduce Britain's dependence on coal—and the nation's vulnerability to the Marxist-led coal miners' union—by replacing coal with nuclear power. She also wanted to establish her scientific credibility as a female prime minister in a largely male political world. Her solution was to raise the alarm about global warming due to fossil-fuel carbon emissions. What's puzzling about this narrative, though, is why Thatcher would hide the need to replace fossil fuels with nuclear behind the issue of climate. The answer is that the scare at the Three Mile Island nuclear plant in the United States (1979) and the nuclear accident at Chernobyl in Ukraine (1986), plus Cold War fears of a nuclear holocaust, meant that public opposition to nuclear power was strongly entrenched. Therefore, Thatcher's solution was to create a smoke-screen for her real intentions. Frighten the people enough with the spectre of warming due to CO_2, she believed, and they would accept nuclear power as an alternative to fossil fuels.[6] As we will see in Chapter 13, this tactic is finally working, albeit after a lag of 20 years—many Greens now prefer nuclear power to carbon power because nuclear generates fewer carbon emissions.

One of the results of Thatcher's interest was the formation in 1988 by the UN Environment Program and the World Meteorological Organization of the IPCC which was, not surprisingly given its political origins, subject to "review by governments." But, then, the person who pays the piper gets to set the tune, and this has been the case with the IPCC—it sings the alarmist song that Thatcher hoped for.

Since the mid-1990s, the politician driving the AGW doomsday bus has been, of course, Al Gore. Gore's political influence can't be understated since, as a senator from 1985-93, he headed science-based senatorial committees and because he was vice-president during the Bill Clinton administration (1993-2001). One college course taken with climatologist Roger Revelle, a pioneer in studying the connection between climate and carbon dioxide, appears to have convinced Gore that he was an expert on climate (even though he got only C's and D's in his science courses at Harvard University[7]). Revelle himself had warned that we should not jump to conclusions about the warming power of CO_2, and that more CO_2 might be beneficial both as a fertilizer and in reducing extremes of weather.[8] However, as a senator and later vice-president, Gore was in a strong position to push the anthropogenic global warming agenda he first wrote about in his 1992 book *Earth in the Balance*. He was, through his political influence as a senator and vice-president, able to direct literally billions of dollars of research funding—the U.S. currently spends more than $5-billion a year on climate issues[9]—in the direction he favored, human-caused warming, and away from a direction he did not favor, natural warming. As physicist John P. Costella notes in his analysis of the Climategate emails:

> The entire industry of "climate science" was created out of virtually nothing, by means of a massive influx of funding that was almost universally one-sided in its requirement that its recipients find evidence for man-made climate change—not investigate whether or how much mankind had caused climate change.[10]

It's worth repeating this: the climate science Gore pushed was biased from the beginning toward the human-warming hypothesis and a climate scientist's chance of getting substantial funding was based on a willingness to accept this hypothesis. Thus, a climate scientist would have to be very brave and/or very foolhardy to risk losing his or her funding—and therefore career—by going against the AGW flow.

While politicians by nature exaggerate issues, few exaggerate as wildly as Gore, perhaps because even politicians realize most issues are considerably more nuanced and therefore considerably less black and white than Gore is prepared to recognize. Gore gets wide public attention with his exaggerations, though, and that attention seems to be what he craves (as we saw in Chapter 11, the "greatest crisis humanity has ever faced" was first, for Gore, nuclear war, then the depletion of the ozone layer, then global warming). As far back as 1992, journalist Gregg Easterbrook noted: "It's worrisome that Mr. Gore, the bright light of political environmentalism, seems increasingly to believe that the only correct stance is to press the panic button on every issue."[11] It's worrisome and unfortunate that the rest of us will be paying for Gore's most recent "save-humanity" crusade.

And we will be paying. The Copenhagen Protocol conference of December 2009 contained a commitment to give developing nations from $75 billion to $100-billion *a year* by 2020 for climate adaptation.[12] That is, the UN believes the developed world should blame itself for inventing industrial civilization and pay the developing world guilt-money to the tune of many billions a year. With this money, explains David Suzuki, "they can adapt to the worst consequences of climate change, reduce their emissions, and benefit from emerging renewable-energy technologies."[13]

Another term for this transfer of wealth from richer countries to poorer is foreign aid, while at the national level it is socialism and, locally, welfare. Yet, like socialism and welfare, direct financial aid to the developing world has almost never worked in the way intended. For example, Dambisa Moyo, a Ghanian economist, notes that the developed world has given Africa more than $1-trillion in aid over the past 60 years. The result? Most African nations continue to be among the planet's poorest. Why? Because to turn aid into a vibrant economy requires several pre-existing conditions. Chief among these conditions, according to economist William J. Bernstein, are property rights (which presupposes the rule of law), scientific rationalism, capital markets and an efficient and extensive transportation network.[14] The failing African nations have few or none of these; instead, they are afflicted by their opposite, political and economic corruption. As Moyo notes:

> The list of corrupt practices in Africa is almost endless. But the point about corruption in Africa is not that it exists, the point is that aid is one of its greatest aids ... Foreign aid props up corrupt governments, providing them with freely usable cash.[15]

Is there any reason to believe the $75-billion to $100-billion a year shipped to developing countries—countries that lack one or more of Bernstein's four pillars of prosperity—will be used by the recipients to develop their economies and fight climate change, whatever that means? Or will the rulers of these countries continue to use this windfall to enrich themselves? The answer is obvious.

Not so obvious, perhaps, is why the United Nations was pushing the Copenhagen climate talks so aggressively, to the point of declaring again and again beforehand, through various world leaders, that if Copenhagen failed we are facing "catastrophe."[16] Since there was no empirical scientific evidence that the planet is facing catastrophe if we didn't get a climate deal in December 2009, we are forced to ask: What is the UN's real agenda here? Since it clearly isn't to propagate scientific truth, the answer is almost certainly politics.

The report by Dr. Edward Wegman on the Michael Mann hockey-stick scandal, discussed in Chapter 4 and in more detail below, notes that the work of the hockey-stick promoters "has been *sufficiently politicized* that *they can hardly reassess their public positions without losing credibility.*"[17] [italics added] In other words, underlying the science of the hockey stick, and therefore underlying the "science" of the IPCC, is a political agenda, and creating mass public fear about global warming is the means to achieve this political agenda. What is this agenda?

The UN's broader agenda seems to be realizing its concept of international social justice, and it is quite willing to lie in the service of this goal. So, UNAIDS, the agency tasked with dealing with the AIDS epidemic, has admitted that it exaggerated the danger of AIDS in order to get public support and, more importantly, international funding. As a result, a disease that causes 3.7 per cent of the world's deaths was getting 25 per cent of the UN's health funding. The UN's excuse, according to one critic of this tactic? It was propagating a "glorious myth," that is, a tale that is "gloriously or nobly false for a good cause."[18] More recently, in October 2009, the UN's World Health Agency was caught exaggerating the threat of H1N1 (swine) flu, declaring it a "pandemic" rather than an "epidemic," for the same reason—the exaggeration was a noble falsehood for a good cause. What cause? Margaret Chan, WHO's director-general, told world leaders that they should use the "devastating impact" of swine flu on poorer countries to push for changes "to distribute wealth" based on the values of "community, solidarity, equity and social justice."[19] Chan also criticized international financial markets for failing to operate "with fairness as an explicit policy objective." While community, solidarity, equity and social justice may be fine things, defining what they mean is a matter of value judgment, not science, and WHO should be promoting just the science when it confronts disease, not the "glorious myth" of a social—i.e., socialist—program of wealth redistribution.

The AIDS and swine flu cases reveal that the UN's top bureaucrats are embarking on an ambitious program to, in effect, bring socialism to the world, and they are quite willing to mislead, exaggerate, or lie to advance their agenda. The same appears to be true of the UN's climate-change initiatives around both the Kyoto Accord and the Copenhagen Protocol. The ostensible aim of these deals is to cope with global warming—and let's call it global warming rather than climate change since climate alarmists aren't worried about global *cooling*. Yet even though the planet was not, in 2010, actually warming, and hadn't for at least a decade, the UN and its followers continued to claim that the global climate situation was becoming *worse*. So, we have *Globe and Mail* columnist Jeffrey Simpson declaring: "Climate-warming predictions of three or four years ago are already out of date. New science suggests an even faster warming than had been thought possible."[20] Simpson was claiming fears of faster warming at a time when the planet was not only not warming but actually slightly cooling. Indeed, much of the CRU Climategate email traffic was about how to "hide the decline" in temperatures in the 21st century from the public. So, what is Simpson's evidence? It can only be computer models, since the actual climate is not co-operating. Again, an agenda other than presenting scientific truth about climate is clearly in operation here.

In other words, political leaders, national and international, are trying to reduce the standard of living of people in the developed nations on the (economically false) zero-sum theory that this will, by definition, increase the standard of living in developing nations. As history has shown time after time, in practice this outright redistribution of income from richer to poorer only works if there is political, economic and social change in the nations receiving the aid, and this sweeping change too often does not occur. Where these changes *do* occur, as in nations like India, Thailand and even China, the people do become better off on average, but foreign aid is at most a peripheral factor. That said, and as we've seen in previous chapters, this lowering of the West's standard of living has been a Green goal for many decades—many environmentalists feel we are using too many resources and that Western civilization is not, therefore, sustainable. The solution, for Greens, is to pull back to a simpler lifestyle that uses fewer resources, reducing our standard of living to, say, that of "a very poor third-world country,"[21] as alarmist eco-journalist George Monbiot puts it in his book *Heat.*

As we saw in Chapter 8, alarmists and moralists—from the Medieval flagellants to the Puritans to today's Islamic Taliban—have been singing the same anti-materialist song for centuries and, at least so far, they've always been wrong. It is possible to have a rising standard of living along with rising populations—indeed, ironically, a rising population seems to be one of the conditions for a rising standard of living. Yet, in July 2009, England's Prince Charles delivered a speech in which he claimed that Armageddon was "94 months" away and that we needed to drastically reduce our standard of living or face disaster. If the Prince has reduced *his* standard of living in response to this perceived crisis, there is no evidence of it. Similarly, IPCC president

Rajendra Pachauri declared in November 2009 that "we have reached the point where consumption and people's desire to consume has grown out of proportion. The reality is that our lifestyles are unsustainable."[22] There is no evidence that multi-millionaire Pachauri is planning to lead by example and reduce his own jet-set lifestyle, which includes a mansion in the most exclusive neighbourhood of New Delhi.[23] Nor, indeed, does Al Gore show any signs of selling his Tennessee mansion for a more "sustainable" shack.

Now, if reducing prosperity is something democratically elected leaders feel is necessary, then naturally they wouldn't want to present this austerity policy directly to their citizens; choosing to be poorer is not exactly a vote-getter if explained honestly. However, if there is some compelling reason why we in the developed world must impoverish ourselves to save the world—and extreme socialism has always made the countries that try it poorer rather than richer—then surely politicians should make this case directly rather than using climate change as a cover story. Similarly, if it is necessary to replace fossil fuels for political and economic reasons, such as the worthy goal of becoming less reliant on Middle Eastern dictatorships, then surely this case should be made directly, not slipped under the public's radar by creating fears over climate. This policy of using climate science and consensus climate scientists as cats' paws for political, social and economic agendas does enormous damage to the integrity of science as a source of truth.

There is another danger when it comes to putting science in the service of an political program to make us poorer rather than richer. The Victorian historian Thomas Macaulay wrote:

> In every experimental science there is a tendency towards perfection. In every human being there is a wish to ameliorate his own condition. These two principles have often sufficed, even when counteracted by great public calamities and by bad institutions, to carry civilisation rapidly forward.[24]

Macaulay is noting that science has always been used to make us materially better off—to "ameliorate" our condition. Yet, just as democratic politicians for the first time in history are trying to make their people poorer, so this is the first time in history that *scientists*—specifically, alarmist climate scientists—are using science to try to make people poorer rather than better off.

In this, alarmist climate science is guilty of an astonishing betrayal of the scientific spirit which, up to now, has been dedicated to material progress, not material regress.

Summing up: alarmist climate research did not arise out of an evolutionary scientific process spurred by a real problem. Comedian Anthony Clark tells of being awakened on an airplane by the flight attendant who says he needs to do up his seat belt because the airplane is about to experience turbulence. "It's a good thing you woke me up," he says to her. "Otherwise, I'd have slept right through that." Similarly, alarmist climatologist Spenser R. Weart has written a book on the history of global warming science in which he makes an intriguing statement:

> If just one of these men [climate scientists like Svante Arrhenius and Roger Revelle who first began to study carbon dioxide] had been possessed by a little less curiosity, or a little less dedication to laborious thinking and calculation, decades more might have passed before the possibility of global warming was noticed.[25]

Similarly, IPCC chairman Rajendra Pachauri, in defending his institution after Climategate, has noted: "If the IPCC wasn't there, why would anyone be worried about climate change?"[26] He's right. No one would be worried. What Weart and Pachauri are saying, although this is not at all what they intended, is that if we didn't have a battalion of climate scientists in the IPCC and politicians like Gore telling us how awful global warming is going to be (reality check: it's not awful so far, is it?), would we even notice? And if we did notice the climate changing, would we be frantic about it, as Gore, the IPCC, et al., wish? As an atmospheric physicist writes:

> Never before in history has it taken a massive publicity campaign to convince the public of a scientific truth. The only reason half the public thinks global warming may be true is the massive amount of money put into global warming propaganda.[27]

If climate change was the *huge*, potentially catastrophic problem that Weart, Pachauri, Gore, Hansen, etc., want us to believe, wouldn't we, the public, have *noticed* something wrong with the world all on our own? Would we have to be terrified by a barrage of propaganda into believing there was a serious issue?

Or would we have "slept right through" the recent, present, and future warmings and coolings, as humanity has coped with climate change, without undue fear, for tens of thousands of years? Instead, alarmist climate science was force-fed by two politicians' egos and agendas, then reinforced by a United Nations policy of promoting international "social justice," i.e., socialism, by scaring people about what would otherwise be a non-issue, climate change. Thus, modern climate science was, from the start, politically motivated and biased toward only one hypothesis—the AGW hypothesis—because that hypothesis supports other, less politically popular goals, like moving to nuclear power (a good idea until something better comes along) and reducing our standard of living in the name of "fairness" (a very bad idea).

PSYCHOLOGICAL ORIGINS OF THE AGW HYPOTHESIS

We looked above at some of the political motivations for propagating a belief in catastrophic global warming. In Chapter 11, we looked at the religious underpinnings of alarmist climate dogmatism. But if alarmist climate science is motivated by political and religious conviction, we still have to explain the source of convictions so strong that many scientists are persuaded to abandon scientific principles in favor of "certainty" about an hypothesis for which there is no empirical proof. To do that, we need to move into the realm of psychological motivation. Specifically, we'll look at three possible *psychological* theories for why alarmist climate scientists are so emotionally attached to the AGW hypothesis. One theory is that humanity is divided into two psychological styles—e.g., optimist and pessimist, liberal and conservative—that affect our approaches to issues like global warming. A second is that pessimism is built into our genetic makeup thanks to natural selection, even when this pessimism is not warranted by the facts. Finally, we will explore whether alarmist climate scientists are victims of groupthink and group polarization.

CONFLICT OF VISIONS

Let's begin with the idea, discussed briefly in Chapter 11, of the importance of one's vision. As American thinker Thomas Sowell notes in *Conflict of Visions*, most human beings ultimately base their actions not on reason but on what Sowell calls their "vision," which is "what we sense or feel *before* we have constructed any systematic reasoning … A vision is our sense of how the world works."[28] Sowell identifies two basic visions, which he calls "constrained" and "unconstrained." Similarly, pioneering psychologist William James (1842-1910) divided humanity into two temperaments, tough-minded (constrained) and tender-minded (unconstrained), and declared long before Sowell that "the history of philosophy is to a great extent that of a certain clash of human temperaments."[29] Neurophysiologist Steven Pinker calls the two world-views Tragic (constrained) and Utopian (unconstrained).[30] The constrained, Tragic vision sees civilization as a necessary curb on a flawed human nature; the unconstrained, Utopian vision sees the flaws in human nature as *caused* by civilization (hence, a belief in the noble savage or the perfection of nature without humanity).[31] The constrained view favors free markets, imperfect as they are; the unconstrained prefers government intervention to make markets fairer (e.g., the United Nations idea of social justice). The constrained view can be reformist, but cautiously (as Thomas Macaulay put it in 1831, "Reform that you may preserve"); the unconstrained view tends to be Utopian and Romantic. The constrained vision favors what Sowell calls "trade-offs"; the more idealistic unconstrained vision seeks total solutions, such as the complete elimination of poverty (or global warming), and anything less than perfection will not do.

In U.S. terms, the constrained vision is Republican, the unconstrained, Democrat; in Canada, the Conservatives tend to the constrained, New Democrats and Liberals to the unconstrained visions. These styles are, of course, the extremes: as Sowell notes, most of us are a mix of the two visions, but usually one or the other is dominant at any given time or situation. As a consequence, most of us build our theories of the world not on what reason tells us but on what our vision tells us. Or, as Bertrand Russell put it, "Political opinions are not based upon reason."[32] Therefore, Sowell observes, what may appear to be conflicts of differing theories or facts are actually conflicts of underlying visions. These "visions" are the unconscious basis for a person's faith or religion—in a way, they *are* one's religion. The vision is how one *sees* the world, at the gut level, before rational thought kicks in.

At some point we may seek facts to test the theory that has arisen from the vision, or we may not. The problem, Sowell notes, is that "when there is a conflict of visions, those most powerfully affected by a particular vision *may be the least aware* of its underlying assumptions—or *the least interested* in stopping to examine such theoretical questions when there are urgent 'practical' issues to be confronted, crusades to be launched, or values to be defended at all costs."[33] [italics added] Sowell's book was published in 1987, but it could be referring to Al Gore's global warming crusade today—for Gore there are always "urgent 'practical' issues to be confronted, crusades to be launched, or values to be defended at all costs." Gore is, in fact, a striking example of the unexamined, unconstrained vision.

In global warming terms, the unconstrained vision battles for climate *mitigation* ("We've got to stabilize the climate," as the Sierra Club puts it, blissfully unaware, apparently, that this is impossible): carbon emissions must be *stopped* at all costs, even if the cost is a wrecked economy and ruined lives. This solution has shades of Marxism's anti-capitalism, another unconstrained vision. Those with the constrained vision believe *adapting* to changing climate is the best fallible human beings can do. The unconstrained vision believes that people can be brought, rationally, to see the enormity of the problem. The constrained vision is not so idealistic, and would prefer to rely on incentives like higher oil prices in a free market to produce rational outcomes. Similarly, those with the constrained vision tend to believe that human beings are not powerful enough to control the climate, either as the cause of this change or the remedy for it, while the unconstrained vision puts human beings and their actions front and centre. Hence, Al Gore and alarmist climate science believe human beings are the "principal" cause of global warming, despite scientific evidence that there are hundreds of factors contributing to climate of which human beings are only a small part.

When it comes to scientists, there is no reason to believe that they, even with their special intellectual training, are immune from this kind of vision-based polarization, visions that distort and even corrupt (*stasis*: see Chapter 7) the scientific discourse on global warming. At the very least, the alarmist, unconstrained climate vision should not be accepted without a high level of scrutiny and skepticism since, as we saw in Chapter 8, the overwhelming majority of alarmist environmentalist visions in the past, all unconstrained, have proved to be wrong.

ARE WE NATURAL-BORN ALARMISTS?

Where do our gut instincts, our visions, right or wrong, come from? Are the alarmist's and skeptic's differing views of climate change based in scientific facts or, perhaps, facts of human psychology such as Sowell's pre-cognitive vision or James's inborn temperaments? Science writer Gregg Easterbrook suggests in *The Progress Paradox* that "natural selection … might have favored early humans who were uneasy, distrustful, inclined to assume the worst about life and one another. … From an evolutionary standpoint, it may be that we are intended to feel unhappy regarding our circumstances."[34] Steven Pinker observes:

> Risk analysts have discovered to their bemusement that people's fears are often way out of line with objective hazards. Many people avoid flying, though car travel is eleven times more dangerous. … Even when people are presented with objective information about danger, they may not appreciate it because of the way the mind assesses probabilities.[35]

That is, human beings appear to be hard-wired to believe the worst—to be alarmists—as an evolutionary survival mechanism. This was undoubtedly a very good trait at a time when dangers were everywhere, as when we were hunter-gatherers and then agriculturalists in a lawless world completely ruled by nature. These fears make little sense in the civilized environment of the advanced countries when, for most of us, the biggest danger we face is getting into a traffic accident. And yet these often irrational fears persist. It is, of course, possible that the fears of global warming will turn out to be justified, but at the moment there is no empirical evidence whatsoever to make global warming "oblivion" inevitable or even plausible. Instead, it is as if human societies need a constant diet of fear to survive. So, if it's not communism or nuclear war or the ozone layer or terrorism that is supposed to terrify us, it's global warming. It is significant, as many thinkers have pointed out, that the political elite began the global-warming scare after the fall of communism in 1989. As historian John Dietrich notes:

> The threat of global warming will eventually recede. The need for an apocalyptic vision, however, will not. The next threat will contain many of the characteristics of the global warming threat. It will predict the end of the world. It will be based on "scientific facts." It

> will require massive counseling for the psychological distress it will cause. It will require the creation of a massive bureaucracy. And it will require the transfer of massive amounts of money to the hypothesized victims of the future crisis.[36]

If this is so then, as Carl Sagan writes: "Part of the duty of citizenship is not to be intimidated into conformity,"[37] not to allow ourselves to be stampeded by these manufactured "crises." Instead, it is the duty of citizens to look closely at what they have been told to fear to see if there is a real problem or if, instead, our genetic propensity to fear is just being manipulated by the political and scientific elite.

GROUPTHINK AND GROUP POLARIZATION

The above hypotheses—unexamined visions and inborn evolutionary pessimism—may seem far-fetched when it comes to explaining the dogmatic alarmism of consensus climate science, or at least the version of climate science presented to the public by Gore, Hansen, Weaver, Schneider, et al. But there is one psychological diagnosis that fits alarmist climate science like a glove: groupthink. With groupthink, we get the best explanation of "how can many, many respected, competitive, independent science folks be so wrong?"

Groupthink was extensively studied by Yale psychologist Irving L. Janis and described in his 1982 book *Groupthink: Psychological Studies of Policy Decisions and Fiascoes*. Janis was curious about how teams of highly intelligent and motivated people—the "best and the brightest" as David Halberstam called them in his 1972 book of the same name—could have come up with political policy disasters like the Vietnam War, Watergate, Pearl Harbor and the Bay of Pigs. Similarly, in 2008 and 2009, we saw the best and brightest in the world's financial sphere come to grief thanks to some incredibly stupid decisions, such as allowing sub-prime mortgages to people on the verge of bankruptcy.[38]

In other words, Janis studied why and how groups of highly intelligent professional bureaucrats and, yes, even scientists, screw up, sometimes disastrously and almost always unnecessarily. The reason, Janis believed, was "groupthink." He quotes Nietzsche to the effect that "madness is the exception in individuals but the rule in groups," and notes that groupthink occurs when "subtle constraints … *prevent a [group] member from fully exercising his critical powers* and *from openly expressing doubts* when most others in the group appear to have reached a consensus."[39] [italics added] Even if the group leader expresses an openness to new ideas, group members value consensus more than critical thinking; groups are thus led astray by excessive "concurrence-seeking behavior,"[40] which, incidentally, is exactly what the Greek term *stasis* refers to (see Chapter 7). Groupthink is "a model of thinking that people engage in when they are deeply involved in a cohesive in-group, when the members' strivings for unanimity override their motivation to realistically appraise alternative courses of action."[41]

The result is what Janis calls "the groupthink syndrome." This consists of three main categories of symptoms:

1. **Overestimate of the group—its power and morality**, including "an unquestioned belief in the group's inherent morality, *inclining the members to ignore the ethical or moral consequences of their actions*."
2. **Closed-mindedness**, including a *refusal to consider alternative explanations* and *stereotyped negative views of those who aren't part of the group's consensus*. The group takes on a "win-lose fighting stance" toward alternative views.[42]
3. **Pressure toward uniformity**, including "a shared *illusion* of unanimity concerning judgments conforming to the majority view"; "direct pressure on any member who expresses strong arguments against any of the group's stereotypes"; and "the emergence of self-appointed mind-guards … *who protect the group from adverse information that might shatter their shared complacency* about the effectiveness and morality of their decisions."[43] [italics added]

It should be obvious that alarmist climate science—as explicitly and extensively documented in the CRU "Climategate" emails—shares *all* of these defects of groupthink, including a huge emphasis on maintaining consensus, a sense that because they are saving the world, climate scientists are beyond the normal moral constraints of scientific honesty ("overestimation of the group's power and morality"), and vilification of those ("deniers") who don't share the consensus. For example, regarding Symptom 1, **overestimation of the group's power and morality**: as we saw in Chapter 8, consensus climate spokespeople like Al Gore, James Hansen, and Stephen Schneider, along with an astonishing *27 per cent* of climate scientists polled (see Chapter 10), feel it's acceptable and even moral to exaggerate global-warming claims to gain public support, even if they have to violate the broader scientific principle of adherence to truth at all costs. Consensus climate science also overestimates the power of humanity to override climate change, whether human-caused or natural, just as U.S. planners overestimated that nation's ability to win the Vietnam War.

Regarding Symptom 2, **closed-mindedness**, this book is filled with cases in which alarmist climate paradigm ignores or suppresses evidence that challenges the AGW hypothesis. Regarding Symptom 3, **pressure toward uniformity**: within consensus climate science there is a "shared illusion of unanimity" (i.e., a belief in total consensus) about the majority view when, as we've seen, this total or near-total consensus has no basis in reality.[44] Climate scientists who dare to deviate from the consensus are censured as "deniers"—a choice of terminology that can only be described as odious. And, as we saw in Chapter 4, the IPCC explicitly aims for "consensus" in its reports—it does not publish minority reports, and yet it is impossible that the alleged more than "2,000 scientists" could completely agree on a subject as complicated as climate. All of these are symptoms of groupthink at work, symptoms that, if pointed out, would undoubtedly be hotly denied by alarmist climate scientists because, of course, alarmist climate science has the (apocalyptic) truth and the situation is too serious (they believe) to tolerate dissent or further debate with those who have the temerity to disagree, especially those outside the alarmist paradigm.

Janis notes one other form of dysfunctional group dynamic that arises out of groupthink and that, in turn, helps create groupthink:

> The tendency for the collective judgments arising out of group discussions *to become polarized*, sometimes shifting toward extreme conservatism and sometimes toward riskier forms of action than the individual members would otherwise be prepared to take.[45] [italics added]

This dynamic is commonly referred to as "group polarization." As a process, "when like-minded people find themselves speaking only with one another, they get into a cycle of ideological reinforcement where they end up endorsing positions far more extreme than the ones they started with."[46] And it is one of the curious qualities of these positions that, because they are so extreme, they are held with extreme ferocity against all criticisms.

Groupthink is not unusual in academic disciplines. For example, Walter Kaufmann, a world-renowned editor of Nietzsche's works, identifies groupthink in his discipline, philosophy, as follows:

> There is a deep reluctance to stick out one's neck: there is safety in numbers, in belonging to a group, in employing a common method, and in not developing a position of one's own that would bring one into open conflict with more people than would be likely to be pleased.[47]

Similarly, as we saw in Chapter 10 in discussing the Climategate emails, Climatic Research Unit director Phil Jones shows this "deep reluctance to stick out one's neck" in writing (July 5, 2005): "The scientific community would come down on me in no uncertain terms if I said the world had cooled from 1998." Keith Briffa laments (Sept. 22, 1999): "I know there is *pressure* to present a nice tidy story as regards 'apparent unprecedented warming in a thousand years or more in the temperature proxy data' but in reality the situation is not quite so simple. ... I believe that the recent warmth was probably matched about 1,000 years ago." [italics added] Elsewhere, Briffa notes (April 29, 2007): "I tried hard to balance the needs of the science and the IPCC, *which were not always the same*. I worried that you might think I gave the impression of *not supporting you well enough* while trying to report on the issues and uncertainties."[48] [italics added] All of the above emails are examples of scientific groupthink—putting the needs and desires of a peer group, in other words, the desire for "consensus," ahead of the scientific facts. We would, undoubtedly, find other examples of alarmist groupthink if we could examine the emails of other promoters of climate alarmism, like the Goddard Institute.

This groupthink isn't at all surprising. After all, alarmist climate scientists go to dozens of conferences with like-minded people (the views of outright "deniers" are not welcome, as the CRU emails clearly reveal) and, in the absence of real debate or dissent, persuade themselves that human beings are the main reason the planet is warming. Why? Because everyone else seems to think so and, in groupthink, consensus is highly valued. The same principles operates strongly, of course, in religion.

Climate alarmists will, of course, angrily dispute that climate science groupthink is as strong as claimed here. However, groupthink is clearly identified in the 2006 Wegman report into the Michael Mann hockey stick controversy. The report was commissioned by the U.S. House Science Committee after Mann's refusal to release to Steve McIntyre and other researchers all the data leading to the hockey stick conclusions. As we saw in Chapter 4, it was McIntrye who revealed that the calculations in Mann's paper had not been checked by the paper's peer reviewers and were, in fact, wrong. The National Academy of Sciences committee, led by Dr. Edward Wegman, an expert on statistics, identified one of the reasons why Mann's paper was so sloppily peer-reviewed as follows:

> There is *a tightly knit group of individuals who passionately believe in their thesis.* However, our perception is that this group has a *self-reinforcing feedback mechanism* and, moreover, *the work has been sufficiently politicized that they can hardly reassess their public positions without losing credibility.*[49] [italics added]

Wegman notes that the Mann paper became prominent because it "fit some policy agendas."[50] The Wegman Report also observed:

> As statisticians, we were struck by the *isolation of communities such as the paleoclimate community* that rely heavily on statistical methods, yet do not seem to be interacting with the mainstream statistical community. The public policy implications of this debate are financially staggering and yet apparently no independent statistical expertise was sought or used.[51] [italics added]

In other words, alarmist climate scientists are part of an exclusive group that talks mainly with itself and avoids groups that don't share the AGW hypothesis or alarmist political agenda. Overall, Wegman is describing with great precision a science community whose conclusions are distorted and polarized by groupthink.

Thanks to the Climategate emails, some consensus climate scientists are beginning to recognize the dangers of groupthink within their discipline. So, climatologist Judith Curry writes:

> In my opinion, there are two broader issues raised by these emails

> that are impeding the public credibility of climate research: lack of transparency in climate data, and "tribalism" in some segments of the climate research community that is impeding peer review and the assessment process.[52]

Similarly, IPCC contributor Mike Hulme writes:

> It is possible that climate science has become too partisan, too centralized. The *tribalism* that some of the leaked emails display is something more usually associated with *social organization within primitive cultures*; it is not attractive when we find it at work inside science.[53] [italics added]

In short, there is no doubt that groupthink—a later, more scientific word for "tribalism"—is strongly at work within alarmist climate science, however much the affected scientists refuse to recognize it. And, as we saw in Chapter 7, it is this tribalism of science that Karl Popper called the "closed society," which leads to "stasis," the abdication of personal responsibility in favor of the group, in contrast to the "open" culture that "expected citizens [of Athens] to be unafraid to speak their minds."[54] Another manifestation of groupthink is over-attachment to a failing paradigm (Chapter 7).

As a result of tribalism (groupthink), alarmist climate science makes assertions that are often extreme (polarized), including the explicit or implicit endorsement of claims that global warming will lead to "oblivion," "thermageddon," mass extinctions, and the like. Indeed, one of the ironies of climate science is that extremist AGW believers like Gore, Hansen, Weaver and Schneider have succeeded it persuading the media and public that those who *don't* make extreme claims, the skeptics, are the extremists. Group polarization is the only rational explanation for these extreme alarmist claims; the empirical scientific evidence is not strong enough to be so confident of these doomsday predictions. It makes more sense to believe that even intelligent, highly educated scientists can be caught in what has been called the "madness of crowds." Indeed, British philosopher Martin Cohen has made this connection explicit:

> Is belief in global-warming science another example of the "madness of crowds"? That strange but powerful social phenomenon, first described by Charles Mackay in 1841, turns a widely shared prejudice into an irresistible "authority". Could it [belief in hu-

> man-caused global warming] indeed represent the final triumph of irrationality?[55]

In other words, there is strong psychological evidence that alarmist fears of climate change are far more the result of groupthink and the group polarization process than scientific evidence, leading to the triumph of irrationality over reason.

PERSONAL AND FINANCIAL GAIN

Finally, in examining how so "many, many respected, competitive, independent science folks [can] be so wrong," another factor must be considered: follow the money. Being an alarmist climate scientist brings a big element of personal gain, both in research money and prestige. As noted in Chapter 10, the U.S. alone spends more than $5-billion a year on climate change research and initiatives, and the vast majority of the research money is earmarked for studies that confirm the anthropogenic global warming hypothesis. *Newsweek* magazine, a strong promoter of the AGW hypothesis, nonetheless wrote in its November 2, 2009, issue:

> Climate change is the greatest new public-spending project in decades. Each year as much as $100-billion is spent by governments and consumers around the world on green subsidies designed to encourage wind, solar, and other renewable-energy markets. The goals are worthy: reduce emissions, promote new sources of energy, and help create jobs in a growing industry. Yet this epic effort of lawmaking and spending has, naturally, also created an epic scramble for subsidies and regulatory favors.[56]

And, of course, an epic scramble by climate scientists for research funding.

By contrast, as we saw in Chapter 7, Greenpeace's "Exxonwatch" website estimates that skeptical climate science has received a paltry $33 million from major funder Exxon oil since 1998.[57] That's just over $3-million a year, compared to the $5-billion a year directed at supporting the AGW hypothesis. It is ludicrous to think that climate scientists like Richard Lindzen, Patrick Michaels, Ian Patterson, Tim Ball and others are skeptics because private industry might be willing to throw a few thousand dollars their way when, by switching sides, they could be partaking of a multi-billion-dollar feast of research money. It is, however, very plausible that many climate scientists have been seduced by the billions coming their way—if they go along with the AGW hypothesis.

Could alarmist climate scientists be so easily bought off? This would be hard to believe if the alarmists didn't so often accuse *skeptical* scientists of being in the pay of corporate interests. These alarmist climate scientists clearly believe that their colleagues can be bought; once we've established that scientists can be bought, it's no stretch to believe that alarmist climate scientists are also susceptible to financial inducements. And, having been bought, whether they believe the AGW hypothesis or not, alarmist climate scientists are wise to keep their heads down, do research that interests them, and not rock the boat. As quoted earlier, paleoclimatologist Donald Prothero calls this the "bandwagon" effect:

> One of the striking sociological features of science is a well-known political and social activity called "jumping on the bandwagon"—latching on to a trendy idea when it is rolling to get a free ride. Politicians do it when a particular candidate, party, or ideology has irresistible momentum to put them in power. *Scientists do it when a particular theory or idea gives one a better change of getting a grant, publishing a paper, or landing a book contract.*[58] [italics added]

Another term for the bandwagon effect is, of course, groupthink. While there is very little, if any, empirical evidence supporting the anthropogenic warming hypothesis, there is a lot of empirical evidence, measurable in dollars, supporting the "follow the money" hypothesis when it comes to explaining the dogmatism of alarmist climate science.

In analyzing the why's of alarmist climate science, we must mention the loss of prestige or, in Asian terms, "face," that alarmist climate scientists would experience if they were to, now, publicly admit doubts about the AGW hypothesis. As we saw above, Edward Wegman alluded to this possible loss of face when he wrote, in his report on the hockey stick, that the work of the hockey-stick promoters "has been sufficiently politicized that *they can hardly reassess their public positions without losing credibility.*"[59] [italics added] In the same vein, Martin Durkin, producer of the skeptical British TV documentary *The Great Global Warming Swindle*, has observed: "Lots of scientists, ... and journalists too, have staked their reputations on this [AGW] theory being true. Many have built their careers on it. I sympathize with them."[60] Many alarmist climate scientists have been hostile and contemptuously dismissive of those offering alternative explanations for climate change. As a result, if the AGW hypothesis is discarded, alarmist scientists will look, in the public's eyes, ridiculous, and rightfully so given their extreme claims and arguments that the science is "settled" when it so clearly isn't. In fact, as we saw in Chapter 10 and will see in Chapter 14, public belief in alarmist climate science is eroding quickly. And so, as the evidence piles up against the AGW hypothesis, dedicated AGW believers become ever more shrill in its defence—"warming is accelerating!"—and ever more out of touch with the climate *reality*—the climate is not currently warming at all.

It didn't have to be this way, and perhaps wouldn't have been this way if it were not for groupthink and group polarization. As noted earlier, the AGW hypothesis is just that—an hypothesis. It has not reached the status of a theory and it is not a scientific fact. As one observer has noted: "The enhanced greenhouse effect remains *a plausible but unproven hypothesis*, with a significant number of question marks hanging over it. The most important question is not whether carbon dioxide warms the earth, but by how much."[61] [italics added] The proper scientific procedure is to present this hypothesis as just that—a tentative explanation—while respectfully considering other hypotheses.

If not so caught in the paranoia of groupthink, the alarmists might also have admitted to the public that their computer models are of limited value in predicting the future of climate. The IPCC *does* admit this, as we've seen, but by the time its conclusions reach the public, highly speculative computer predictions become apocalyptic scientific "facts." This is not how science ought to work, nor is it ethically acceptable for scientists to use exaggeration, misleading arguments and, sometimes, outright falsehoods to get the public's support and, not incidentally, feather their own research nests. The proper scientific approach is to rely on *empirical* evidence. A warming Arctic and melting glaciers, for example, are not evidence that humans are at fault—we are in an *interglacial* after all, and in interglacials, ice melts by definition; other, natural factors are also involved in warming and are almost certainly far more important than anything humans could do. Yet alarmists do everything they can to ensure that the public does not learn this alternative point of view. This blinkered approach was condemned by physicist Richard Feynman, who wrote:

> You should not fool the layman when you're talking as a scientist. … I'm talking about a specific, extra type of integrity that is … bending over backwards to show how you are maybe wrong, that you ought to have when acting as a scientist. And this is our responsibility as scientists, certainly to other scientists, and I think to laymen.[62]

This "extra" type of integrity is rare in writings for laypeople by alarmist climate scientists, who prefer instead the Doctrine of Certainty.

WHAT WE KNOW

Alarmist climate science presents, to the public at least, an air of utter certainty that the planet is warming dangerously and that humans are primarily at fault. In the process, evidence that this *cannot* be the case—for example, the lack of correlation both in the geological past and at present between temperature levels and carbon-dioxide levels, not to mention the CO_2 saturation effect—is dismissed and/or ignored. Criticisms of the IPCC process, computer models and conclusions—such as the IPCC models' inability to capture the behaviour of clouds when clouds represent roughly 20 per cent of the greenhouse effect—are also dismissed and ignored. Instead, skeptical criticisms of consensus climate science normally attract *ad hominem* attacks on the critics, rather than a substantive response based on science (see Chapter 7's discussion of the alarmist *ad hominem* assault on Bjorn Lomborg). Undoubtedly, most alarmist climate scientists sincerely believe the evidence for their case is "overwhelming."[63] That they believe so strongly when the empirical evidence is so weak indicates the influence—the *strong* influence—of factors outside science: political pressure, personal gain, ideological agendas, a genetic predisposition to alarmism and unexamined psychological factors such as groupthink.

British philosopher David Hume (1711-1776) sums up quite well the *psychological* and *religious* factors that might lead to global warming alarmism. In his essay "Of Superstition and Enthusiasm," Hume writes:

> The mind of man is subject to certain *unaccountable terrors and apprehensions*, proceeding either from the unhappy situation of private or public affairs, from ill health, from a gloomy and melancholy dis-

position, or from the concurrence of all these circumstances. In such a state of mind, *infinite unknown evils are dreaded from unknown agents*; and *where real objects of terror are wanting*, the soul, active to its own prejudice, and fostering its predominant inclination, *finds imaginary ones, to whose power and malevolence it sets no limits.*

As these enemies are entirely invisible and unknown, the methods taken to appease them are equally unaccountable, and consist in ceremonies, observances, mortifications, sacrifices, presents, or in any practice, however absurd or frivolous, which either folly or knavery recommends to a blind and terrified credulity.[64] [italics added]

Everyone supports reducing our production of greenhouse gases—until they are told how much it will cost the economy. Maybe that's why they're never told.

—**Roy W. Spencer**[1]

Only when people appreciate how precisely their lives will change, what those changes will cost and when the changes must be implemented, can a reasonable measure be taken of public opinion. ... At no time has the Chrétien government spoken truthfully to Canadians about what lies ahead.

—**Jeffrey Simpson**[2]

People are not the enemy of the environment. Nor is affluence the enemy. Affluence does not inevitably foster environmental degradation. Rather, affluence fosters environmentalism. As people become more affluent, most become increasingly sensitive to the health and beauty of their environment. And gaining affluence helps provide the economic means to protect and enhance the environment.

—**Jack M. Hollander**[3]

Chapter 13

WHAT IS TO BE DONE—OR NOT DONE?

In *Storm Warning*, her 1999 book about the potential catastrophic consequences of global warming, science journalist Lydia Dotto compares being prepared for global warming to taking out an insurance policy. Her point is that regardless of whether global warming fears are justified or not, we can't do *nothing*. After all, she notes, we don't really believe that our houses will burn down when we buy insurance; in buying protection against the financial risks of fire we are just being prudent. Similarly, although Dotto acknowledges that we don't have proof that global warming is human-caused or that it will be catastrophic[4]—and in this respect it's worth noting that in 2010 the global temperature is lower than it was when Dotto published her book in 1999—we still need to take out insurance like the Kyoto Accord and other carbon-reducing measures just in case. She writes:

> The interesting thing about insurance is that we buy it *despite the considerable uncertainties that the disasters they protect us against will ever happen*. We do not expect *proof* that our house is going to

> burn down, that we'll be in a car crash, or that we'll die early of a heart attack before we buy insurance. ... It's precisely because the consequences are potentially devastating that we try to protect ourselves against such events *even if we believe the probability of their happening is very low.*[5] [italics the author's]

This seems a reasonable position—until we begin to look at how much the climate insurance Dotto wants us to buy will cost and whether it will adequately protect us against the (possibly slim) risks we anticipate. At that point, like many analogies when faced with the reality, Dotto's breaks down. In the case of climate change, it breaks down quite badly.

For a start, the "premium" that Dotto and the other global warming alarmists want the developed world to pay isn't just expensive, as insurance coverage can be, although most house insurance coverage is still a small fraction of the house's market value. The global warming premium is ruinously expensive: billions of dollars a year just for Canada alone with a total bill that could cripple the economies of both the developed and developing world. For example, a report by Environment Canada in 2007 estimated that reducing CO_2 emissions by 6 per cent below 1990 levels, as sought by the then-Stephane Dion-led Liberals, would cost 275,000 jobs, increase electricity prices 50 per cent, increase gasoline prices by 60 per cent, and cost each Canadian family $4,000 a year in lost income.[6] Another Environment Canada report the same year, explaining why the Conservative Stephen Harper government would not be implementing Kyoto, said the following:

> The Government's analysis, broadly endorsed by some of Canada's leading economists, indicates that Canadian Gross Domestic Product (GDP) would decline by more than 6.5% relative to current projections in 2008 as a result of strict adherence to the Kyoto Protocol's emission reduction target for Canada. This would imply a deep recession in 2008, with a one-year net loss of national economic activity in the range of $51-billion relative to 2007 levels. By way of comparison, the most severe recession in the post-World War II period for Canada, as measured by the fall in real GDP, was in 1981-1982. Real GDP fell 4.9% between the second quarter of 1981 and the fourth quarter of 1982.[7]

These figures are low compared to some other estimates. In 2002 the Canadian Manufacturer's Association predicted that implementing the Kyoto Accord would cost 450,000 jobs. Whatever the final tab—and nobody knows exactly what the cost of Kyoto would have been or what the cost of its successors might be—we can be sure the bill will be high, with a major negative impact on the Canadian economy, which means a negative impact on individual Canadians and their families.

We in British Columbia have already had a taste of what is in store if a carbon tax comes into being. In July 2008, British Columbia was the first province to add a carbon tax to the price of gasoline: 2.4 cents a litre, with another 1.17 cents added in 2009, and more increases coming in future years. At the same time, the effects of $138 US-a-barrel oil were being felt across North America and Europe. Gasoline prices rose to almost $1.50 in Victoria, with, again, 2½ cents of that being a tax added to the market price, and to $4 US a gallon in the U.S. As a direct result, Victoria's Meals on Wheels program ended, in part, the organizers said, because of rising gasoline costs. Meals on Wheels, which had been in operation for 35 years in Victoria, said it would need $2,000 a week just to cover the rising cost of gasoline—an extra $100,000 a year that simply wasn't there.[8] The price of food and most other goods increased because of higher fuel costs.[9] Tourism was down in Victoria and across Canada, in part because many Americans didn't feel they could afford to drive here. Taxi drivers in many Canadian cities got a considerable fare increase to cover the cost of gas.[10]

I could go on, but the point is that rising fossil-fuel costs in the summer of 2008 gave North Americans a taste of the effects of a high carbon tax—exactly the type of tax many global warming alarmists want. Multiply these hardships by several times and we have some idea of the size of the "insurance premium" that alarmists like Dotto would have us pay to protect ourselves against a warming that may or may not occur and, if it did occur, could be minimal and even benign.

And what would we get for our very expensive pains? As we saw in Chapter 9, Andrew Weaver admits that Kyoto, even if fully implemented, would have "zero effect" on global warming. Weaver also acknowledges that meeting the Kyoto standards would have been "extraordinarily difficult":

> The only way to meet them is through changing technology. I support Kyoto *not because it will have an immediate climatic effect*, but

> because in order to meet it one needs to develop new technologies and change our conception of energy and how we get it.[11] [italics added]

Thus, for Weaver and many other climate alarmists, the Kyoto Accord was only "the first step." Weaver writes:

> There's a danger that people will believe that Kyoto is the solution to the problem. But it's only a first step. The required reductions to actually stabilize levels of greenhouse gases at four times pre-industrial—a level that hasn't been seen since dinosaurs roamed the Earth—means reducing emissions globally by much more than 50 per cent of 1990 levels.[12]

If the economy-killing Kyoto was only a "first step," one can only reflect with horror on what the "next steps" might be. And yet, whatever we do will have "zero effect" on global warming. The same can undoubtedly be said about anything the 2009 Copenhagen meetings might have produced.

But, then, the expected costs of Kyoto have never been clearly explained to the Canadian public, as Dotto admits.[13] She quotes from a 1997 column by *The Globe and Mail's* Jeffrey Simpson in which Simpson writes:

> Only when people appreciate how precisely their lives will change, what those changes will cost and when the changes must be implemented, can a reasonable measure be taken of public opinion. For now, people have only the haziest idea, if they have any at all, about how their lives must change.
>
> And change they must. Emissions must decline by about one-fifth in the period 2000 to 2010 for Canada to meet its Kyoto target. No reduction of that incredible magnitude *can be achieved without wrenching changes. At no time has the [Jean] Chrétien government spoken truthfully to Canadians about what lies ahead.*[14] [italics added]

More than a decade later, Canada's politicians still aren't speaking truthfully to Canadians about the costs of cutting carbon emissions deeply enough to make any difference to global warming (assuming, again, that the planet is still warming and assuming that cutting carbon would actually reduce warming). For example, in a 2008 commentary article in the *National Post*, then Liberal leader Stephane Dion promised a "richer, greener, fairer Canada" that will "unleash the power of the free market, not stifle it."[15] Dion didn't explain how artificially loading carbon taxes onto market-set prices for fossil fuels qualifies as unleashing a "free market." The public is being told that all these anti-carbon measures will prepare Canadian business to "thrive" in the "new economy"—a thriving economy that is in fact nothing but a government fiction based on unproven fears of global warming. Anyone with even the slightest political understanding knows that Dion's promises of a better, greener future through higher taxes are total moonshine. Fortunately, many Canadians do appear to have that understanding, because they sent the Dion Liberals a strong anti-green message in the 2008 federal election, cutting the Liberal's seats to 77 from the 103 they had in 2006 and leading to Dion's ouster as leader.

It is telling that the Liberal governments of Jean Chrétien and Paul Martin (Dion was Martin's Environment Minister) did nothing concrete to meet the Kyoto commitments after Chrétien signed the accord in 2002, although Ottawa somehow managed to spend a reported $4-billion or more in the process of doing nothing. While Ottawa fiddled, carbon emissions went up 24 per cent by 2006, when they were supposed to go below 1990 levels by 2012.[16] Why didn't Ottawa want to take the necessary carbon-cutting steps? Because, as Simpson pointed out in his column, the Liberal government knew the Kyoto measures would cripple the Canadian economy—otherwise, why not act? And the Liberals knew that once Canadians realized how much Kyoto would actually cost them, individually and as families, support for Kyoto and for the political party that brought in Kyoto would collapse.

Every government knows, or should know, that these "green" measures, which are sold by the environmental movement as profit generators, are money-losers. That's why "green" industries need hefty subsidies; if green economic measures were profitable, private industry would leap on them. Dion was sugar-coating the real costs that the proposed Liberal carbon tax would have imposed on Canadians because, again, if Canadians knew the truth, they would be horrified. Or, as American meteorologist Roy Spencer puts it, echoing Simpson: "Everyone supports reducing our production of greenhouse gases—until they are told how much it will cost the economy. Maybe that's why they're never told."[17] The public would be even more horrified if it knew that all these sacrifices will be in vain—even if we were able to bring CO_2 emissions to a halt, there would be little effect on warming. Why? Because the actual behavior of the climate in the past decade shows that carbon dioxide is simply not as powerful a greenhouse gas as alarmist climate models predict. Thus, reducing carbon emissions will have little or no effect on climate. The utter futility of Kyoto is not a secret, but it's not exactly trumpeted about by Kyoto supporters, either.

We can expect no better from Kyoto's successor, the Copenhagen Protocol, and here is one example. Journalist Christopher Monckton has done a detailed calculation showing that if the Copenhagen targets were all met, and even using the IPCC's inflated equations for CO_2 warming—in other words, the IPCC's worst-case warming scenario—by 2020 the amount of warming prevented would be .02°C.[18] To prevent this minuscule amount of warming, the developed world would be crippling its economy to a tune of billions of dollars and sending at least $75-billion a year to the developing world, or $750-billion over the decade. This isn't an "insurance policy," it's economic insanity.

In short, as Dotto points out, we may well run risks in not doing anything about carbon emissions that, in her view, may lead to catastrophic warming. What she doesn't point out is that we may be running even more risks if we take hasty action, including the strong likelihood of a wrecked economy, which means thousands and maybe even millions of wrecked livelihoods and therefore wrecked lives. In other words, the cost of Dotto's "insurance" is too high without a lot more proof that we are, in fact, heading for catastrophe. So far, as we've seen, that proof exists only in computer models (our "only tool").

To put this another way: the worst thing that could happen to the world right now is not a degree or two or three Celsius of warming. A prosperous world economy can easily cope with that, just as the world and humanity coped perfectly well during the Holocene Maximum of 7,000 years ago, which was up to 3°C warmer than today. The worst thing that could happen is the collapse or crippling of the economic system upon which the planet's six-billion-plus people depend and which gives the developed world and increasingly the developing world enough wealth to cope with warming, if that is what the climate does. If Dotto and her fellow alarmists have their way, we in the developed world will take a major and *self-created* hit to our standard of living. As noted in Chapter 12, this will be *the first time in history that democratic governments and scientists have deliberately tried to reduce the prosperity of their people.*

As we saw in Chapter 9, developed nations have far better environmental policies than developing nations because well-off people don't mind spending some of their surplus income on protecting nature. They don't mind precisely *because* they have surplus income. It seems counter-intuitive, but the best course of action in coping with global warming (assuming that global warming is driven primarily by human carbon emissions, which it almost certainly isn't, but assuming) is to let the carbon emissions continue until the developing nations have reached the West's standard of living. At that point, these nations, too, will begin to protect the environment, including, perhaps, reducing their carbon emissions if that is still considered necessary. In other words, David Suzuki has it exactly backwards when he states, in *The Sacred Balance*: "Some people believe that a clean environment is only affordable when the economy is strong, but in fact, it's the other way around; the biosphere is what gives us life and a living."[19] Recent history has shown that it is the strong economies that work hardest to protect the biosphere. Encouraging poor countries to become as well off as we are is probably the most useful—and humane—policy the developed world could adopt in responding to global warming. And if, as a bonus, the extra CO_2 emissions actually have little warming effect, for reasons we discussed in Chapter 3 (e.g., carbon dioxide saturation), we will not have wasted billions of dollars to no purpose.

In terms of Dotto's insurance analogy, Kyoto and Copenhagen aren't like paying $500 a year for insurance on a $500,000 house, but like spending a tenth of your annual salary and then getting almost no protection because Kyoto and it successors would have "zero effect" on warming. No matter. For Dotto and those who support her position, no amount of money is too much to protect the planet. And as for the little problem of no proof—and Dotto states quite categorically, "There is *no proof* that global warming will cause adverse, even catastrophic damages around the world"[20] [italics the author's]—well, she has an answer to that, too: it's not up to the alarmists to provide proof of harm, it's those who claim that warming won't cause catastrophe—the skeptics—who need to provide the proof. In fact, Dotto seems surprised and even offended that "opponents of greenhouse gas cuts have so successfully placed the burden of proof on the scientific community, environmental activists, and politicians who advocate immediate action to curb global warming."[21]

Dotto is here invoking the Precautionary Principle. In its most extreme form, and this is the form that strong environmentalists prefer, if there is even the remotest chance that some human behavior or product could cause harm, even if there is no proof whatsoever, then that behavior or product should be banned. The precautionary principle has become extraordinarily powerful and has led to the banning of many products, such as plastic water bottles, that might, maybe, perhaps, possibly cause harm. The burden of proof is always on the product's producer, not the accuser, unlike, say, the court system where the accused is innocent until proven guilty.

Based on the precautionary principle, but without (as Dotto admits in 1999 and as was still the case in 2010) any conclusive proof, the global-warming scare is to be accepted as *de facto* true because, if it were true, it would be disastrous. Thanks to the precautionary principle, and despite a complete lack of evidence that carbon curbs will do any good, the onus of proof is, incredibly, on those who are skeptical of the global-warming hypothesis. And while there might be a case for expecting manufacturers to prove their products safe, there is no logical justification for requiring the same of global warming skeptics—it's not the skeptics who are making extreme claims.

Instead, the alarmist claim that global warming will be catastrophic is presented to the public as scientific fact, rather than an untested, unfalsifiable, and therefore unscientific hypothesis. And because the public has been led to believe the AGW hypothesis is a "fact," it's difficult for governments to approve any other course of action than a full-scale assault on the presumed cause of warming—human industrial activity. And so, thanks to the control of the media by the global-warming alarmists, the public is being given only two choices, both disastrous: the global warming catastrophe that *might* occur, maybe, perhaps, if we don't curb carbon emissions, and the global economic catastrophe that will *almost certainly occur* if we do severely curb carbon emissions. In short, we appear to be between a rock and a very hard place. Is there another alternative? Yes: how about doing nothing until we have stronger evidence that catastrophe is coming, given that "much of the belief that we are in the midst of an environmental crisis depends on computer models"?[22]

DOING NOTHING

Doing nothing is what Christopher Monckton, a former British political advisor, bluntly suggested to the climate conference in Bali in December 2007: "There is no climate crisis. The correct policy response to a non-problem is to have the courage to do nothing."[23] Of course, thanks to overreliance on the precautionary principle, "nothing" is one of the most difficult things a democratic government can do when someone cries wolf. Thanks to the litigious nature of modern society, governments and institutions try to protect themselves against any legal eventuality, however unlikely. Furthermore, politicians don't want to appear anti-environment lest they lose votes. For example, in 2001, a British Columbia environmentalist group claimed that hunting threatened grizzly bear populations in the province. Therefore, the B.C. government shut down grizzly hunting for three years (a new government reinstated the hunt six months later) to study what turned out to be a non-existent problem—in fact, the grizzly bear population is increasing to the point of becoming a hazard to people.[24]

So it is with global warming. The mere possibility that global warming may cause problems isn't dealt with rationally and scientifically. The rational response is to recognize that there's very little we can do to stop warming (or cooling) because nature is, as a rancher friend of mine told me, "a lot bigger than anything humans can throw at it." Nor, as we've seen, is the relationship between increasing carbon dioxide and increasing temperature as clear-cut as the global-warming alarmists would have the public to believe. However, in the Western world we're programmed to think that we have to *do* something about perceived problems, even if they are beyond our control and even if the proposed solutions, like the Kyoto Accord and Copenhagen Protocol, make no sense. Yet here's why doing nothing, at least from the governmental point of view, may be the right policy.

For a start, as we saw in Chapter 3, the Peak Oil believers, who are often global warming alarmists as well, tell us that fossil fuels, and particularly oil, will run out in the next few decades. What does Peak Oil mean for carbon emissions? Quite simply, that at some point in the next 40 years or so a major source of carbon emissions—oil—will stop being significant. It's believed natural gas will run out in 60 years, coal in 130 years.[25] No government regulation will be required to make this happen, and no draconian carbon curbs are needed.

Humanity has been using fossil fuels extensively for the past 100 years. In that time, the IPCC tells us, CO_2 levels have risen 120 ppm, from 280 ppm to almost 400 ppm, and, undoubtedly, humanity has to take some of the blame for this increase. For the purposes of this argument, let's assume that humans are responsible for all of this 120 ppm increase. We aren't, but let's make that assumption. Given that fossil-fuel use is now at or close to the Peak—that is, we've used about half of what is said to be available—then all things being equal, using up the remaining fossil fuels should add roughly the same amount of CO_2 as the previous 100 years, that is, about 120 ppm. That would take CO_2 levels to about 520 ppm. Given that population may go up by three billion in the next 50 years, let's add another 200 ppm to be generous, for a total of 720 ppm of carbon dioxide in the atmosphere.

For a start, we can see immediately that Andrew Weaver's fear of carbon dioxide levels four times pre-industrial levels (280 ppm), or almost 1,200 ppm, seems unwarranted—we couldn't possibly burn up that much carbon in a century.[26] Second, as we saw in Chapter 3, thanks to carbon dioxide saturation (the more CO_2 put into the atmosphere the less warming effect each additional amount has), 720 ppm would probably produce a very slight additional warming of, at most, 1-2°C. At the same time, 720 ppm would be extremely beneficial to the planet's plant life. Doing nothing would also be extremely beneficial to humanity's economic life.

PRIVATE INDUSTRY'S RESPONSE TO PEAK OIL

What about private industry? How will it respond to the threat of global warming and/or peak oil? Well, we can expect the oil companies, in particular, to move heaven and earth to find another source of energy so they can stay in business when the oil becomes too expensive for average consumers. That means spending some of their profits from today's high fossil-fuel prices for research and development of exactly the alternate energies the Greens want. To believe this is not naïve right-wing optimism—it's how a market economy works. It's how cities moved from horse power to horsepower. Similarly, the auto companies will be turning themselves inside out to keep cars on the road when there's no gasoline to run them. They'll do this without government regulation, spurred on by the desire to survive and make a profit, which is the engine of economic prosperity.

In general, government intervention is not conducive to economic prosperity. Faced with an increase in demand, government's solution, as Canadians have seen in the case of socialized health care, is to ration what is available. Business's solution is to try to produce and sell more of the product or service in greater demand. With the government approach, citizens get less. With a business approach, consumers get more, and almost always at cheaper prices. For example, in the early 1980s my Apple II with 4 k of RAM cost about $5,000, and that's at 1980s prices; today, an Apple iPhone with computing power the Apple II user could have only dreamed of costs about $400. The palm-sized iPhone also uses fewer natural resources. Thus, there's no need for heavy-handed and usually futile government regulations to make adaptation to global warming happen—in fact, government just gets in the way. As geologist Ian Plimer notes:

> Carbon and emissions trading schemes are a God-given opportunity to plunder. New legislation on pie-in-the-sky emissions will result in the public paying more for everything. Trading schemes will be based on a mythical commodity [carbon dioxide]. Such schemes have not stood the test of time and will require constant amendment. The opportunities for fraud are breathtaking. ... Governments just cannot resist an opportunity to raise more taxes, to increase the bureaucracy and impose more regulations.[27]

However, if governments can restrain their greed for the extra income that carbon taxes will bring them, it makes sense for them to step back, "do nothing" (while keeping a close eye on developments) and let markets handle the problem of what to do about global warming, if warm it does over the next few decades. And if we must have a carbon tax for political reasons, then the suggestion by economist Ross McKitrick makes perfect sense: base the size of the carbon tax on the amount of warming that actually occurs. If the planet warms as alarmist climate science predicts, then bring on the carbon taxes, increasing the tax as warming increases. If the planet doesn't warm, as is currently the case, then there's no need for a carbon tax at all.[28]

Here is the basic problem that needs to be faced head-on: the developed economies are too dependent on fossil fuels to cut back enough to make a difference to carbon emissions. Buying fluorescent light bulbs won't do it; driving a hybrid car won't do it; curbside recycling won't do it. As an article in *Maclean's* magazine noted:

> If every home in the U.S. put in one compact fluorescent light bulb, the savings in greenhouse gas emissions would be wiped out by fewer than two medium-sized coal plants. The kind of plant that is being built in China at a rate of one a week.[29]

The 2008-2009 economic slump was created by a drop in GDP in the United States of 6.2 per cent.[30] The Great Depression of the 1930s, the worst economic disaster in recent times, saw a 25 per cent cut in production[31]; that was enough to put one in five Americans out of work. The draconian economic measures proposed by the global-warming alarmists—whose strategy is always based on pulling back rather than moving forward—invite a repetition of this 1930s economic catastrophe.

Why is the alarmist negative economic strategy bad? Because, as already noted, a crippled economy is less capable of responding to the real problems of climate change, whether human-caused or natural, whether warming or cooling. If the climate warms, as it will do in an interglacial, sea levels will rise. The wealthy nations can easily cope with this by spending a few billion dollars on dikes—Bjorn Lomborg estimates dikes might cost about .4 per cent of GDP,[32] which is much less than the IPCC's estimate of three per cent of GDP to "stabilize" CO_2 at about 500 ppm.[33] If the oceans rise, some people may have to move. *I* may have to move—my home in Victoria, B.C., is on the ocean. As Lomborg notes, however, the IPCC believes

> the sea-level rise between now and 2100 will be somewhere between six inches and two feet—not 20 feet—with most estimates around one foot. Now, we have already seen a foot of sea-level rise over the last 150 years, so it will be a bit faster by 2100. But ... was the 20th century marked particularly by the fact that the sea level rose? ... That doesn't mean it isn't a problem, but it's a problem we can deal with.[34]

Climate change is not a catastrophe, although it will require adaptation, the kind of adaptation that human beings have been making for tens of thousands of years. Relocations in the face of rising ocean levels may be necessary in the next few centuries—not decades as the alarmists claim. A strong economy can easily handle this kind of change. A weak economy, with millions of unemployed, cannot. And it is this weak economy that the global warming alarmist policies are aiming us toward.

WHAT WE *SHOULD* DO

While we should do nothing about global warming directly, at least until we have actual empirical evidence—not just the predictions of computer models—that warming is going out of control, we will have to do something about declining fossil fuels. In the next few decades, fossil fuels will become increasingly difficult to find and extract, and may therefore become too expensive for everyday use (I say "may" because pessimists have been predicting this scenario for decades). Thus, every dollar spent developing alternative energy sources is a dollar well-spent, while every dollar spent trying to stop climate change—an impossibility since climate always changes—is a dollar wasted. And yet, the evidence so far indicates that the Greens' preferred alternative-energy solutions—wind and solar power—cannot come even close to supplying the power humanity will actually need in the next century if we are not all—all nine billion of us—to sink back into poverty. Until wind and solar power have the necessary capacity, or some other, perhaps still unknown energy source comes available, the only feasible alternative to fossil fuels is nuclear power. And here again, we have a problem in that Greens have usually been the most vocal opponents of this form of power.

The environmentalist reasons for opposing nuclear power are couched in both scientific and religious/moral terms. On the scientific front, environmentalists note that nuclear power is more of a danger to surrounding areas if it goes out of control than other forms of energy. For example, climatologist Stephen Schneider asks if the "small risk [of nuclear technology] is justifiable, given that other alternatives—including conservation—are available."[35] As an example of the risks, Greens point to the Soviet Union's Chernobyl disaster in 1986 and, in the United States, the Three Mile Island near-meltdown in 1979. Chernobyl was the worst but also the *only* nuclear-plant disaster in the 55-year history of nuclear power. It was also a limited rather than apocalyptic disaster. Although hundreds of thousands of deaths have been claimed due to Chernobyl—a Greenpeace study alleges 93,000 fatal cancer cases[36]—a study by the International Atomic Energy Agency, the UN Development Program and the World Health Organization found fewer than 50 deaths. This study estimated that an additional 4,000 might die from radiation exposure.[37] Further, the problem at Chernobyl was brought about by Soviet incompetence and corruption, not nuclear power as such. And while Three Mile Island could have been a disaster, the leak was fully contained and there were no deaths or injuries among either plant workers or the surrounding community. The accident also led to new safety regulations and procedures to prevent a repetition.[38]

Environmentalists argue that disposal of nuclear wastes is an insurmountable problem. As we'll see in the next paragraph, safe radioactive disposal is quite feasible. Greens also point to the possibility of nuclear fuel being used for terrorism, which is a valid point. But if we adopt that argument, then we wouldn't do anything out of fear of terrorists. Finally, Greens argue that moving to nuclear will delay the change to truly earth-friendly sources of power.[39] And, as usual when science fails them, Greens fall back on morality and regard nuclear energy as yet another human attempt to gain god-like powers.

That said, some of the more realistic environmentalists are beginning to favor nuclear, in part because nuclear power will reduce CO_2 emissions. In other words, for these environmentalists, nuclear is now the lesser of two evils. For example, Mark Lynas, arch-alarmist author of *Six Degrees*, has moved to a pro-nuclear stance, arguing that it is a non-carbon-emitting, stopgap measure "while truly clean energy systems are developed," and therefore better than the warming he believes will follow from continued reliance on fossil fuels.[40] Similarly, while still opposed to nuclear power in principle, George Monbiot, author of the ultra-alarmist *Heat*, has also softened his position on nuclear, writing: "As the likely effects of climate change have become clearer, nuclear power, by comparison, has come to seem less threatening." What convinced him? A Finnish report showing that nuclear wastes can be contained, a report by the Sustainable Development Commission showing that nuclear power is safe, and the knowledge that ionizing radiation from coal plants is actually higher than from nuclear plants.[41]

Support for nuclear power also comes from the unlikely quarter of arch-alarmist James Lovelock, creator of the Gaia hypothesis, who writes:

> Renewable energy sounds good, but so far it is inefficient and expensive. It has a future, but we have no time now to experiment with visionary energy sources: civilization is in imminent danger and has to use nuclear energy now, or suffer the pain soon to be inflicted by our outraged planet. ... Nuclear energy is merely the medicine that sustains a steady secure source of electricity to keep the lights of civilization burning until clean and everlasting fusion, the energy that empowers the sun, and renewable energy are available.[42]

While nuclear may not be Lovelock's preferred power source, he also produces figures showing that nuclear plants, far from being risky, are in fact far safer than any other type of energy generation. So, for example, from 1970-1992, coal had 342 deaths per terawatt year (twy) of energy, almost entirely coal workers. Natural gas had 85 deaths per terawatt year, to members of both the public and workers. Hydroelectric power had 883 deaths per twy, mostly to members of the public due, for example, to bursting dams. Deaths in nuclear plants? Eight workers per terawatt year—a mere one per cent of the deaths caused by non-carbon-emitting, renewable hydroelectric power.[43] As physicist Jack M. Hollander notes:

> Over four decades of nuclear plant operations in the United States, not a single documented fatality involving radiation from nuclear plant accidents or waste materials has occurred, while thousands of fatalities have resulted from accidents related to other energy sources.[44]

In short, the nuclear safety issue, which looms so large in the mind of the public and Greens, is a fabrication of the environmental movement. These fears have no scientific basis, unless you assume power generation can be 100 per cent risk-free, which isn't possible for any human endeavor. However, nuclear power comes closer to zero risk than any other source of power.

As fossil fuels decline and the world moves toward greater use of nuclear power—France is the leader in this respect with more than 75 per cent of its power coming from nuclear plants[45]—humanity will, or at least should, be using money that might have gone to fight climate change to improve the lives of the planet's people. This is the approach suggested by Bjorn Lomborg in his books[46], articles and interviews. Lomborg *does* think we need to do something about global warming, but what we do should be smart, not stupid, wasteful, and irrelevant. For example, Lomborg notes in an interview:

> When I first started in the global-warming debate, I was struck by the fact that the world was going to pay $180-billion a year for a protocol [Kyoto] that could at best reduce the temperature by 0.3 degrees Fahrenheit by the end of the 21st century. The UN estimates that for less than half that amount, we could provide clean drinking water, sanitation, and basic health care and education to every single human being on the planet. The same warped sense of priorities will continue to bedevil us this December [2009] in Copenhagen.[47]

Lomborg is not opposed to a carbon tax, but he doesn't think it will have much impact. He feels the same about carbon cap-and-trade systems, with their potential for abuse by industry. Instead, he says:

> Everybody seems to be saying, let's make carbon-emitting fossil fuels so expensive, nobody will want to use them. But that is bound to fail. So rather than making fossil fuels so expensive, we should try to make green energy so cheap that everybody will want to use it. ...
>
> You can use the money raised [by a carbon tax] to help finance research and development into green alternatives. That is the treaty I would like to see in Copenhagen. Every nation would promise to spend 0.05% of GDP on research and development on non-carbon-emitting energy technologies. This would be about $7-billion for the U.S. It would be about $30-billion for the entire world. We could easily get everybody on-board because it is a fairly low amount. And it would have a much greater chance of dealing with climate change in the long run, because it will focus on making alternative technology so cheap that everybody would want to use it.[48]

Thus, we need two approaches to climate change, whether warming or cooling. One, we need to come up with an alternative to fossil fuels in the next few decades, and that can only be nuclear power. Second, and looking beyond nuclear, we need to invest in research and development of non-nuclear alternatives, such as solar panels, tidal power, geothermal power, etc., that are, currently, not economical on a mass scale. And beyond that,

Lovelock believes that humanity's best and probably final power source will be nuclear fusion, "the ultimate clean and everlasting energy source." How do we know fusion is possible? Because fusion is the source of the sun's energy. If the sun can do it, human ingenuity can do it, and as Lovelock notes, we have already made steps in that direction.[49]

Whatever course we choose, trying to "stop climate change" should not be one of them, because it can't be done. As the many scientific sources referred to in this book have tried to show, climate change is a fact of life. We can't stop it any more than King Canute could stop the tides. Further, as this book has tried to show, while humans have some role in climate, it is not a primary role; we are bit players compared to the sun and oceans. Nor are higher levels of carbon dioxide a threat: as we've seen, temperature and carbon dioxide are not as closely linked as the climate alarmists would have the public believe—this close linkage exists only in alarmist computer models, not in the real world. Finally, in the next 100 years we will stop using fossil fuels anyway, whether we wish to or not. Fossil-fuel carbon emissions are a *short-term* problem. We need long-term solutions.

Putting this another way: while climate caution is always called for, as Lomborg writes: "Climate change is not an imminent planetary emergency that will bring down civilization."[50] Alarmists like Lydia Dotto are asking us to buy insurance (in the form of lost productivity due to carbon curbs) at an exorbitant rate for a risk that so far exists only in computer models and may not exist at all.

THE REAL THREATS

This doesn't mean there aren't real threats facing us as a planet and a civilization. One continues to be toxic pollution—but *real* pollution to the air, ground and oceans, not carbon dioxide "pollution." As we've seen, the developed nations have greatly reduced toxic pollution and will continue to do so, as will the developing nations when they have the resources to do so. Another threat is overpopulation. A third is poverty. But, really, the two problems of overpopulation and poverty come down to one: poverty. As Hollander notes in his book *The Real Environmental Crisis*: "The essential prerequisites for a sustainable environmental future are a global transition from poverty to affluence, coupled with a transition to freedom and democracy."[51]

The IPCC-inspired climate policies do nothing to address either overpopulation or poverty, and will actually make the problem worse by creating more poverty rather than less.

The United Nations estimates the world's population will reach 9-10 billion by 2050 before beginning to fall. However, the belief that this number will overwhelm the world's resources is not true. As Hollander notes:

> Typical of today's environmental pessimism, these doomsday pronouncements contain grains of truth embedded in a sea of exaggeration. ... Such broad-brush statements mislead the public and are scientifically inaccurate. For example, they usually represent environmental quality as deteriorating, which is not the case. They usually represent the earth's productive capacity as rapidly diminishing, which is not the case. They usually present population growth as a global threat, which is not the case.[52]

Feeding three billion extra people over the next 40 years will be a huge challenge, but humanity has the technological ability to meet this challenge. In fact, "overpopulation" is not an objective demographic reality; it is a relative term. Whether a region is "overpopulated" depends on many factors that have little to do with the actual number of people living there and everything to do with the region's ability to feed, house, clothe and employ its people, i.e., its level of prosperity. If population density is, in itself, bad, Gregg Easterbrook asks,

> why would the densely populated Netherlands be prosperous and reasonably clean while the Sudan, sparsely populated, is impoverished and shows numerous signs of environmental distress? Why is densely populated Switzerland prosperous and squeaky-clean while sparsely populated Mozambique is poor and has terrible water pollution? Why do the densely populated countries of Western Europe have the world's highest life expectancies?[53]

Singapore and Hong Kong are among the most crowded cities on earth, but they are not "overpopulated" because these cities can afford to feed and house their large populations. Indeed, for Singapore and Hong Kong, large numbers of workers are a productive asset since these cities lack natural resources—people are, literally, their "capital." North America is not "overpopulated" because this region can feed and house most of its people in some comfort. Many developed parts of the world, such as Canada, not only aren't "overpopulated" but would be declining in population without high levels of immigration. The parts of the world we perceive as "overpopulated" are the countries and regions that are poor, such as Pakistan, Bangladesh, and many African states—nations that cannot adequately feed and house their many people.

So, yes, there will be problems feeding these three billion extra people in the "overpopulated" (i.e., impoverished) part of the developing world. That said, thanks to the invention of effective birth control techniques and much-improved health care, having many children—perhaps more children than a family can feed—is now primarily a *cultural* choice; there is no part of the world any longer where families are *forced* by biology to have large families. Population, not climate, is an issue the world *should* be talking about because a less-populated world would have fewer difficulties coping with inevitable climate change—warming or cooling. And any climate policy that hinders the growth of prosperity—as the alarmist anti-carbon and therefore anti-industrial policies would do—will also make worse the problems of "overpopulation" and poverty. On the other hand, because wealthier societies can far more easily cope with climate change, warmer or cooler, our task in the develped world should be to help, not hinder, the creation of wealth. As Hollander writes:

> For the affluent nations to assist people in the developing world is socially responsible and morally right. But from an environmental perspective the issue is more than ethical. It is pragmatic as well, since the environmental self-interests of the affluent would be well served by the eradication of poverty.
>
> This idea disturbs those who fear that people emerging from poverty will inevitably become "wasteful" consumers like ourselves and will only exacerbate the globe's environmental damage as they pursue the trappings of the good life. This fear is understandable, but the conclusion is wrong. Without doubt, people tasting affluence will embrace consumerism and become proud owners of property, vehicles, computers, cell phones, and the like. But they will also pursue education, good health, and leisure for themselves and their families. And *they will become environmentalists*.[54] [italics the author's]

To repeat: the alarmist climate policies will not address either poverty or population pressure—only prosperity can do that. Instead, the alarmist climate policies will reduce prosperity and therefore, in the long run, do even greater damage to the environment.

WHAT WE KNOW

Climate alarmists would have the Western world give up its prosperity to meet a climate threat for which there is no evidence outside of computer models. That is, the rationale for this impoverishment is not empirical evidence—as Dotto admits, there is no empirical proof behind fears of global warming—but the precautionary principle: if a disaster *could* happen, we must respond as if it *will* happen. In the process, the usual requirements for scientific evidence are tossed aside. Abandoning the scientific method is not the way to develop a sensible policy on anything, much less climate. As we saw in Chapter 7, environmentalists and others have been predicting disaster for centuries. So far, they have always been wrong. There is no reason to believe the apocalypticists will be right when it comes to global warming, either.

If there was a *real* global warming problem, we, the public, would surely have seen signs of it by now; we would not have to rely on the precautionary principle. Until we are sure we have a problem, the best policy is to watch carefully but *do* nothing directly. Why? Because impulsive, poorly thought out action will almost certainly make the situation worse. In particular, the alarmist climate policies will make the *real* and most pressing global problem—poverty—worse, rather than better, by damaging the prosperity of both the developed and developing world. And, we should also be aware that many environmentalists have another agenda apart from a cleaner environment: a socialist or communistic political and economic system. Both are historically tried and tested ways of destroying prosperity, political freedom, and, ironically, the environment—it is communist countries, not capitalist ones, that have been among the most polluted places on earth. Writing about the failures of socialism, economist Friedrich Hayak notes:

> One of the most influential political movements of our time, *socialism*, is based on demonstrably false premises, and despite being inspired by good intentions and led by some of the most intelligent

> representatives of our time, endangers the standard of living and the life itself of a large proportion of our existing population.[55] [italics added]

Now, replace the word "socialism" with "global warming alarmism":

> One of the most influential political movements of our time, *global warming alarmism*, is based on demonstrably false premises, and despite being inspired by good intentions and led by some of the most intelligent representatives of our time, endangers the standard of living and the life itself of a large proportion of our existing population.

This is a perfect description of both the extremist environmental movement and global warming alarmism.

Dotto writes: "When the history of global climate change is finally written, it will show that the debate was badly derailed by three simple words: proof, uncertainty, and risk."[56] When the history of climate change is written, let us hope that *taking seriously* the words "proof," "uncertainty" and "risk"—and especially "proof"—will save us from spending vast amounts of money and ruining millions of lives to achieve nothing in terms of climate while wrecking the global economy.

Oh what a tangled web we weave, When first we practice to deceive.
—Sir Walter Scott

The great tragedy of Science—
the slaying of a beautiful hypothesis by an ugly fact.
—Thomas H. Huxley[1]

When people come to know what the truth is,
they will feel deceived by science and scientists.
—Physicist Syun-Ichi Akasofu[2]

Climate change science is in a period of "negative discovery"—the more we learn about this exceptionally complex and rapidly evolving field the more we realize how little we know. Truly, the science is NOT settled.
—Letter to UN Secretary General from 141 scientists

The [London Science] Museum is abandoning its previous practice of trying to persuade visitors of the dangers of global warming. It is instead adopting a neutral position, acknowledging that there are legitimate doubts about the impact of man-made emissions on the climate.
—London Times[3]

Chapter 14

THE BLACK SWANS OF ALARMIST CLIMATE SCIENCE

A scientific hypothesis or theory is only as strong as its weakest link. This is not just a figure of speech; it takes only one decisive disproof to eliminate a theory from scientific contention. As Einstein wrote: "No amount of experimentation can ever prove me right; a single experiment can prove me wrong."[4] He applied this principle to his own general theory of relativity:

> The chief attraction of the theory lies in its logical completeness. *If a single one of the conclusions drawn from it proves wrong, it must be given up*; to modify it without destroying the whole structure seems to be impossible.[5] [italics added]

This is, of course, Karl Popper's falsification criterion and a well-known analogy is white swans. The "consensus" hypothesis states that all swans are white. Everywhere one sees white swans. The evidence that all swans are white is "overwhelming." That all swans are white is "certain." The science of white swans is "settled." But all it takes is a sighting of one black swan and the theory is falsified. And, indeed, the real white-swan hypothesis, widely held in Europe for thousands of years, was falsified when black swans were discovered in Australia in the 17th century.

Naturally, scientists don't give up cherished hypotheses easily, especially if their research grants and reputations depend on their particular paradigm being dominant. So, they will claim a white swan has been, say, painted black by that devious black-swan industry—in other words, the apparently falsifying data is wrong. They will accuse white-swan skeptics of being in the pay of the black-swan industry. They will claim that the white-swan deniers have an "unscrupulous determination to deny the facts," and that they are trying to "distort and undermine" the white-swan science. They will claim that belief in black swans is "irresponsible" and "immoral." They will point to the "consensus" that all swans are white. They will cite computer programs proving that black swans can't exist. They will refuse to debate with black-swan theorists to avoid giving the black-swan theory "credibility," and try to stifle the voices and publications of those who believe in black swans. Anyone trying to get an article supporting black swans into a peer-reviewed journal will find obstacles not faced by the white-swan believers. But if a black swan is sighted, and then a second, and a third, and so on, eventually the white-swan paradigm crumbles and a new hypothesis emerges, leading to a new paradigm.

Of course, all scientific hypotheses and theories have black swans—no theory is perfect, which is why Darwin's theory of evolution is still called a theory despite vast amounts of empirical confirmation. But alarmist climate science may be unique in claiming, at least to the public, that it has *no* black swans within its major hypothesis. Therefore, it says, there's no need for further questioning of the science by pesky skeptics. However, in this book I've tried to reveal some of the black swans that alarmist climate science ignores or suppresses. Let's have a look at some of them.

THE ANTHROPOGENIC WARMING HYPOTHESIS

In Chapter 1 is an example of the consensus hypothesis from climate scientists R.A. Warrick, E.M. Barrow and T.M.L. Wigley. Let's look at it again. Their hypothesis reads as follows:

> The potential rates and magnitudes of the GHG [greenhouse gas]-induced change ... give rise to legitimate concerns about the future. These concerns include the following:
>
> - first, that humankind may now be a potent factor in causing unidirectional global changes which could dominate over natural changes on the decade-to-century time scale;
> - secondly, that, in terms of recent human experience, changes in climate and sea level could accelerate to unprecedented rates;
> - thirdly, that human tinkering with the global climate system could have unforeseen catastrophic consequences (e.g., "runaway" warming or sea level rise from strong positive feedbacks); and
> - finally, that the quickened rates of change could exceed the capacity of natural and human systems to adapt without undue disruption or cost.[6]

While these would indeed be "legitimate concerns" if true, we've seen that at least so far human carbon emissions have *not* been a "potent" factor in causing climate change—the IPCC models appear to overstate the warming power of CO_2 by a factor of up to three and perhaps more, and *human* carbon emissions are a fraction—less than five per cent—of natural carbon emissions each year. Also, the atmospheric band that traps carbon dioxide, and therefore the earth's radiated heat, is almost certainly saturated, so further warming beyond a degree Celsius or two from increased carbon emissions is unlikely. This is a fact of physics but climate alarmists give the public the impression that more carbon dioxide will cause *accelerated* warming. We've also seen that climate change over the past hundred years, when carbon emissions have increased, is not "unidirectional." Instead, the climate has gone up and down and, as of at least the late 1990s, stopped warming and may even have entered a spell of global cooling. Also, levels of carbon dioxide eight times today's levels didn't keep the planet from starting to cool 50 million years ago, taking us into the current ice age. Therefore, the climate is not (and never was and never will be) "unidirectional," and carbon dioxide (much less the small human contribution) is not a "potent" factor compared to the other natural forces that caused the 21st century non-warming and are almost certainly responsible for the late 20th century warming as well. Indeed, we should be grateful for whatever small warming our CO_2 emissions can produce if they delay the onset of the inevitable next glacial. Our planet has been in an ice age for two million years and the last thing humanity needs in an ice age is more cooling.

The three authors fear that climate change "could accelerate to unprecedented rates." But, as we've seen, today's rates of climate change are far from "unprecedented" and they were not "accelerating" as of 2010. The Younger Dryas cooling of 12,000 years ago saw temperature changes of 5-10ºC in a few decades. Abrupt warming and cooling was common even within the glaciations of the past million years (the Dansgaard-Oeschger cycles, Chapter 5). And, as former Climatic Research Unit head and Climategate email writer Phil Jones has admitted, warming was no more "accelerated" from 1975-1998 than it was in previous warming periods of the past 150 years, 1860-1880 and 1910-1940.[7]

Are we "tinkering" with the climate system through our industrial technology? Perhaps, but how could we not "tinker" given that humans no longer live in caves and most of us do not wish to return to them? We are always "tinkering," we are always, as one climatologist put it (albeit with an alarmist spin), "playing with fire."[8] Playing with fire *is*, by definition, civilization—if humans hadn't "tinkered" with fire there'd have been no civilization. Fortunately, the empirical evidence indicates that so far our tinkering has not been disastrous, and there is no evidence outside of computer models that it will be disastrous.

What about the "unforeseen catastrophic consequences" of sea-level and temperature rise? As we've seen, temperatures and sea levels (not to mention carbon dioxide levels) in the past have been both much higher and much lower than today's without creating "catastrophe" or "oblivion." As long as the planet is in an interglacial there will continue to be "background" sea level and temperature increases; these would occur whether humans were here or not. At worst, our activities might enhance this warming by a percentage point or two.

Finally, will "the quickened rates of change … exceed the capacity of natural and human systems to adapt without undue disruption or cost"? First, as Jones has admitted, from 1995 until 2010 the rates of change were not "quickening" at all: instead, the temperature has showed no "statistically significant" warming and even slight cooling.[9] Second, the "undue disruption" we should fear most is that of the economic system through "tinkering" with various forms of carbon curbs to deal with a probably non-existent climate problem. If humanity continues to enjoy increasing prosperity, as the UN predicts if we did nothing about warming,[10] we will have all the resources we need to cope with sea level and temperature increases, quickened or not. Any "catastrophic" consequences will be the result of poverty and the resulting "overpopulation," not climate itself. Wealthy nations can easily handle climate change.

If not human activity, what *is* the "principal" (to use Al Gore's term) cause of global warming? There isn't one principal cause, and if there was, human carbon emissions certainly wouldn't be it. The two main sources of climate change are solar activity and the oceans. Almost all of the other climatic factors—clouds, water vapor and, yes, greenhouse gases—are the product of a combination of solar and oceanic factors. Within the climate system, water vapor is a far more potent greenhouse gas than carbon dioxide—most scientists agree that water vapor accounts for up to 90 per cent of the greenhouse effect. In addition to fossil-fuel emissions, human activities like cities (the Urban Heat Effect) and agriculture undoubtedly contribute to planetary warming, and likely far more than our fossil-fuel additions. However, nobody except the most extreme deep-ecology environmentalists is suggesting we abandon cities and agriculture, so some warming from these activities is inevitable.

IGNORING THE BLACK SWANS

Why does the IPCC so deliberately ignore the black swans in its science? As we've seen, the IPCC was created as a *political* body—its summaries are subject to "review by governments"—with a *political* agenda. Originally, the IPCC's political agenda was to reduce the public's resistance to nuclear energy. More recently, as shown by the Copenhagen Protocol, the agenda is to transfer vast sums—$75-billion to $100-billion a year—to the developing world, ostensibly to help poorer nations to cope with global warming while creating "social justice." And, of course, some groups use the IPCC to advance their own particular political, religious and environmental programs.

For example, as noted in Chapter 4, consensus climatologist Judith Curry observes that the IPCC is

> framed in the context of the UNFCCC (United Nations Framework Convention on Climate Change). That's who they work for, basically. The UNFCCC has a particular policy agenda—Kyoto, Copenhagen, cap-and-trade, and all that—so the questions that they pose at the IPCC have been framed in terms of the UNFCCC agenda. That's caused a narrowing of the kind of things the IPCC focuses on. It's not a policy free assessment of the science.

The IPCC's political agenda became crystal clear in November 2009 thanks to the Climategate emails, followed by further revelations that the IPCC had used non-peer-reviewed and inaccurate data on glacier melting, weather disasters, Amazon forest destruction, the possibility of drought in Africa due to warming—and these are just a few of the errors—to bolster its case. These errors were allowed into the IPCC reports because they supported the AGW hypothesis. Given this, any claims that the IPCC is an unbiased source of climate analysis are clearly ridiculous, to the point where, as we saw in Chapter 10, even some alarmist climate scientists have disowned the organization.

The IPCC's close connection with Al Gore is another sure sign that it is off the scientific rails. As we saw in Chapter 6, almost every "inconvenient truth" in Gore's books and movie is misleading, exaggerated or false, yet alarmist climate scientists have not only not criticized Gore's many errors, but they actually endorsed Gore to the point of accepting a Nobel Peace Prize with him. Gore may have the details wrong, they say, but overall he's got the right "message." But given that an hypothesis is built up out of details, it's hard to understand how Gore's details can be wrong but his message right. A British judge even noted, in finding Gore's movie *An Inconvenient Truth* riddled with errors: "It is now common ground that [*An Inconvenient Truth*] is not simply a science film—although it is clear that it is based substantially on scientific research and opinion—but that *it is a political film*."[11] [italics added] In accepting a Nobel with Gore, the IPCC endorsed Gore's extreme *political* message, and thereby further damaged its credibility as a *scientific* organization.

Finally, the Climategate emails showed that *top* IPCC contributors had countenanced and encouraged the manipulation of data to meet the IPCC's political message (i.e., actions perilously close to scientific fraud if not outright fraud), tried to conceal information about planetary cooling from the public and even other researchers ("hide the decline"), and manipulated the peer-review process. Nor has the IPCC repudiated the scientists who wrote these emails, as other scientific disciplines would if confronted with similar misdeeds. Instead, as we saw in Chapter 10, the IPCC has fully supported the Climategate scientists. Overall, the evidence is overwhelming that the IPCC is not the objective scientific organization it claims to be. Therefore, nothing the IPCC has told us in the past or, without major reforms, could tell us in future reports can be trusted.

As we saw in Chapter 8, the IPCC and many of its climate scientists have joined the long list of prophets of eco-doom, from Malthus to Marx to Paul Ehrlich to David Suzuki. The eco-doomsayers have always been wrong in the past; why would they suddenly be correct now? The eco-apocalyptics of the last 200 years have been wrong because the capitalist system they detest is the source of the wealth that makes it possible for humanity to protect, rather than simply exploit, the environment, while also cushioning humanity from the inevitable disasters, such as earthquakes, that nature throws at us. Indeed, the 20th and 21st centuries are the first time in history that humanity has tried to preserve species from extinction, with some success, just as, for the first time in history, we are trying to preserve eco-systems rather than merely exploit them. But environmental protection comes at a huge cost; only a wealthy, developed country can afford that cost. The result has been, in the developed nations at least, a dramatic *decrease* in air, water and soil pollution along with improvements in health and longevity unlike anything humanity has experienced before. Yet, environmentalists continue to claim that the environment is deteriorating, in part to keep donations flowing to their organizations.

Free markets are one pillar of Western prosperity. Another is science. Science, for the first time in earth's 4.5-billion-year history, has freed a species—us—from utter subservience to nature. Many in the environmental movement do not like this development; they want us to return to nature's ways. And they have a point: to the extent that humans are part of nature—as we are—we need some connection with the natural world to feel truly alive. But this connection need not include, as was unavoidable until very recently, an early death due to starvation, helplessness in the face of environmental calamity and what are today easily preventable diseases. It is science, as both the source of and outcome of economic prosperity, that has given us this freedom from the casual brutality of nature. Yet, AGW believers are working to destroy the integrity of science. For the scientist truth should be sacred—the basis of science is an unswerving, unblinking adherence to empirical truth, no matter how unpopular or politically incorrect that truth may be. Yet, as this book has demonstrated again and again, alarmist climate science has betrayed truth in favor of swaying the public toward a political and/or environmentalist agenda. In other words, "truth is sacrificed to strong conviction." And so, leading global-warming promoters offer "scary scenarios," an "over-representation of factual presentations," "emphasis on extreme scenarios" and "hide the decline" to persuade the public that the planet is warming (when, at the moment anyway, it isn't) and that warming will be the evil they predict but cannot prove.

How do we know alarmist climate science cannot prove its contentions? Because it falls back so often on the Precautionary Principle and the appeal to consensus and authority rather than just presenting and debating the facts. Alarmists then get frustrated because much of the public doesn't "get it," i.e., demands actual evidence that humanity is at fault for what are mainly natural climatic phenomena. If *empirical* evidence for anthropogenic warming existed, the public would accept that evidence, as non-scientists accept myriad other scientific findings, and governments would not have to spend tens of millions of dollars on global warming propaganda. Furthermore, if the evidence was as "settled" and "overwhelming" as the alarmists claim, there would be no need for deception of the sort revealed in the Climategate emails and IPCC reports. As philosopher Walter Kaufmann observes: "What distinguishes knowledge is not certainty but evidence."[12] Consensus climate science has certainty—indeed, a Doctrine of Certainty, a dogma of certainty. What it doesn't have is the evidence—the empirical proof—to back up that certainty, although it believes it has and claims to the public, politicians and the media that it has.

Part of the problem for climate science is that, unlike physics or chemistry, it is not the "hard," predictive science it wishes the public to believe it is. In fact, alarmist climate science doesn't meet any of the generally accepted requirements for hard science, including testable, falsifiable hypotheses (because "you can't put earth in a test tube"), and the ability to make accurate predictions. Prediction is the essence of hard science but not one of the IPCC's climate models predicted the non-warming of the start of the 21st century. Therefore, as meteorologist Neil Frank notes:

> Skeptics believe that climate models are grossly overpredicting future warming from rising concentrations of carbon dioxide. We are being told that numerical models that cannot make accurate 5- to 10-day forecasts can be simplified and run forward for 100 years with results so reliable you can impose an economic disaster on the U.S. and the world.[13]

Even worse: alarmist climate science as revealed in the Climategate emails and IPCC reports has besmirched the reputation of all scientists to the point where the head of the U.S. National Academy of Science, Dr. Ralph Ciccerone, noted in an interview with the BBC:

> There is a feeling that scientists are suppressing dissent, stifling their competitors through conspiracies. ... Public opinion polls are showing that the answers to questions like: 'how much do you respect scientists?' or 'are they behaving in disinterested ways?', have deteriorated in the last few months.[14]

Since top alarmist climate scientists *are* suppressing dissent and stifling their competitors, as well as manipulating data to meet political and other agendas, this public distrust is, sadly, well-merited.

And yet, even after their deceptions are exposed, alarmist climate scientists continue to cling to the AGW hypothesis, while refusing to admit to the public, openly and honestly, that anthropogenic global warming *is* just an hypothesis, not a scientific fact that is "certain" and "settled." Rather than admit its lack of certainty, and as the Climategate emails reveal clearly, alarmist climate science resorts to bluster, propaganda, concealment of evidence and intimidation of critics in order to persuade the public that its views are true. Thus, thanks to the Climategate emails, as economist Robert P. Murphy writes:

> Those of us who are not experts on climate models now have proof that the official line that "the science is settled" was a bluff. Of course it's still possible that the IPCC projections may turn out to be ac-

curate when all is said and done, but the *confidence* we should right now place in their modeling is much lower than what their biggest enthusiasts have been assuring us for years.[15]

WHEN SCIENCE AND RELIGION CONVERGE

How did this corruption of climate science occur? There are several reasons, but three major ones are, first, climate science's attachment to a political agenda, which we've discussed above, second, the infusion of the environmentalist religion into climate science and, third, groupthink.

There is little doubt that environmentalism is primarily a religion. While environmentalism has done much good work in helping to make the planet a cleaner and healthier place, it also has, as we saw in Chapter 11, every one of the features of a religion. Disturbingly, alarmist climate science also includes many if not all of environmentalism's religious features, including dogmatic certainty, original sin (even though a carbon footprint is unavoidable), the belief that having too many material possessions is morally wrong (the term used is the scientific sounding "unsustainability"), attacks on those who refuse to accept the dogma (the "deniers"), a belief in apocalypse if we sinners don't repent our ways, prophecy (computer predictions), religious rituals (e.g., Kyoto and Copenhagen, which even alarmist climatologists admit would have had "zero effect" on global warming) and, for some alarmist scientists, a belief in a supreme earth being (Gaia, Mother Nature) who may seek "revenge" upon us.[16] In other words, alarmist climate science has abandoned the value-free search for truth in favor of value-laden moral, environmental and/or religious "truths," such as the perceived "good of the planet."

Why have alarmist climate scientists been so susceptible to this merging of religious belief and science? Human beings act from many unconscious motives and scientists are no exception. One of the largely unconscious forces in most human beings, including scientists, is a need for harmony with their fellows. In politics, in economics, in philosophy, in religion, in all human activities including the sciences, this psychological need tends toward groupthink. As we saw in Chapter 12, groupthink is "a model of thinking that people engage in when they are deeply involved in a cohesive in-group, when the members' strivings for unanimity override their motivation to realistically appraise alternative courses of action."[17] In the desire for group consensus, even scientists sometimes cease to "realistically appraise alternative courses of action." Another result of groupthink is group polarization, in which a group's views are more extreme than views that individuals might have arrived at, rationally, on their own. Both problems, groupthink and group polarization, are strongly present within alarmist climate science, including the keen desire for consensus and beliefs that are, rationally and objectively considered, extreme and even absurd. The public utterances of alarmist climate promoters like James Hansen, Al Gore, Stephen Schneider, and many others reveal extreme apocalyptic views. This is group polarization at work, based in groupthink, and, while very common in all walks of life, groupthink is inimical to good science.

The extremist conclusions of alarmist climate science have led the world to the brink of a series of potentially disastrous economic measures. Carbon-tax and carbon trading schemes will damage the livelihoods of millions of people to no useful end—that is, even if we make these sacrifices, they will not meaningfully reduce global warming. Meanwhile, the emphasis on *stopping* climate change—a quixotic quest that is, for the rational person, insane—will take attention and billions of dollars away from issues that really do need attention, most notably global poverty and the *real* toxic pollution (carbon dioxide is not a pollutant) that continues to exist. Therefore, the most sensible course is to do nothing about global warming, at least until we have some actual empirical evidence that warming is the problem the alarmists say it is *and* that this problem is caused *primarily* by humans. Or, if we must do something then, as Bjorn Lomborg suggests, let's spend our money smartly, where it will do the most good: in developing alternative, energy-efficient, non-carbon-emitting technologies so inexpensive we will all *want* to use them. This way, coercion by a huge, expensive global climate bureaucracy as envisaged by the Copenhagen Protocol will not be necessary. And let's spend our money smartly to reduce poverty. That doesn't mean throwing money at the poor countries, as the Copenhagen Protocol would do, but helping them to gradually develop the institutions and infrastructure that have created prosperity in the developed world. The solution to the world's climate and environmental problems isn't less prosperity, as some environmentalists and alarmist climate scientists believe, but *more* prosperity. Unfortunately, the anti-carbon climate policies proposed by the IPCC and UN will only make more of us poorer while, if history is any guide, making the bureaucracy creating the decline richer.

TELLING THE TRUTH ABOUT CLIMATE

There are, of course, some *white* swans in alarmist climate science. The anthropogenic global warming theorists may have the correct "message"; that is, their "intuition" may be right even if they don't have the empirical facts to back it up. Our carbon emissions might cause a catastrophic failure of the earth's biological systems, and therefore of our human systems. In this book I am not primarily arguing that the alarmists are wrong. Although my "intuition" tells me they are wrong, I don't know this for sure and—and this is my point—*neither do the alarmists*. There are no empirical grounds whatsoever for claiming "certainty" that a disaster will happen and that humans are primarily to blame, something even the alarmists admit in their more honest moments as we saw in Chapter 10.[18] The planet heated up .6ºC over the 20th century—was *that* a disaster? Not at all—in fact, this warming helped feed the billions who came into the world during that century by allowing longer growing seasons and increased agricultural yields. Yet if the climate alarmists had been in existence at the start of the 20th century, they would have argued that the climate then, in 1900, was the "perfect" climate and tried to create fears over warming that was, we in 2010 now know, largely beneficial. Therefore, instead of panicking and throttling back our economic prosperity, we should adopt a cautious, wait-and-see approach until we have actual evidence of impending climate disaster. However, if past apocalyptic predictions are any guide, the latest prophecies of eco-doom are yet another false alarm among many.

What could climate scientists have done differently? From the start, instead of claiming "certainty," a concept that has no place in science, climate scientists could have explained to the public that their models are tentative. They could have admitted in 2007, when the non-warming of the 21st century became clear, that the planet wasn't warming, rather than attacking those who were simply stating the truth. Instead, the IPCC's 2007 report declared that warming of the 21st century was "unequivocal," an astonishing conclusion but quite in line with the IPCC's agenda of creating climate fear rather than climate reason.

Climate scientists could have educated the public on the geological history of climate. They didn't, because the geological facts make it clear that the planet's warming, today and in the past, is due to a quite normal interglacial trend. Instead, they have allowed the public to believe that today's global warming is somehow unusual and unnatural.

Climate scientists could have admitted that their spectres of doom are highly speculative, and that the scenario they prefer, "oblivion," is only one of a range of possible climate outcomes, some of them benign. Instead, they have chosen to present only the "scary scenarios" to get public support for their hypothesis—and to ensure research funding, of course. They even made an alliance with Al Gore whose frantic claims that we are facing a "planetary emergency" are, quite simply, science fiction.

Finally, alarmist climate science could have welcomed criticism and debate, as is proper in science. Instead, the AGW believers have declared "the science is settled," while doing their best to keep the media and scientific journals from publishing views that might suggest the science isn't settled. However, as a letter to *The Globe and Mail* on skeptical climate blogs suggested:

> Instead of shunning their critics, these [alarmist] climate scientists should be happy to have the debate joined by bloggers. If the evidence is truly balanced one way or the other, it will survive controversy and debate from all sides. Engaging the public in a meaningful discourse will allay the fear and distrust that so many have surrounding the issue.[19]

A few in the alarmist camp are also coming to this conclusion. For example, James Lovelock actually has good things to say about climate skeptics:

> I think you have to accept that the sceptics have kept us sane — some of them, anyway. They have been a breath of fresh air. They have kept us from regarding the science of climate change as a religion.[20] It had gone too far that way. There is a role for sceptics in science. They shouldn't be brushed aside. It is clear that the angel side wasn't without sin.[21]

IPCC climatologist Mike Hulme suggests: "Scientists need to be prepared to argue about their 'considered opinions', to embrace consensus but without closing down argument or suggesting that matters are settled."[22] And consensus climatologist Judith Curry, who has been trying to repair the damage caused by Climategate, has written:

> Ignoring skeptics coming from outside the field is inappropriate; Einstein did not start his research career at Princeton, but rather at a post office. I'm not implying that climate researchers need to keep defending against the same arguments over and over again. Scientists claim that they would never get any research done if they had to continuously respond to skeptics. The counter to that argument is to make all of your data, metadata, and code openly available. Doing this will minimize the time spent responding to skeptics; try it! If anyone identifies an actual error in your data or methodology, acknowledge it and fix the problem.[23]

Even John Beddington, the British government's chief science advisor and a believer in human-caused warming, has stated:

> I don't think it's healthy to dismiss proper skepticism. Science grows and improves in the light of criticism. *There is a fundamental uncertainty about climate change prediction* that can't be changed.[24] [italics added]

In other words, alarmist scientists could have acknowledged to the public the problems with climate science right from the start and accepted, rather than contemptuously dismissing, the challenges to their science by skeptics. They didn't. Instead, alarmist climate science has created a kind of Potemkin Village of global warming—sorry, "climate change"—a façade that, they hope, will carry them past at least a decade of non-warming that has left their hypothesis in ruins and made it clear the climate models don't predict as advertised. The awareness of failure is very clear in the Climategate emails, along with a near-frantic desire to hide this fact. Meanwhile, backed into a corner by their arrogance and hubris, *leading* alarmist climate scientists—not alarmist mavericks—have exaggerated the facts, misled the public, media and politicians, and, if necessary, told outright lies, such as the claim by the head of the IPCC, Rajendra Pachauri, and others that warming has "accelerated" when in fact the planet is slightly cooling. The result? As of 2010, the disconnect between the AGW theory and the climate reality has become so pronounced that alarmist climate science is rapidly losing whatever credibility it may have had. For example, a Gallup poll published in March 2010 found that 48 per cent of Americans now believe global warming fears are exaggerated, up from 41 per cent less than a year earlier.[25] And, as we've seen, the errors and, worse, the dogmatism of alarmist climatologists are damaging the reputation of science as a whole.

WHAT IS CERTAIN: CLIMATE SCIENCE IS IN TROUBLE

Therefore, while this book doesn't claim certainty about our climate future, it is certain of one thing: there is something seriously wrong with alarmist climate science. As Bertrand Russell has written,

> One of the greatest benefits that science confers upon those who understand its spirit is that it enables them to live without the delusive support of subjective certainty. *That is why science cannot favour persecution.*[26] [italics added]

But, alarmist climate science *does* claim subjective certainty and *does* favour persecution against AGW skeptics by denying them research grants and trying to keep their papers out of peer-reviewed journals. I personally know three fine scientists who fear that their careers would be jeopardized if they revealed themselves as climate skeptics. Lawrence Solomon reports that some of the skeptical scientists in his *National Post* series "The Deniers" were angry at being "outed"—but why should a scientist be worried about stating publicly what he or she believes to be true? And the Climategate emails make this campaign of persecution against AGW non-believers abundantly clear.

A truly sad aspect of this issue is how thoroughly alarmist climate science has bought into the doom-and-gloom mentality about Western industrial civilization, when Western civilization is the most successful economic, political and social experiment in history. No other society, ever, has managed to feed, house and clothe so many of its citizens in such comfort, with such freedom, while still protecting the environment. Science has been the major driving force behind this achievement. And yet, this astonishing record of scientific and social progress means nothing to the alarmists, who prefer to focus almost exclusively on the negative. As columnist Robert Fulford has written (quoted earlier), this pessimism and alarmism is a triumph of ideology over observation.[27]

Where does the public stand in all of this? In discussing climate with friends and acquaintances, I am often told, "I don't know what to believe." This, in itself, is surprising since the climate propaganda is overwhelmingly on the side of the AGW believers. Yet, a recent Rasmussen poll found that more than half (52 per cent) of Americans believe the science is not "settled," and 59 per cent believe (I think correctly given what is revealed in the Climategate emails) that "it's at least somewhat likely that some scientists have falsified research data to support their own theories and beliefs about global warming."[28] A British poll taken after the Climategate revelations found 46 per cent of respondents did not believe humans were "largely responsible" for warming, and only one in four thought global warming was a serious problem.[29] Polls show Canadians are less skeptical than Americans but, overall, Canadians' fears of global warming are on the decline. A November 2009 survey found only 26 per cent consider global warming a major concern.[30] In other words, a significant number of Canadians continue to have doubts about the AGW hypothesis, often, I suspect, without clearly understanding why. And no wonder; they're not given the full spectrum of scientific opinion in most of the Canadian media. And so, for many, the AGW hypothesis "just feels wrong," as they might put it. That the AGW hypothesis "feels wrong" was the starting point for this book, back in June of 2007, and after more than two years of intensive research, I still believe the AGW hypothesis feels wrong, and that the alarmist science that produced it also feels wrong—very wrong.

In this book I have tried to present the arguments for why the AGW hypothesis not only *feels* wrong but quite possibly *is* wrong. It is a book for the citizen who wants information from both sides of the climate debate—information that the national media apart from the *National Post* has so far not been willing to provide, and certainly that the alarmist climate orthodoxy is not willing to provide. However, with at least some knowledge of the *total* spectrum of opinion and research in climate science, this book's readers can hopefully make an informed decision. If they decide to support the AGW hypothesis, with its potentially damaging economic measures, at least they will be aware of more of the facts. If they decide not to support the AGW hypothesis until there is actual empirical evidence that disaster is looming, then they will have some of the intellectual ammunition they need to counter the alarmist propaganda barrage.

My personal conclusion? We don't know what the climate is going to do, period. Those who say they know, that they are "certain," are deliberately not telling the public the truth, the whole truth and nothing but the truth, or are deluded by groupthink. That said, from my research I know for a fact that many alarmist climate scientists have put the "message"—that is, a political agenda, environmental and religious values, and their personal "intuitions"—ahead of the bare, empirical facts. This is a very bad sign for climate science and by extension for all science since, once the public's good opinion of science has been lost, it may be lost (like Mr. Darcy's good opinion in *Pride and Prejudice*) forever. As the Japanese scientist Dr. Kiminori Itoh has written, warming fears are the "worst scientific scandal in history," and he quotes physicist Syun-Ichi Akasofu as follows: "When people come to know what the truth is, they will feel deceived by science and scientists."[31] This would be a tragedy because science is humanity's best hope for a prosperous and happy future for all of us, not just those in the developed world. In the case of alarmist climate science, however, this loss of trust and public suspicion of deception is wholeheartedly deserved; I, personally, find it difficult to believe a word that alarmists like Gore, Hansen, Schneider, Mann, or Weaver might have to say on this topic. And I again put to them the question, asked by French thinker Jean-François Revel: "How is it possible for a theory, which is false in its component parts, to be true as a whole?"[32]

For me, the most telling argument against the AGW hypothesis is this: If the alarmist case is so "overwhelming," if the evidence is so "certain" and "settled," *why is there the need for so much misrepresentation, exaggeration and, sometimes, outright falsehood*, as documented in this book and by other skeptical observers? Why the need to try to hide the climate facts—"hide the decline" of 21st century temperature, for example? Why the refusal to admit to the public that there are problems—black swans—with the AGW hypothesis? Why the continued claims of "certainty" when there is so much uncertainty—even a "maddening degree of uncertainty," according to alarmist Stephen Schneider?[33] Why the "scary scenarios" when the public has shown, in the past, that it is quite willing to accept scientific conclusions if those conclusions are presented with reason, sincerity, and evidence? Why the literal persecution of those who offer alternative hypotheses?

So, unlike alarmist climate scientists, I can't be "certain" about what the climate will do. It could suddenly turn warm, even very warm—in which case, we could be just experiencing another case of the abrupt climate change that is "inevitable," according to climate scientists themselves.[34] Or the planet could continue to cool for a few more years, which should finish off the AGW hypothesis for good.

However, based on almost three years of research I *am* certain that fears of anthropogenic global warming and its supposed catastrophic consequences are more the product of political, environmental and religious agendas and scientific groupthink than empirical scientific evidence. Look at the facts. Is the planet warming? Perhaps, but that's natural in an interglacial, and the planet has not warmed, on average, for at least a decade. Are humans the principal drivers of warming? We undoubtedly have a role, but a small role. Are we facing a disaster "whose ultimate consequences could be second only to global nuclear war"?[35] There is no evidence whatsoever to support these fears, given that the planet has been much warmer and more carbon-dioxide rich in the past without creating "oblivion." In other words, much of what we—the public, politicians and the media—have been told about global warming is misleading, exaggerated, or just plain wrong. However sincere the climate alarmists may be as private individuals and as scientists in their belief that we are creating a future catastrophe with our carbon dioxide emissions, based on the evidence (or lack of it) so far, the conclusion that makes the most sense is: false alarm.

It would be impossible for any investigative reporter ... to objectively delve into global warming and conclude that the science was settled.

—Lawrence Solomon[1]

[Warming alarmism is] the worst scientific scandal in history.

—IPCC contributor Kiminori Itoh[2]

You can fool some of the people all of the time, and all of the people some of the time, but you can not fool all of the people all of the time.

—Abraham Lincoln

Concluding Unscientific Postscript:

CANADIAN JOURNALISM AND CLIMATE SCIENCE

In March of 2002, two Greenpeace activists met with the Victoria *Times Colonist* editorial board, including myself, with a warning about global warming. At that time I was interested in climate change but I hadn't done any in-depth research into the topic. However, as the meeting went on, it became obvious that the Greenpeacers were offering a rather dubious theory, given the number of assumptions it contained, and so I wrote a column on our meeting entitled "A shaky tower of assumptions."[3] In particular, I noted that the global warming hypothesis was based on four assumptions, each less probable than the one before it.

The first assumption was that the planet is on a permanent ("unidirectional") warming trend. Everyone, including climate skeptics, agrees the planet has been, overall, warming since the end of the last glaciation about 15,000 years ago, and also warming since the end of the Little Ice Age in 1850. In 2002 we still weren't aware that the planet had stopped warming, but I knew even then that the planet had been warmer as recently as a thousand years ago (the Medieval Warm Period) and that future cooling was just as likely as future warming.

The Greenpeacers' second assumption was that human carbon emissions are the *primary* source of late 20th century warming. I thought in 2002 that this was a dubious assumption and the lack of 21st century warming (and perhaps earlier) indicates that the Greenpeace assertion is probably wrong.

The third assumption, even less probable, was that we could stop warming if we signed onto the Kyoto Accord. As we saw in Chapter 11, even Andrew Weaver believes that Kyoto, even if fully implemented, would have "zero effect" on warming.

The fourth Greenpeace assumption was that signing on to Kyoto would help our economy by encouraging new technologies and the more efficient use of energy. I agree totally that we need new technologies and should be efficient as possible in our energy use. However, I don't think crippling the carbon economy, which pays for the research that will give us these new, greener technologies, is the way to go. Put all four assumptions together and you end up with a theory that is highly improbable. Therefore, the column concluded:

> Market forces are already pushing us toward more energy efficiency and use of alternative power sources. That push will continue as the fossil fuels run out, whether we endorse Kyoto or not; crippling the market in favor of dubious social engineering may actually hamper the change to more efficient forms of technology.
>
> Environmentalists point to the "precautionary principle" when they can't marshal enough facts to prove global warming: if global warming could happen, they say, we should behave as if it will. Before this country takes steps, like signing Kyoto, that could wreck or at least cripple its economy, Canadians should demand at least some certainty. No sensible nation would make major policy decisions based on a tower of assumptions as shaky as the global-warming hypothesis.

Again, this column was written in 2002 based on common sense and some general reading into climate change but without, as noted above, the benefit of the more than two years of in-depth research, much less the knowledge that the planet isn't behaving as the climate models predicted (i.e., the planet isn't currently warming). Having now done this research, I continue to endorse the column's conclusions. What puzzles and disturbs me, though, is why so few other Canadian journalists have also taken a critical stance when faced with improbable, extreme and sometimes patently absurd claims by alarmist climate scientists and environmentalists.

For example, most journalists still accept the myth that the Arctic polar bears are endangered by global warming when even the Inuit, who have the most to lose if polar bears go extinct, say the bears are, so far at least, doing fine and even thriving.[4] Another example: In 2008, the *National Post* dutifully reported that a "record-breaking meltdown" was expected in the Arctic and quoted the Canadian Ice Service as saying that the opening of the Northwest Passage that year was "unprecedented."[5] However, a quick Internet search quickly reveals that the Passage has been ice-free several times just in the past hundred years,[6] and undoubtedly many times before that during the planet's various warm periods. As for the melting Arctic being due to increased human-caused warming: if you leave ice cubes on the kitchen counter, they will melt. This doesn't mean the room temperature is *increasing*. Similarly, we are in an interglacial; during interglacials, ice melts. (It's worth noting, though, that in March 2010, Arctic sea ice had returned to the levels of a decade earlier—another blow to the AGW hypothesis.[7]) Finally, most of the Canadian media continue to regard Al Gore's claims as if they actually made sense when, in fact, even a small amount of research quickly reveals that Gore goes far beyond even the IPCC's wildest imaginings.[8]

WHY JOURNALISM FALLS SHORT IN SCIENCE REPORTING

One reason for this failure of criticism is that most journalists aren't scientists. Therefore, they have little or no training in the methods and philosophy of science, and they are often not knowledgeable about the specialty of the scientist they are interviewing or whose research they are reporting. In particular, journalists rarely know the *critical* questions to ask when interviewing a climate scientist. Instead, alarmist climate comments are reported as if they make perfect sense simply because they come from "experts," without critical examination. And, yes, skeptical climate comments should be treated with the same critical eye.

Of course, experts should be respected. But if their claims are extreme, if they appear to be in the grip of ideological and/or political agendas, as leading climate alarmists rather obviously are ("review by governments"), and particularly if the experts are clearly not telling the public the truth, the whole truth and nothing but the truth, as is so often the case with alarmist climate scientists (e.g., Climategate), these claims should be investigated closely and critically before being reported as gospel. For example, any good journalist should be looking at extreme claims—a "sixth extinction," for example, or the assertion that we are facing "oblivion" if temperatures rise more than two degrees Celsius—with considerable caution regardless of where these claims come from.

Instead, most of Canada's print and broadcast media present only one view on climate change, the "consensus" (alarmist) view. The skeptical view of climate science is marginalized and even suppressed. I know this from personal experience. At the time of writing, the main newspaper in my city, the Victoria *Times Colonist*, had an explicit if unwritten policy against publishing articles skeptical of the alarmist climate "consensus." As a result, several of my skeptical opinion articles were rejected, and efforts to meet with the *Times Colonist* editor-in-chief to discuss this alarmist censorship were politely rebuffed. In other words, the *Times Colonist's* editorial mind is made up and not open to contrarian facts.

I have a subscription to *The Globe and Mail*; I rarely if ever see an article critical of the consensus climate position there apart from columnists Margaret Wente, Neil Reynolds, and Rex Murphy (who has left *The Globe* for the more congenial *National Post*, reportedly because *The Globe*'s new editor-in-chief disagreed with Murphy's skeptical climate views). I have a subscription to *Maclean's* magazine; apart from columnist Mark Steyn, its view of consensus climate science is largely unquestioning although, recently, an article by *Maclean's* columnist Andrew Coyne lamented that Canadians knew too little about the climate issue to make an informed decision.[9] This is not surprising. If the Canadian media don't present all sides of this vital issue, where is the average Canadian to go for information?

So far, to my knowledge and as of April 2010, not one of the major Canadian TV networks has staged a major public debate on the science of climate change; if there'd been a full-dress debate on climate on any of the networks, I think I'd have been aware of it.[10] So far, to my knowledge, no major Canadian television network has run the British skeptical documentary *The Great Global Warming Swindle* although, to its credit, in 2007 the CBC did air a skeptical documentary entitled *Global Warming Doomsday Called Off*. In Britain, until the Climategate scandal caused it to reconsider, the state-run BBC had an explicit policy against airing skeptical views[11]; perhaps Canada's networks have the same policy. The only major Canadian newspaper, indeed, as far as I am aware, the only major Canadian news outlet of any kind, that consistently presents more than one side of the climate issue is the *National Post*.

As for politicians—in Canada they are all on the AGW bandwagon, or pretend to be.[12] Even Conservative Prime Minister Stephen Harper, who knows better, has bowed to the Kyoto and Copenhagen gods and tries to make his government appear green while, sensibly and like the "green" Jean Chrétien and Paul Martin Liberal governments before him, doing as little as possible to avoid economic damage. Politicians could have tried to educate the voters about the problems with alarmist climate science, rather than timidly accepting the alarmist climate smokescreen. However, given the power of the environmental lobby, that course is clearly politically too risky.

TAKING ON THE ALARMIST CLIMATE INDUSTRY

The media's reluctance to take on the alarmist climate science industry is understandable. Alarmist climate science is a multi-billion-dollar Goliath and attacking Goliath is a formidable task, not to be undertaken lightly. The alarmists are *scientists*, after all; they are the self-proclaimed climate experts. And they've said the science is completely settled and anyone who dares to argue otherwise is in the pay of the oil industry, mischievous, scientifically illiterate or simply deluded. In short, and despite claims by Weaver, Suzuki, and others that the media offer too much "balance" to skeptics, alarmist climate science rules in Canada's print and airwaves. What journalist would be foolish enough to contest this alarmist juggernaut? And, indeed, I have at times questioned my own sanity as, time after time, my research reveals how often alarmist climate science is misleading, exaggerated and sometimes plain wrong in what it tells the public. Yet, these are the *experts*, aren't they?

At least, this is how the situation appears if one accepts what the alarmist climate science industry *says*. However, anyone who actually critically *investigates* alarmist climate science—which, again, few journalists or politicians in Canada have been inclined to do—discovers quite a different story. Far from being a Goliath, alarmist climate science is more like the Great and Terrible Wizard of Oz. There is a lot of fire and fury out front. But behind the scenes, as the Climategate emails showed very clearly, is a group of rather petty men and women who, far from being great and terrible, are mostly terrified. Terrified of what? Terrified that the public will find out that they have no empirical evidence to back up the climate fears they have created. And terrified that they will have to admit, publicly, what they acknowledge privately—that the world is not warming at the moment, as their models predict, and that, therefore, their computer models don't work. To again quote James Lovelock (from Chapter 10):

> The great climate science centres around the world are more than well aware how weak their science is. If you talk to them privately they're scared stiff of the fact that they don't really know what the clouds and the aerosols are doing. They could be absolutely running the show. We haven't got the physics worked out yet.

In short, the anthropogenic hypothesis is falling apart and alarmist climate scientists are desperate that the public not know it—this was a consistent theme throughout the Climategate emails. I am certain that, if we had emails from the other alarmist climate centres, like James Hansen's Goddard Institute, they would tell the same story.

The Great and Terrible Climate Oz is, as Dorothy found when *she* did some behind-the-scenes investigation, a fraud. It's a well-meaning fraud, to be sure—most alarmist climate scientists appear to sincerely believe that we are facing a human-caused disaster. But it is a fraud in scientific terms nonetheless because there is no *empirical evidence* to support this hypothesis, and a great deal of empirical evidence (the black swans) to undermine it. And so, to mix children's stories, the Wizard is desperately afraid that someone in the media will point out that the Emperor has no clothes. Thus, the alarmist climate Wizard tries to silence all opposition with his claim that the evidence is "overwhelming" and therefore "the science is settled" and "certain."

The result, as anyone (like myself) who cared to investigate can *easily* find out, is a tissue of evasions, misleading pronouncements, exaggerated figures, and outright lying, all in the service of a higher cause—the "glorious myth" of Saving the Planet, even though there is so far *no* empirical evidence that the planet needs saving now, any more than it has for most of the past, say, 250 million years during which temperatures and carbon dioxide levels have been much higher than today's. Seeing these evasions, it's difficult to avoid the conclusion that the alarmist climate establishment—the so-called "experts," including the IPCC—cannot be trusted for objective, unbiased and accurate information on climate. The Climategate emails only reinforce this impression of a science that puts ideology—the alarmist "message"—ahead of scientific facts, which are considerably less alarming.

What alarms me, personally, is that, thanks to the lack of balanced coverage on climate change in the Canadian media, I wouldn't have known about the flaws in alarmist climate science if it hadn't been for Lawrence Solomon's "The Deniers" series in the *National Post*, since made into a book.[13] A review of Solomon's book describes it as follows:

> This short book shows how the mainstream view of man-made global warming *has become an orthodoxy that cannot be questioned*, its much-publicized "scientific consensus" sustained in part by pressure, *the sidelining of even authoritative dissent*, and the *politically-motivated machinations* of the government-led Intergovernmental Panel on Climate Change.[14] [italics added]

Without Solomon's series, I, like so many in the Canadian public, media and politics, would have meekly accepted the alarmist climate gospel while anthropogenic warming believers dismantled the Western economy with their anti-carbon ideology and legislation.

I began my research into climate and climate science in June 2007. The more I learned, the more astonished and concerned I became that the public, media and politicians weren't being told the whole truth about climate and climate science. But I was also mystified. Why weren't other journalists doing the same very basic research and coming up with the same skeptical conclusions, when the problems with alarmist climate science are so obvious?

I also began to explore the blogosphere, which is almost the only place where the average person can get both the alarmist and skeptical points of view. I found that the alarmist sites, like *RealClimate, Climate Progress* and *DeSmog Blog,* tended to be bombastic, reliant on *ad hominem* attacks (especially *DeSmog Blog*) and dismissive of criticism. The skeptical sites, like Anthony Watts's *Watts Up With That*, Steve McIntyre's *Climate Audit*, and the blogs of partial skeptics Roger Pielke, Jr. and Sr., to name just a few, struck me as far more sane, reasoned, and balanced. The skeptical blogs offer a wealth of climate *data*—not ideology—to reach conclusions quite different from the alarmists'. Prior to Climategate, very little of this non-alarmist information was reported in Canada's media. Indeed, I suspect one reason conventional media is in decline is because it has lost the critical edge that is essential to good journalism. It has certainly lost that critical edge when it comes to reporting on climate. As a result, I am grateful to bloggers like Watts, McIntyre, the Pielkes and many others for trying to keep climate science honest, even though they are vilified for it by the alarmists.

BECOMING A SKEPTIC

I became a confirmed climate skeptic in the same way that Solomon did: I decided to actually *investigate* alarmist climate science claims, as any good journalist should, to see if they were true. (When Bjorn Lomborg did the same with environmentalist claims, he became the "skeptical environmentalist.") The claim Solomon first investigated was that the AGW theory had *complete consensus* within the climate science community, all settled, no need for discussion. This claim is patently absurd but, for some reason, continues to be widely accepted by the Canadian media, politicians and public. Instead, Solomon found that many top-ranked climate scientists didn't agree with the "consensus." In other words, the highly touted "consensus" is a smokescreen for the alarmists' lack of actual evidence. Based on his experience, Solomon has written: "It would be impossible for any investigative reporter … to objectively delve into global warming and conclude that the science was settled."[15] I could not agree more. Investigation of the sort the media ought to have been doing reveals very quickly that very few alarmist climate claims can stand up to critical, objective, empirical scrutiny.

ONE 'CRISIS' AFTER ANOTHER

The global warming "crisis" is not new. We've had these "crises" before—in fact, there seems to be a new crisis every few years (see Chapter 8). Remember the Y2K scare in 1999? The impending meltdown of the world's computers? Chaos and the fall of technological civilization? Absolutely *nothing* happened. The population bomb and mass starvation? Not yet, anyway. AIDS pandemic? Not so far. Nuclear war? Thankfully, avoided so far. DDT poisoning the bird life? Wrong again. I quoted historian John Dietrich in Chapter 12 and would like to repeat his observation here:

> The threat of global warming will eventually recede. The need for

> an apocalyptic vision, however, will not. The next threat will contain many of the characteristics of the global warming threat. It will predict the end of the world. It will be based on "scientific facts." It will require massive counseling for the psychological distress it will cause. It will require the creation of a massive bureaucracy. And it will require the transfer of massive amounts of money to the hypothesized victims of the future crisis.[16]

How many times do journalists have to be caught up in these phony crises—these apocalyptic visions—before they see the repeating pattern and say, "No, I'm not going to be fooled again. I will do more research and ask more questions"?

For a journalist, alarmist climate science is a rich source of investigative material—it is the Watergate scandal of our time, but far, far bigger. At the very least, it is worth critically examining alarmist climate claims. Perhaps, after a thorough *critical* examination of alarmist climate science, some journalists will conclude there is nothing wrong. After all, many journalists—from Lydia Dotto to Ross Gelbspan to the *New York Times*' Andrew Revkin—have been convinced by the IPCC's alarmist climate science and reported it as gospel. However, with a bit more digging it becomes clear that the IPCC is, again, the curtain that the Great and Terrible Oz spreads before his operations. The key facts in the alarmist climate puzzle won't be found in the IPCC reports but in the politics and ideology that is behind them.

I have no doubt that alarmist climate scientists will disagree vehemently with what I've written in this book. However, following Climategate, several "consensus" climatologists, most notably Judith Curry and Mike Hulme, have come out with criticisms of alarmist climate science that confirm many of the conclusions I arrived at independently and long before the emails were revealed. Even Andrew Weaver has acknowledged that the IPCC is too politicized. Yet, in my view, these are conclusions that any sensible person, much less journalist, would reach after investigating alarmist climate science. The Climategate emails simply confirmed that the skeptical concerns about alarmist climate science were legitimate.

This book has tried to report on what is behind the alarmist climate curtain. It is not a pretty sight. What is revealed is tainted science, dogmatism, propaganda, coercion, and political and scientific manipulation. What is *not* revealed is any cause for climate alarm. Surely it's time more Canadian journalists also looked behind the curtain. If they do, I predict they will be, as I was, shocked and even alarmed that so much of what they've been told by the alarmist climate scientists is misleading, exaggerated, or just plain wrong.

NOTES

Introduction

1 Rex Murphy, "Global warming more a cause than a science." *Globe and Mail*, June 8, 2007.

2 Al Gore, *The Assault on Reason*. New York: Penguin, 2007, p. 23.

3 Samuel Butler, *The Way of All Flesh*. Toronto: Rinehart, 1951, chapter 84, p. 377.

4 Sharon Begley, "The Truth About Denial." *Newsweek*, Aug. 13, 2007, pp. 20-29.

5 *Newsweek Online*. Available at http://www.msnbc.msn.com/id/20123112/ site/newsweek/ #anc_nwk_070804_global_warming.

6 "Worse than Nazis?" Letter to the editor, *The Globe and Mail*, May 11, 2007, p. A16.

7 Quoted in Tom Zoellner, "Nuclear power gets its swagger back." *Globe and Mail*, March 14, 2009, p. F5.

8 Richard Lloyd Perry, "Act now on climate change or face oblivion, warns UN chief." *Times Online*, Dec. 12, 2007.

9 Quoted by Elizabeth May, "Don't shoot the messenger." *National Post*, March 11, 2010.

10 "Q&A: Professor Phil Jones." *BBC*, Feb. 13, 2010.

11 For example, former Liberal Party leader Stephane Dion wrote in the *National Post* on June 6, 2008: "Climate change is real, it is man-made and unless something is done, it will damage the planet and our way of life." Previously, though, one of Al Gore's most common phrases was that "global warming is real."

12 Edward Aguado and James E. Burt, *Understanding Weather & Climate*. Upper Saddle River, NJ: Pearson Education, 2004, p. 479. *Understanding Weather & Climate* is used in university geography courses; its views are scientific mainstream, not controversial or skeptical.

13 For a very good, searchable overview of these emails see "Climate Cuttings 33," Bishop Hill blog, Nov. 20, 2009. For a detailed analysis of the emails and their implications for

climate science, see John P. Costella, "Climategate Analysis," Jan. 20, 2010. Both are available online.

Chapter 1: Consensus Climate Science's 'Doctrine of Certainty'

1 John Stuart Mill, "On Liberty," Chapter 2. *On Liberty and Other Essays*. John Gray, ed. Toronto: Oxford Univ. Press, 1998 (1991), p. 22.

2 Bertrand Russell, "What Is an Agnostic?" *The Basic Writings of Bertrand Russell: 1903-1959*. New York: Simon and Schuster, 1961, p. 584.

3 Walter Kaufmann, *Critique of Religion and Philosophy*, Princeton, NJ: Princeton Univ. Press, 1990 (1958), p. 107.

4 Richard P. Feynman, *The Pleasure of Finding Things Out: The Best Short Works of Richard P. Feynman.* Cambridge, MA: Persius Books, 1999, p. 111.

5 Eric Hoffer, *The True Believer*. New York: Harper & Row, 1951, p. 76.

6 Mark Warawa, "Conservative Action on Climate Change," broadcast, *The Nation's Business*, CBC, April 24, 2007.

7 Brian Fagan, *The Journey from Eden: The Peopling of Our World*. London: Thames and Hudson, 1990, p. 13.

8 Fagan, *The Journey from Eden*, p. 214.

9 Lawrence Solomon's "Denier" series, which profiles global warming skeptics who are also top-ranked scientists, is a good start for anyone wondering about just how unanimous the so-called global warming "consensus" is. The entire series can be found on the *National Post*'s website at nationalpost.com.

10 Andrew Weaver, "Climate change is no conspiracy," letter to the *Times Colonist*, May 24, 2007, p. A15.

11 My blog is at http://paulmacrae.com.

12 Bob Unrah, "31,000 scientists reject 'global warming' agenda." *WorldNetDaily*, May 19, 2008.

13 Christopher Essex and Ross McKitrick, *Taken by Storm: The Troubled Science, Policy and Politics of Global Warming*. Toronto: Key Porter, 2002, p.17.

14 Essex & McKitrick, p. 25.

15 Andrew Weaver, *Keeping Our Cool: Canada in a Warming World*. Toronto: Viking Canada, 2008, pp. 22-23.

16 Andrew Weaver, "The climate tells the story." *The Ring*, Jan. 23, 2003.

17 Barbara Kay, "David Suzuki vs. Michael Crichton." *National Post*, Feb. 21, 2007.

18 Al Gore. *Earth in the Balance: Ecology and the Human Spirit*. Toronto: Penguin Books, 1993 (1992), p. 39.

19 Stephen Schneider, *Laboratory Earth: The Planetary Gamble We Can't Afford to Lose*. New York: Basic Books, 1997, p. 67.

20 Lydia Dotto, *Storm Warning: Gambling with the Climate of Our Planet*. Toronto: Doubleday Canada, 1999, p. 280.

21 Dotto, *Ibid.*, p. 273.

22 Quoted in Robert C. Balling, Jr., *The Heated Debate: Greenhouse Predictions Versus Climate Reality*, San Francisco: Pacific Research Institute for Public Policy, 1992, p. xxvii.

23 Gore, *Earth in the Balance*, p. 37.

24 Al Gore, *An Inconvenient Truth.* Emmaus, PA: Rodale, 2006, p. 261.

25 Gore, *An Inconvenient Truth*, p. 262.

26 R.A. Warrick, E.M. Barrow & T.M.L Wigley, *Climate and Sea Level Change: Observations, Projections and Implications.* New York: Cambridge Univ. Press, 1993, p. 3. I've converted the paragraph into bullets for easier reading.

27 Warrick, et al., p. 3.

28 *20/20*, ABC television, August 30, 2006. Quoted in Amy Menefee, "Life is convenient when you define "truth'." *Business and Media Institute*, Jan. 31, 2007.

29 Kyle L. Swanson & Anastasios A. Tsonis, "Has the climate recently shifted?" *Geophysical Research Letters,* 36, March 31, 2009. See also "Could we be in for 30 years of global COOLING?" *Daily Mail Online,* Jan. 11, 2010.

30 The data for this graph comes from the Hadley Centre for Climate Prediction and Research. The graph was prepared by meteorologist Anthony Watts and can be seen at http://wattsupwiththat.wordpress.com/2008/02/19/january-2008-4-sources-say-globally-cooler-in-the-past-12-months.

31 Andrew Weaver, *Keeping Our Cool*, p. 84.

32 William F. Ruddiman, *Earth's Climate: Past and Future*. New York: DH Freeman, 2001, p. 171.

33 Environment Canada puts the human contribution to carbon dioxide levels at a mere two per cent. See www.ec.gc.ca/pdb/ghg/about/gases_e.cfm. Other sources say four to five per cent.

34 From the 800s to the 1300s, "a retreat of glaciers both on Iceland and southern Greenland … had opened up a considerable amount of rich land for farming." Roy A. Gallant, *Earth's Changing Climate*. New York: Four Winds Press, 1979, p. 128. A recent textbook on climate refers to Greenlanders' sheep feasting "on grasses flourishing in the long summer days" (Edward Aguado and James E. Burt, *Understanding Weather and Climate*. Upper Saddle River, NJ: Pearson Education, 2004, p. 475).

35 Steven Mithen, *After the Ice: A Global Human History*. Cambridge, MA: Harvard Univ. Press, 2003, pp. 507-508.

36 Bjorn Lomborg's *How to Spend $50 Billion to Make the World a Better Place* (Cambridge: Cambridge Univ. Press, 2006) has valuable suggestions about where our climate-change money could be more effectively spent. Another useful discussion in this vein can be found in Jack M. Hollander's *The Real Environmental Crisis* (Univ. of California Press, Berkeley, 2003).

37 "Q&A: Professor Phil Jones." BBC, Feb. 13, 2010.

38 IPCC 2007 Summary for Policymakers, p. 5.

39 Elizabeth Nickson, "The Kyoto Accord's hidden menace." *National Post*, Sept. 6, 2002, p. A20.

40 Marilyn Berlin Snell, "Cool heads tackle our hottest issue." Sierra Climate Exchange, Sierra Club, May/June 2007.

41 Richard Somerville, "A response to climate change denialism." *Scripps Institute of Oceanography*, Jan. 14, 2010.

42 Dan Gardner, *Risk: Why We Fear the Things We Shouldn't—and Put Ourselves in Greater Danger*. Toronto: McClelland & Stewart, 2008, p. 122.

43 Morris Cohen, *Reason and Nature: An Essay on the Meaning of Scientific Method.* Glencoe, IL: Free Press, 1931. Quoted in Carl Sagan, *The Demon-Haunted World: Science as a Candle in the Darkness*. New York: Ballantine Books, 1996, p. 251.

44 Sagan, *The Demon-Haunted World*, pp. 427, 434.

45 Steven Yearly, *Sociology, Environmentalism, Globalization: Reinventing the Globe.* London: Sage, 1996, p. 118. Quoted in Greg Gerrard, *Ecocriticism*. London: Routledge, 2004, p. 168.

46 Greg Gerrard, *Ecocriticism*, p. 169.

47 John Stuart Mill, "On Liberty," p. 21.

48 Thomas B. Macaulay, "Southey." *Reviews, Essays and Poems.* London: Ward Lock, 1896, p. 150.

49 At the time of writing, Pachauri was under pressure to resign after a series of IPCC errors.

50 Elaine Dewar. *Bones: Discovering the First Americans*. New York: Carrol & Graf, 2001, p. 6.

CHAPTER 2: GLOBAL WARMING: THE GEOLOGICAL BACKGROUND

1 Edward Aguado & James E. Burt, *Understanding Weather & Climate*. Upper Saddle River, NJ: Pearson Education, 2004, p. 479.

2 John & Mary Gribbin, *Ice Age*. Toronto: Penguin Books, 2001, p. 1.

3 John & Mary Gribbin, *Ice Age*, pp. 1-2.

4 Aguado & Burt, p. 479.

5 James Zachos, Mark Pagani, Lisa Sloan, Ellen Thomas, and Katharina Billups, "Trends, Rhythms, and Aberrations in Global Climate 65 Ma to Present." *Science*, 2001, Vol. 292 (5517): pp. 686–693.

6 Figure 1.2 from Jouzel, J. et al. (2004). EPICA Dome C Ice Cores Deuterium Data. IGBP Pages/Word Data Centre for Palaeoclimatology Data Contribution Series # 2004-038. Boulder, CO: NOAA/NGDC Palaeoclimatology Program. Found in Nicholas Schneider, *Understanding Climate Change*. Vancouver: Fraser Institute, 2008, p. 20.

7 IPCC 2007, *Summary for Policymakers*, "A Paleoclimatic Perspective," p. 9. "The last time the polar regions were significantly warmer than present for an extended period (about 125,000 years ago), reductions in polar ice volume led to 4 to 6 metres of sea level rise."

8 Aguado & Burt, p. 481. See also Jonathan T. Overpeck, Bette L. Otto-Bliesner, Gifford H. Miller, Daniel R. Muhs, Richard B. Alley, Jeffrey T. Kiehl, "Paleoclimatic Evidence

for Future Ice-Sheet Instability and Rapid Sea-Level Rise." *Science*, March 24, 2006, Vol. 311. no. 5768, pp. 1747-1750.

9 Harm de Blij, *Why Geography Matters: Three Challenges Facing America.* Toronto: Oxford Univ. Press, 2005, p. 82.

10 Brian Fagan, et. al. *The Complete Ice Age: How Climate Changed the World.* London: Thames & Hudson, 2009, p. 67.

11 Bjorn Lomborg, *The Skeptical Environmentalist: Measuring the Real State of the World.* Cambridge, UK: Cambridge Univ. Press, 2001, p. 292. Aguado and Burt, p. 483. Figure 1.3 from IPCC 1990.

12 "Southern Greenland's average temperature probably was as much as four degrees Celsius … higher than it is today [1979]." Roy A. Gallant, *Earth's Changing Climate.* New York: Four Winds Press, 1979, p. 127.

13 Figure 1.4: Temperature reconstruction by C.R. Scotese; CO_2 reconstruction after R.A. Berner & Z. Kothavala, "Geocarb III: A revised model of atmospheric CO_2 over Phanerozoic time." *American Journal of Science*, Vol. 301, Feb. 2001, pp. 182-204.

14 Aguado & Burt, p. 479.

15 Aguado & Burt. P. 477. Of these seven ice ages spread out over billions of years, our series of glacials and interglacials is just the latest.

16 Aguado & Burt, p. 479.

17 Tim Flannery, *The Eternal Frontier: An Ecological History of North America and its Peoples.* London: Vintage, 2002, p. 84.

18 Donald R. Prothero, *Eocene-Oligocene Transition: Paradise Lost.* New York: Columbia Univ. Press, 1994, pp. 22, 238,

19 Prothero, p. 35.

20 Flannery, p. 84.

21 Source: Ice Age Now, available at http://www.iceagenow.com/Extent_of_Previous_Glaciation.htm.

22 "Glacial-Interglacial Climate Cycles—Summary." *CO_2 Science*. Available at http://www.co2science.org/subject/c/summaries/glacialcycles.php.

23 Gordon Jaremko, "Causes of climate change varied: poll." *Edmonton Journal,* March 6, 2008.

24 Quoted in Alexander Cockburn, "Dissidents against dogma," June 9, 2007. This does not mean, of course, that there aren't geologists who do believe in human-made global warming.

25 From two million to one million years ago, glacial and interglacial cycles occurred on a 41,000-year cycle and were not as extreme as the million years just before our time.

26 Martin Bell & Michael J.C. Walker, *Late Quaternary Environmental Change: Physical and Human Perspectives*. Toronto: Pearson Education, 2005 (1992), p. 83.

27 Ian Plimer, *Heaven and Earth—Global Warming: The Missing Science*. London: Quartet, 2009, p. 170.

28 William F. Ruddiman, *Plows, Plagues, and Petroleum*. New York: DH Freeman, 2001, p. 41.

29 Aguado & Burt, p. 479; Bell & Walker, *Late Quaternary Environmental Change,* p. 83.

30 Ruddiman, *Plows, Plagues, and Petroleum*, p. 20.

31 The previous million years had seen cold and warm cycles on a 41,000-year cycle, matching the cycle of the earth's tilt.

32 John & Mary Gribbin, pp. 88-89.

33 IPCC 2007, Chapter 6, Paleoclimate, Section 6.1, p. 449.

34 Ruddiman, *Earth's Climate: Past and Future*. New York: DH Freeman, 2001, p. 171. Ruddiman's hypothesis is that human agriculture, starting 8,000 years ago, put enough carbon dioxide and methane into the atmosphere to delay an interglacial that should have already begun. Again, in Ruddiman's view, temperature is highly sensitive to changes in CO_2 levels.

35 Aguado & Burt, p. 486.

36 Fagan, et al., *The Complete Ice Age*, p. 61.

37 Intergovernmental Panel on Climate Change (IPCC), Third Assessment Report (2001), Section 14.2.2.2, page 774.

38 Several peer-reviewed scientific papers in particular suggest we may be entering a decade or two of cooling. They include Kyle L. Swanson and Anastasios A. Tsonis, "Has the climate recently shifted?" *Geophysical Research Letters,* 36, March 31, 2009; and N. S. Keenlyside, M. Latif, J. Jungclaus, L. Kornblueh & E. Roeckner, *Nature* 453, pp. 84-88 (1 May 2008). Climate alarmist Fred Pearce of the *New Scientist* quotes Latif as suggesting there will be a decade or two of cooling (available at http://www.newscientist.com/article/dn17742-worlds-climate-could-cool-first-warm-later.html).

39 Greg Roberts, "Revealed: Antarctic Ice Growing, Not Shrinking." *The Australian*, April 18, 2009.

CHAPTER 3:
IS CARBON DIOXIDE THE GLOBAL WARMING VILLAIN?

1 Paul Sheehan, "Beware the climate of conformity." Interview with Australian geologist Ian Plimer. *Sydney Morning Herald*, April 13, 2009.

2 Aaron Wildavsky, *But Is It True? A Citizen's Guide to Environmental Health and Safety Issues*. Cambridge, MA: Harvard Univ. Press, 1995, p. 446.

3 Intergovernmental Panel on Climate Change (IPCC), Third Assessment Report (2001), Section 14.2.2.2, page 774.

4 Paul N. Pearson & Martin R. Palmer, "Atmospheric carbon dioxide concentrations over the past 60 million years." *Nature*, vol. 406 (August 17, 2000), pp. 695-699. The authors' research shows early Eocene carbon dioxide levels as high as 2,000 parts per million, although other researchers regard this as too high an estimate. Today's levels are about 388 ppm.

5 N. Shaviv & J. Veizer, "Celestial driver of Panerozoic climate?" *Geological Society of America,* vol. 13 (2003), pp. 4-10. Cited in S. Fred Singer and Dennis T. Avery, *Unstoppable Global Warming: Every 1,500 Years*. Lanham, MD: Rowman & Littlefield, 2007, p. 198.

6 Environment Canada, "Greenhouse gases." Available at http://www.ec.gc.ca/pdb/ghg/about/gases_e.cfm.

7 Charles W. Rice & Richard Nelson, "What is Carbon and the Carbon Cycle?", Table 1. Montana State University. Most other sources show lower percentages of human-caused carbon dioxide.

8 IPCC, 2001, Chapter 3, Section 3.1, p. 187.

9 See Christopher Monckton, "Climate sensitivity reconsidered." *American Physical Society Forum on Physics & Society*, July, 2008.

10 Patrick J. Michaels & Robert C. Balling, Jr., *The Satanic Gases: Clearing the Air about Global Warming*. Washington, DC: Cato Institute, 2000, p. 25.

11 Lomborg, *The Skeptical Environmentalist.* New York: Cambridge Univ. Press, 2001, pp. 270-273.

12 Roy W. Spencer, "Clouds cool the climate system—but amplify global warming?" *Global Warming*, April 14, 2009.

13 IPCC 2001, Technical Summary, p. D1. Available at http://www.ipcc.ch/ipccreports/tar/wg1/022.htm.

14 Volker Mrasek, "Cleaner air means a warmer Europe." *Spiegel Online International.* April 14, 2008.

15 Aguado & Burt, pp. 500, 485.

16 Intergovernmental Panel on Climate Change (IPCC), Third Assessment Report (2001), Section 14.2.2.2, page 774.

17 As we saw in Chapter 1, Stanford climatologist Stephen Schneider has written: "Computer modeling is our *only available tool* to perform what-if experiments such as the human impact on the future." [italics added] *Laboratory Earth: The Planetary Gamble We Can't Afford to Lose*, p. 67.

18 Al Gore. *Earth in the Balance: Ecology and the Human Spirit.* Toronto: Penguin Books, 1993, p 96.

19 Judith Curry, "On the Credibility of Climate Research, Part II: Towards Rebuilding Trust." Feb. 24, 2010. Available online.

20 *IPCC Climate Change 07*, Summary for Policymakers, p. 12. This and most other IPCC documents cited are available at the IPCC website, www.ipcc.ch/pub/reports.htm.

21 Quoted in Peter Foster, "The Weather Exploiters." *National Post*, Oct. 15, 2009.

22 Ian Plimer, "Vitriolic atmosphere in academic hothouse." *The Australian*, May 29, 2009.

23 "Carbon dioxide in greenhouses." Fact Sheet. Ontario Ministry of Agriculture, Food & Rural Affairs, Available at http://www.omafra.gov.on.ca/english/crops/facts/00-077.htm.

24 IPCC 2001, Chapter 3, Section 3.2.2.4, p. 195.

25 Sherwood B. Idso, "Carbon dioxide and global change." *Rational Readings on Environmental Concerns*, Jay H. Lehr, ed. New York: Van Nostrand Reinhold 1992, p. 422.

26 Robert M. Carter, "Wong's climate paper clouded with mistakes." *The Age*, July 29, 2008.

27 R.F. Sage, "Was low atmospheric CO_2 during the Pleistocene a limiting factor for the origin of agriculture?" *Global Change Biology,* Vol. 1, issue 2, April, 1995, pp. 93-106. Sage writes: "In the late Pleistocene, CO_2 levels near 200 ppm may have been too low to support the level of productivity required for successful establishment of agriculture."

28 Lawrence Solomon, "The ice-core man." *National Post*, May 4, 2007. Available on the *National Post* website as part of the "Deniers" series.

29 Tim Patterson, appearing as an expert witness before Canadian House of Commons Committee on Environment and Sustainable Development, Feb. 10, 2005. Available at http://cmte.parl.gc.ca/Content/HOC/committee/381/envi/evidence/ev1623904/enviev18-e.htm.

30 "Climate science leader Rajendra K. Pachauri confronts the critics." Interview with *Science*, Jan. 29, 2010.

31 IPCC 2007, Summary for Policymakers, p. 12.

32 John Bluemle, "Some thoughts on climate change." *North Dakota State Geological Survey Newsletter,* Vol. 28 (2001), pp. 1-2. Quoted in Singer and Avery, p. 179.

33 William F. Ruddiman, *Earth's Climate: Past and Future*. New York: D.H. Freeman, 2001, p. 134.

34 For a discussion of carbon saturation and temperatures using the same approach but a more sophisticated formula, see the website "Cold Facts About Global Warming" at http://brneurosci.org/co2.html. This more complex formula finds the maximum increase of temperature to be between 1.02 and 1.85°C.

35 Steven Malloy, "The *Real* 'Inconvenient Truth'." *JunkScience.com*, August 2007.

36 At the start of the Eocene, 55 million years ago, the earth came close to a runaway greenhouse when volcanic eruptions ignited vast amounts of *methane*, not CO_2. For details, see Tim Flannery, *The Weather Makers.* Toronto: HarperCollins, 2005, pp. 51-53. And a series of massive volcanic eruptions that spewed millions of tonnes of carbon dioxide into the atmosphere helped cause the disastrous Permian-Triassic warming of 250 million years ago. But, as noted elsewhere, there is no prospect of this kind of geological catastrophe occurring at present.

37 Union of Concerned Scientists, "How natural gas works." *Clean Energy.*

38 BP Statistical Review of World Energy June 2008, pp. 10, 26, 32. The time left for a given energy source is described as the R/P (Reserves to Production) ratio.

39 Keith Jackson, "The Effect of Finite Fossil Fuel Reserves on Climate Change Potential." HM (British) Government Treasury, 2006. This article appears to have been removed from the British Treasury site.

40 Jackson, *Ibid*, p. 10.

41 Plimer, *Heaven and Earth*, p. 293.

42 Goddard Institute for Space Studies, "NASA Study Illustrates How Global Peak Oil Could Impact Climate." Sept. 10, 2008.

43 Patrick J. Michaels & Robert C. Balling, Jr., *Climate of Extremes: Global Warming Science They Don't Want You to Know*. Washington, DC: Cato Institute, 2009, p. 2.

44 Ruddiman, *Earth's Climate*, p. 95.

45 Arthur C. Clarke, "Presidents, experts, and asteroids." *Science*, June 5, 1998.

46 Roger Pielke, Sr., "Summary Of Roger A. Pielke Sr's View Of Climate Science." *Climate Science* website, April 7, 2009.

47 Roger Pielke, Sr., "Roger A. Pielke Sr.'s Perspective On The Role Of Humans In Climate Change." *Climate Science*, March 31, 2008.

CHAPTER 4: THE IPCC'S RUSH TO JUDGMENT

1 Albert Einstein, "Philosophy and Religion." In *Ideas and Opinions*. New York: Dell Publishing, 1954, p. 56.

2 David S. Landes, *The Wealth and Poverty of Nations*. New York: W.W. Norton, 1999, p. 517.

3 IPCC, "16 Years of Scientific Assessment in Support of the Climate Convention," December, 2004, p. ii.

4 Keith Briffa, Climategate email to Michael Mann, April 29, 2007.

5 IPCC, *Ibid.,* p. 4.

6 IPCC, *Ibid.*, p. ii.

7 IPCC, *Ibid.*, p. ii.

8 Paul Feyerabend, "How to defend society against science," *Radical Philosophy 2* (Summer 1975), pp. 4-8.

9 IPCC, *Climate Change 90*. Quoted in Vincent Gray, *The Greenhouse Delusion: A Critique of Climate Change 2001*, p. 7. Gray says this position was relaxed, after protests, in later reports. See also Gray, "Climate Change 95: An Appraisal," which is available online.

10 Mike Hulme, "The IPCC, consensus and science." Feb. 19, 2010. Available online.

11 Fred Guteri, "It's Gettin' Hot in Here: The Big Battle Over Climate Science." *Discover*, April 2010.

12 IPCC, *Climate Change 2007*, Summary for Policymakers, p. 9.

13 "In Europe as a whole, about 200,000 people die from excess heat each year. However, about 1.5 million Europeans die annually from excess cold. That is more than seven times the total number of heat deaths." Bjorn Lomborg, *Cool It: The Skeptical Environmentalist's Guide to Global Warming*. New York: Alfred A. Knopf, 2007, p. 17.

14 Shawn McCarthy, "Arctic holds 13 per cent of world's undiscovered oil." *Globe and Mail*, May 29, 2009.

15 Michael Crichton, *State of Fear*. New York: Avon Books, 2004, p. 272.

16 Al Gore, *An Inconvenient Truth*, p. 261.

17 Kate Gibson, "Al Gore calls on world business leaders to push for climate deal." McClatchy Tribune News Service, May 24, 2009.

18 Al Gore, *Earth in the Balance,* p. 40.

19 Andrew Brod, "It's Not Too Late—Yet—to Counteract Global Warming." *Greensboro News & Record*, April 29, 2007.

20 Quoted in Peter Menzies, "Concerned about the cost of Kyoto." *Calgary Herald,* Dec. 14, 1998, p. A12.

21 Quoted in Robert James Bidinotto, "Environmentalism: Freedom's Foe for the 90s." *The Freeman*, vol. 40, no. 11, November 1990.

22 Quoted in Dixy Lee Ray, *Trashing the Planet*. New York: Harper Perennial, 1990, p. 167.

23 IPCC, *Climate Change 07*, Working Group I: The Physical Basis of Climate Change, Chapter 9, p. 704.

24 S. Fred Singer, Letter to IPCC (Working Group I) Scientists, available at http://web.archive.org/web/19980629122454/http:/www.sepp.org/ipcccont/ipccflap.htm.

25 Singer, Letter to IPCC (Working Group I) Scientists.

26 "Climate debate must not overheat." *Nature*, June 13, 1996 (Vol. 381, Issue 6583), p. 539.

27 Paul N. Edwards and Stephen H. Schneider, "The 1995 IPCC Report: Broad Consensus or 'Scientific Cleansing'?" *Ecofable/Ecoscience*, vol. 1, no. 1 (1997), pp. 3-9.

28 For a listing of these changed and deleted passages, see "Changes to 1995 IPCC report." *Reclaiming Climate Science*, Feb. 26, 2009. Available at http://www.greenworldtrust.org.uk/Science/Social/IPCC-95-Ch8.htm.

29 IPCC, 2007 Summary for Policymakers, Figure SPM.5, p. 14. The bottom line, in the IPCC graphic, is a scenario in which CO_2 concentrations are kept at 2000 values. By coincidence, the bottom line can also represent the lack of warming since 2001.

30 Bjorn Lomborg, *The Skeptical Environmentalist: Measuring the Real State of the World.* Cambridge, UK: Cambridge Univ. Press, 2001, p. 266.

31 Roy W. Spencer, "A Layman's Explanation of Why Global Warming Predictions by Climate Models are Wrong." *Global Warming* website, May 29, 2009.

32 Christopher Monckton, "Climate sensitivity reconsidered." *American Physical Society Forum on Physics & Society*, July, 2008.

33 IPCC, *Climate Change 07, Physical Basis of Climate Change*, Chapter 8, Climate models and their evaluation, p. 601.

34 Albert Einstein, *Ideas and Opinions*, p. 56.

35 Quoted in Fred Pearce, "World's climate could cool first, warm later." *New Scientist*, Sept. 4, 2009.

36 Quoted in Vincent Gray, *The Greenhouse Delusion: A Critique of Climate Change 2001.* Brentwood, Essex: Multi-Science Publishing, 2002, pp. 70-71. All these IPCC comments can be found in the various reports put out by the IPCC and published online.

37 IPCC, *Climate Change 01*, Chapter 14, Advancing Our Understanding, Section 14.2.2.2.

38 Kevin Trenberth, "Predictions of climate." *Climate Feedback*, June 4, 2007. Trenberth continues to believe, however, that human carbon emissions are the main cause of warming: "A consensus has emerged that 'warming of the climate system is unequivocal,' to quote the 2007 IPCC Fourth Assessment Working Group I Summary for Policy Makers, and the science is convincing that humans are the cause. Hence mitigation of the problem: stopping or slowing greenhouse gas emissions into the atmosphere is essential. The science is clear in this respect."

39 Vincent Gray, *Greenhouse Delusion*, p. 70.

40 IPCC, *Climate Change 07*, Chapter 8, p. 600.

41 IPCC, "Climate Models and Their Evaluation." *Climate Change 2007*: Chapter 8. Working Group I to the Fourth Assessment Report of the Intergovernmental Panel on Climate Change, p. 601.

42 John Manzi, "Prediction Time: Global-warming 'truths' are not as certain as some claim them to be." *National Review Online*, March 20, 2007.

43 Quoted in Steven D. Levitt & Stephen J. Dubner, *Superfreakonomics: Global Cooling, Patriotic Prostitutes and why Suicide Bombers Should Buy Life Insurance*. Toronto: Harper Collins, 2009, p. 181.

44 Michael Mann, R.S. Bradley & M.K. Hughs, "Global-scale temperature patterns and climate forcing over the past six centuries." *Nature*, 392, vol. 6678, April 23, 1998, pp. 779-787.

45 Details of this intellectual battle can be found in Essex and McKitrick, *Storm Warning*, pp. 154-174. See also Christopher Monckton, "Hockey Stick? What Hockey Stick?" available at http://scienceandpublicpolicy.org/monckton/what_hockey_stick.html, and Bishop Hill, "Caspar and the Jesus paper," available at http://bishophill.squarespace.com/blog/2008/8/11/caspar-and-the-jesus-paper.html. Steve McIntyre's excellent blog site, *Climate Audit*, discusses the hockey stick in detail at http://www.climateaudit.org/?page_id=354.

46 IPCC, 2007 Summary for Policymakers, "A Paleoclimatic Perspective," p. 9. The IPCC's discussion of the "hockey stick" controversy can be found in IPCC 2006, Chapter 6, Paleoclimate, Section 6.6.1.1. The hockey stick graphic can be found, among other places, in IPCC 2001, Chapter 2, Section 2.3.3, Figure 2.20.

47 M. Bell & M.J.C. Walker, *Late Quaternary Environmental Change: Physical and Human Perspectives*, 2nd ed. Toronto: Pearson Education, 2005 (1992), p. 90.

48 Steve McIntyre, "Some Thoughts on Disclosure and Due Diligence in Climate Science." *Climate Audit,* Feb. 15, 2005.

49 Vincent Gray, *The Greenhouse Delusion*, p. 58.

50 IPCC, *Climate Change 2001:* Model Evaluation, "What is Meant by Evaluation?", Section 8.2.2.

51 Mike Hulme, "The IPCC, consensus and science." Feb. 19, 2010.

52 Nigel Lawson, *An Appeal to Reason: A Cool Look at Global Warming.* New York: Duckworth Overlook, 2008, pp. 107-108.

53 Landes, *The Wealth and Poverty of Nations*, p. 517.

54 Michael D. Lemonick, "Freeman Dyson Takes On The Climate Establishment." *Environment 360*, Yale School of Forestry and Environmental Studies, June 4, 2009.

55 IPCC, *Climate Change 2007,* IPCC Working Group II: Impacts, Adaption and Vulnerability, Chapter 10, Section 6.2.

56 David Rose, "Glacier scientist: I knew data hadn't been verified." *Daily Mail Online,* Jan. 24, 2010.

57 Rose, "Glacier scientist."

58 IPCC, 2007, Working Group II: Impacts, Adaption and Vulnerability, Chapter 1, Section 3.8.5.

59 Jonathan Leake, "UN wrongly linked global warming to natural disasters." *Sunday Times Online*, Jan. 2, 2010.

60 Indur Goklany, "The IPCC: More sins of omission—telling the truth but not the whole truth." *Watts Up With That*, Jan. 25, 2010.

61 Donna Laframboise, "More dodgy citations in the Nobel-winning IPCC report." *There Is No Frakking "Scientific Consensus" on Global Warming.*" Jan. 23, 2010.

CHAPTER 5: THE *REAL* GLOBAL WARMING SUSPECTS

1 Aaron Wildavsky, *But Is It True? A Citizen's Guide to Environmental Health and Safety Issues*. Cambridge, MA: Harvard Univ. Press, 1995, p. 374.

2 Henrick Svensmark & Nigel Calder, *The Chilling Stars: A New Theory of Climate Change.* Cambridge: Icon Books, 2007, p. 30.

3 IPCC, *Climate Change 2007,* Summary for Policymakers, pp. 4-5, 10.

4 National Aeronautics and Space Administration, "NASA study finds increasing solar trend that can change climate."March 20, 2003. See also Richard A. Willson and Alexander V. Mordvinov, "Secular total solar irradiance trend during solar cycles 21–23." *Geophysical Research Letters*, Vol. 30, No. 5 (2003).

5 U. Cubash et al., "Simulation of the influence of solar radiation variations on the global climate with an ocean-atmosphere general circulation model." *Climate Dynamics*, Vol. 13, No. 11, pp. 757-767. Cited in Bjorn Lomborg, *The Skeptical Environmentalist: Measuring the Real State of the World.* Cambridge, UK: Cambridge Univ. Press, 2001, p. 276.

6 Max Planck Institute, "How Strongly Does the Sun Influence the Global Climate?" Press release, Aug. 2, 2004. The press release added that without some other contributing factor, this solar change was not enough to explain current temperatures, although earlier fluctuations of temperature were correlated with solar variability. One possible augmenting factor might be cosmic rays, discussed below.

7 Cindy Robertson, "Global warning? Controversy heats up in the scientific community." *Carleton University Magazine*, Spring 2005.

8 From Eigil Friis-Christensen & K. Lassen, "Length of the solar cycle: An indicator of solar activity closely associated with climate." *Science*, Nov. 1, 1991, pp. 698-700. Carbon dioxide concentration added by Friends of Science website.

9 The alarmist explanation is that increased sulfate particles, called aerosols, put into the atmosphere from the 1940s to the 1970s by industry had a counteracting cooling effect by creating more cloud cover. After the 1970s, the theory goes, anti-pollution measures reduced these particles and the planet began warming again. A simpler and more logical explanation is that the planet just got colder in the 1940s, and warmer in the 1970s, for natural (including solar) reasons.

10 John A. Eddy, "The Case of the Missing Sunspots." *Scientific American*, May, 1977. Quoted in Roy A. Gallant, *Earth's Changing Climate*. New York: Four Winds Press, 1979, p. 180.

11 James Lovelock, *Gaia*: *A New Look at Life on Earth.* Toronto: Oxford Univ. Press, 1979, p. 20.

12 Kate Ravilious, "Mars melt hints at solar, not human, cause for warming, scientist says." *National Geographic News*, Feb. 28, 2007.

13 Phil Chapman, "Sorry to ruin the fun, but an ice age cometh." *The Australian*, April 23, 2008.

14 Andrew Weaver, *Keeping Our Cool*: *Canada in a Warming World.* Toronto: Viking Canada, 2008, pp. 167-169.

15 Weaver, *Keeping Our Cool,* p. 194-198.

16 S. Fred Singer & Dennis Avery, *Unstoppable Global Warming: Every 1,500 Years.* Toronto: Rowman & Littlefield, 2007, p. 22.

17 In R.M. Carter, "The myth of dangerous human-caused climate change." AUSIMM New Leaders' Conference, May 2-3, 2007, p. 69. Source of data: P.M. Grootes, et al. "Comparison of oxygen isotope records GISP2 and GRIP Greenland ice cores." *Nature,* 366, 1993, pp. 552-554.

18 Svensmark & Calder, p. 13.

19 Singer & Avery, p. 198,

20 Anthony Watts, "Where have all the sunspots gone?" *Watts Up With That.* Feb. 13, 2008. Other postings on Watts's site have updated details on the sunspot cycle.

21 IPCC, 2007 Summary for Policymakers, p. 5.

22 Svensmark & Calder, p. 79.

23 IPCC, Climate Change 01, Summary for Policymakers, p. 46.

24 Svensmark & Calder, pp. 73-74.

25 Svensmark & Calder, pp. 99-131.

26 Svensmark & Calder, pp. 132-155. For a shorter description of the cosmic ray hypothesis, see Singer & Avery, pp. 195-197.

27 Quoted in Lawrence Solomon, "From chaos, coherence." *National Post,* August 15, 2007.

28 Singer & Avery, p. 189.

29 Kyle L. Swanson and Anastasios A. Tsonis, "Has the climate recently shifted?" *Geophysical Research Letters,* 36, March 31, 2009.

30 The hypothesis that oceans (not humans) are a key factor in 20th century and 21st century climate gets support from climatologist Judith Curry, who notes in an interview with *Discover* magazine (Fred Guteri, "It's Gettin' Hot in Here: The Big Battle Over Climate Science." April 2010): "[One] issue is these big ocean oscillations, like the Atlantic Multidecadal Oscillation and the Pacific Decadal Oscillation, and particularly, how these influenced temperatures in the latter half of the 20th century. I think there was a big bump at the end of the 20th century, especially starting in the mid-1990s. We got a big bump from going into the warm phase of the Atlantic Multidecadal Oscillation. The Pacific Decadal Oscillation was warm until about 2002. Now we're in the cool phase. This is probably why we've seen a leveling-off [of global average temperatures] in the past five or so years. My point is that at the end of the 1980s and in the '90s, both of the ocean oscillations were chiming in together to give some extra warmth. If you go back to the 1930s and '40s, you see a similar bump in the temperature records. That was the bump that some of those climate scientists [Climategate] were trying to get rid of [in the temperature data], but it was a real bump, and I think it was associated with these ocean oscillations. That was another period when you had the Pacific Decadal Oscillation and the Atlantic Multidecadal Oscillation chiming in together. These oscillations and how they influence global temperature haven't received enough attention, and it's an important part of how we interpret 20th-century climate records. Rather than trying to airbrush this bump in the 1940s and trying to get rid of the medieval warm period—which these hacked e-mails illustrate—we need to understand them."

31 Weaver, *Keeping Our Cool,* p. 5.

32 Weaver, *Keeping Our Cool*, p. 202.

33 Graph is from IPCC 2007, Summary for Policymakers, p. 11.

34 R.A. Warrick, E.M. Barrow, & T.M.L. Wigley, *Climate and Sea Level Change: Observations, Projections and Implications*. New York: Cambridge Univ. Press, 1993, p. 3.

35 Weaver, *Keeping Our Cool*, pp. 116, 218.

36 Robert B. Gagosian, "Abrupt Climate Change: Should We Be Worried?" Woods Hole Oceanographic Institute paper for panel on abrupt climate change at World Economic Forum, Davos, Switzerland, January 27, 2003.

37 Richard B. Alley, et al. *Abrupt Climate Change: Inevitable Surprises*. Washington: National Academy Press, 2002, p. 27.

38 Alley, et al., *Abrupt Climate Change*, p. 1.

39 This was told to me by an archeologist at the University of Victoria.

40 Geologist John McClenney's "The Sky is Falling, or, Revising the Nine Times Rule" argues, based on geological data, that sea levels have been, on average, 30 metres (100) feet higher than today's levels in previous interglacials.

41 IPCC, 2007 Summary for Policymakers, p. 5.

42 Gregory Murphy, "Interview: Dr. Nils-Axel Morner: Claim that sea level is rising is a total fraud." *Executive Intelligence Review,* June 22, 2007.

43 Warrick, et al., p. 3.

44 From John D. Cox, *Climate Crash: Abrupt Climate Change and What It Means for Our Future*. Washington, DC: Joseph Henry Press, 2005, p. 115.

45 Cox, *Climate Crash*, p. 115. This graph also appears in *Abrupt Climate Change: Inevitable Surprises*, p. 37.

46 Cox, p. 4.

47 Roger Pielke, Sr., "Main conclusions." *Climate Science.* Available at http://climatesci.org/main-conclusions. Pielke's views are expanded in *Radiative Forcing of Climate Change: Expanding the Concept and Addressing Uncertainties*, published by The National Academies Press in 2005, which offers a detailed analysis of these other anthropogenic forcings.

48 As seen in Chapter 2, note 34, climatologist William Ruddiman believes that prehistoric agriculture, and particularly rice cultivation in China, produced enough carbon dioxide and methane, and therefore warming, to delay the start of the next glaciation, which, if past interglacials averaging 15,000 years are any measure, should have started by now. There is, however, no general scientific acceptance of Ruddiman's hypothesis. See Ruddiman, *Plows, Plagues and Petroleum,* p. 100

49 Svensmark & Calder, p. 30.

CHAPTER 6: AL GORE: GOOD SCIENCE, BAD SCIENCE, AND BOGUS SCIENCE

1 Richard Feynman, *The Character of Physical Law*. Cambridge, MA: MIT Press, 1965, p. 156.

2 Karl Popper, *Unended Quest: An Intellectual Biography*. Glasgow: Fontana/Collins, 1976, p. 41.

3 Jean-François Revel, *Neither Marx nor Jesus*. New York: Laurel, 1971, p. 15.

4 Popper, *Unended Quest*, p. 41.

5 Popper, *Unended Quest*, pp. 24, 38.

6 Al Gore, *Earth in the Balance: Ecology and the Human Spirit*. Toronto: Penguin Books, 1993 (1992), pp. 57-60.

7 For example, a series of massive volcanic eruptions in the Deccan area of what is now India may have caused or contributed to the death of the dinosaurs 65 million years ago, regardless of whether an asteroid hit the earth. A series of huge volcanic eruptions in Siberia may have caused the massive Permian-Triassic extinction of 250 million years ago.

8 Thomas J. Crowley & Gerald R. North, *Paleoclimatology*. New York: Oxford Univ. Press, 1991, p. 105. If there is a global climate effect, as with the Mount Pinatubo eruption in the Philippines in 1991, the effect is temporary.

9 See Hubertus Fischer, et al., "Ice Core Records of Atmospheric CO2 Around the Last Three Glacial Terminations." *Science*, March 12, 1999 (Vol. 283, No. 5408), pp. 1712-1714. Also, Nicolas Caillon et al., "Timing of Atmospheric CO2 and Antarctic Temperature Changes Across Termination III." *Science*, March 14, 2003 (Vol. 299, No. 5613), pp. 1728-1731.

10 Edward Aguado and James E. Burt, *Understanding Weather & Climate*. Upper Saddle River, NJ: Pearson Education, 2004, p. 495.

11 Paul MacRae, "The $8 billion global-warming swindle." *Times Colonist*, May 22, 2007.

12 Martin Golder, "Columnist is up for another award," letter to the *Times Colonist*, May 24, 2007, p. A15.

13 Andrew Weaver, "Climate change is no conspiracy." *Times Colonist,* May 24, 2007, p. A15.

14 John Houghton, "The Great Global Warming Swindle." John Ray Initiative, March 8, 2007.

15 Al Gore, *Earth in the Balance,* pp. 66-67.

16 Tim Flannery, in *Weather Makers* (Toronto: HarperCollins, 2005), p. 44, regards the idea that the Medieval Warming was global as "bunk." Gore, for one, seems to think the warming extended at least as far as Central America. Yet there is ample evidence that the Medieval Warming was also experienced as far away as China. This evidence is summarized in Fred S. Singer and Dennis T. Avery's *Unstoppable Global Warming*, p. 106. In other words, it's likely the Medieval Warming and Little Ice Age were not localized to Europe but global, much as alarmists like Flannery would like to believe otherwise to support their theories.

17 Gore, *Earth in the Balance*, p. 64.

18 Brian Fagan, *The Little Ice Age*. New York: Basic Books, 2000, pp. 15-16. This book catalogues some of the horrors of a globe plunged into cold weather. On the plus side, Fagan suggests that the cold did prompt human ingenuity in seeking survival, including, perhaps, the Scientific Revolution. Humankind seems to make its most radical

evolutionary and technological leaps when faced with colder climates. Unfortunately, these cold snaps also cause a lot of deaths.

19 Harm de Blij, *Why Geography Matters: Three Challenges Facing America.* Toronto: Oxford Univ. Press, 2005, p. 53.

20 Gore, *Earth in the Balance*, p. 95. Other sources indicate this warming began 12,000 or 13,000 or 15,000 years ago.

21 Gore, *Earth*, p. 91.

22 Gore, *Earth*, p. 96.

23 Weaver letter, *Times Colonist,* May 24, 2007.

24 Weaver, *Keeping Our Cool: Canada in a Warming World.* Toronto: Viking Canada, 2008, p. 110.

25 William J. Broad, "From a rapt audience, a call to cool the hype." *New York Times*, March 13, 2007.

26 Broad, *Ibid.*

27 Lawrence Solomon, "The hurricane expert who stood up to UN junk science." *National Post*, Feb. 2, 2007.

28 Gore, *An Inconvenient Truth.* Emmaus, PA: Rodale, 2006, p. 261.

29 Patrick Michaels, *Meltdown: The Predictable Distortion of Global Warming by Scientists, Politicians, and the Media.* Washington, DC: Cato Institute, 2004, p. 9. Michaels goes on: "We don't have everything to do with it [global warming]; but we can't stop it, and we couldn't even slow it down enough to measure our efforts if we tried."

30 Naomi Oreskes, "The Scientific Consensus on Climate Change." *Science*, December 3, 2004, p. 1686. Interestingly, the author said in her original published paper that she used the search terms "climate change," which would have produced more than 12,000 documents and, therefore, likely much more disagreement. The article was later corrected with the more limiting search terms she actually used, "global climate change," which produced the smaller number of articles she cited in her study.

31 Benny Peiser, "The letter *Science* magazine refused to publish." *Benny Peiser Homepage.* Jan. 4, 2005. Alarmists have quibbled with some of Peiser's categorizations, but the basic point is clear enough: the "100 per cent consensus" is a myth.

32 J. Lighthill et al. "Global Climate-Change and Tropical Cyclones." *Bulletin of the American Meteorological Society*, Vol. 75, issue 11 (November 2994), pp. 2147-2157.

33 R.L. Keeney, "Establishing Research Objectives to Address Issues of Climate-Change." *Socio-Economic Planning Sciences*, Vol. 28, issue 1 (January 1994), pp. 1-8.

34 The list of abstracts explicitly disagreeing with the "consensus" can be found at "Peiser's 34 abstracts," Deltoid: Peter Lambert's weblog, http://timlambert.org/2005/05/peiser.

35 Peiser, "The letter…".

36 Dennis Bray & Hans von Storch, *The Perspectives of Climate Scientists on Global Climate Change.* Geesthacht, Ger: Institute for Coastal Research, 2007, p. 62.

37 The preliminary results of the 2008 Bray-Storch poll move the goalposts considerably closer to Gore's "100 per cent" consensus. In this survey, only 16 per cent of 375 respondents now say they are either neutral on or disagree with the question of whether humans are causing "most" of recent or near-future climate change. However, an

additional 17 per cent expressed only lukewarm support for the proposition that humans are the main cause of climate change, for a total of 33 per cent—one-third of the climate scientists polled—who either don't believe that humans are the main force warming the climate or believe it's possible but are not certain. Also, the 2008 survey had only 375 respondents, or two-thirds of the 2003 poll. Nonetheless, one-third of the climate scientists polled in 2008 expressed skepticism or only lukewarm agreement on the question of the human origins of global warming, which still falls short of either "100 per cent" agreement or "no debate." It is curious that this increase in anthropogenic warming believers occurs at a time when the planet isn't warming, and leads one to suspect the bandwagon effect at work. That said, Gore made his claims before the results of the 2008 survey were known. See Dennis Bray and Hans von Storch, "A response to RealClimate concerning a new survey of climate scientists," Roger Pielke, Sr., website Prometheus, Oct. 13, 2008.

38 Camille Paglia, "Real inconvenient truths." *Salon* magazine, April 11, 2007.

39 "Polar bear worries unproven, expert says." CBC News, May 15, 2006.

40 Gore, *Earth,* p. 79.

41 Gore, *Earth,* p. 28.

42 Bjorn Lomborg, *The Skeptical Environmentalist.* New York: Cambridge Univ. Press, 1998, p. 252.

43 Norman Myers, "Extinction rates past and present." *Bioscience*, Vol. 39, No. 1 (Jan. 1989), pp. 39-41.

44 Myers, "Specious: On Bjorn Lomborg and species diversity." *Grist,* December 12, 2001.

45 "Largest mass extinction in 65 million years underway, scientists say." *Wikinews*, March 8, 2006.

46 Nassim Khadem, "Earth faces mass extinction." *The Age*, March 16, 2006.

47 These figures can be found at http://www.iucnredlist.org. I am indebted to Jeffrey Foss's *Beyond Environmentalism* (Hoboken, NJ: Wiley, 2009, p. 36) for bringing the Redlist to my attention.

48 Jeffrey E. Foss, *Beyond Environmentalism*, p. 37.

49 Cited in Lomborg, p. 251. Lomborg's estimate is about one species lost per year.

50 Stephen Schneider, in his book *Laboratory Earth: The Planetary Gambit We Can't Afford to Lose*. (New York: Basic Books, 1997, p. 102-103), cites biologist E.O. Wilson's *The Diversity of Life* to explain how the calculations going into the extinction models work.

51 The judge's full ruling can be found at http://www.bailii.org/ew/cases/EWHC/Admin/2007/2288.html.

52 For a comprehensive listing of the errors in *An Inconvenient Truth* see Marlo Lewis, *Al Gore's Science Fiction: A Skeptic's Guide to An Inconvenient Truth.* Also highly recommended is Christopher Monckton, *35 Inconvenient Truths: The Errors in Al Gore's Movie.* Both are available online.

53 Simon, *Hoodwinking the Nation.* New Brunswick, NJ: Transaction Publishers, 1999, pp. 87, 89.

54 Gore, *Earth,* p. 256.

55 Aristotle, *History of Animals*, Book 2, Part 3.

56 Thomas B. Macaulay, "Lord Bacon." *Reviews, Essays and Poems*. London: Ward Lock, 1890, p. 408.

57 Macaulay, "Lord Bacon," pp. 416-417.

58 Francis Bacon, *Novum Organum*. Indianapolis: Bobbs-Merrill Educational Publishing, 1983, pp. 88-89 (Aphorisms Book One, No. 89).

59 Bacon, *Novum Organum*, p. 89 (Book One, section 89).

60 Gore, *Earth,* pp. 220-224.

61 Gore, *Inconvenient Truth*, pp. 306-310.

62 George Monbiot, *Heat: How to Stop the Planet from Burning.* Toronto: Doubleday Canada, 2006, p. xv.

63 "Al Gore's Personal Energy Use Is His Own 'Inconvenient Truth'," Tennessee Center for Policy Research, Feb. 26, 2007. Granted, Gore uses part of his home as an office and reportedly has, since the publication of this information, incorporated many "sustainability" features in his home. In May, 2010, it was revealed that Gore had purchased a 1.5-acre California "mega-mansion" with a swimming pool, spa, five bedrooms, six fireplaces and nine bathrooms for almost $9 million.

64 For details on Gore's "green" investments that are making him the first "carbon billionaire," see John M. Broder, "Gore's dual role: Advocate and investor." *New York Times*, Nov. 9, 2009.

65 Jean-François Revel, *Neither Marx nor Jesus.* New York: Laurel, 1971, p. 15.

CHAPTER 7: 'OPEN' SCIENCE AND ITS ENEMIES: THE GLOBAL WARMING PARADIGM

1 Melanie Phillips, "The global warming fraud." *Daily Mail,* January 12, 2004. Phillips is a British newspaper columnist and writer.

2 Karl R. Popper, *The Open Society and Its Enemies,* Vol. 1. Princeton, NJ: Princeton Univ. Press, 1971 (1943), p. 4.

3 Steve Fuller, *Kuhn vs. Popper: The Struggle for the Soul of Science*. Cambridge: Icon Books, 2006 (2003), p. 222.

4 "Sees what he believes" is a loose translation of St. Augustine's maxim *crede ut intelligas*, "I believe in order to understand."

5 Albert Einstein, *Ideas and Opinions*. New York: Dell, 1954, p. 54. From a 1941 symposium entitled "Science, Philosophy and Religion, in their Relation to the Democratic Way of Life."

6 Bertrand Russell, *The Problems of Philosophy*. Indianapolis: Hackett Publishing, 1912, p. 121.

7 Thomas Kuhn, *The Structure of Scientific Revolutions,* 2nd ed. Chicago: Univ.of Chicago Press, 1970 (1962), pp. 17-18.

8 Kuhn, p. 16.

9 Kuhn, p. 24, calls "normal science" this effort "to force nature into the reformed and relatively inflexible box that the paradigm supplies."

10 Karen Armstrong, *The Battle for God.* New York: Alfred A. Knopf, 2000, p. 70.

11 Kuhn, p. 24.

12 Kuhn, p. 19.

13 Kuhn, p. 24.

14 Popper, *Unended Quest: An Intellectual Biography*. Glasgow: Fontana/Collins, 1976, p. 51.

15 Kuhn, p. 24.

16 Kuhn, p. 11. He writes: "Men whose research is based on shared paradigms are committed to the same rules and standards of scientific practice. That commitment and the apparent consensus it produces are prerequisites for normal science."

17 Fuller, *Kuhn vs. Popper,* p. 16.

18 Popper, *Open Society,* p. 173.

19 Popper, *Open Society*, pp. 9, 172.

20 Fuller, p. 26.

21 Popper, *Open Society,* p. 1.

22 Popper, *Open Society*, p. 175.

23 Karl Popper, *Conjectures and Refutations: The Growth of Scientific Knowledge*. New York: Routledge & Kegan Paul, 2007 (1963), p. 35.

24 Bertrand Russell, "What Is an Agnostic?" *The Basic Writings of Bertrand Russell: 1903-1959.* New York: Simon and Schuster, 1961, p. 584.

25 Fuller, p. 105.

26 Fuller, p. 107.

27 Donald R. Prothero, *The Eocene-Oligocene Transition: Paradise Lost*. New York: Columbia Univ. Press, 1994, p. 89.

28 Hugh W. Ellsaesser, "Response to Kellogg's Paper," *Global Climate Change, Human and Natural Influence,* ed. Fred Singer, Paragon House, NY, 1989, p 72.

29 Email from Phil Jones to Michael Mann dated July 8, 2004. The texts of the Climategate emails are available at "Climate Cuttings 33," *Bishop Hill* blog, Nov. 20, 2009, and John P. Costella, "Climategate Analysis," Jan. 20, 2010.

30 Email from Tom Wigley to Timothy Carter dated April 24, 2003.

31 Christopher Essex & Ross McKitrick, *Taken by Storm: The Troubled Science, Policy and Politics of Global Warming*. Toronto: Key Porter, 2002, p. 50.

32 Richard Lindzen, "Global Warming: The Origin and Nature of the Alleged Scientific Consensus." Cato Institute, *Regulation*, Vol.15, No. 2, Spring 1992. Quoted in Essex and McKitrick, p. 45.

33 Austin Jenkins, "Scientists Say Cascade Snowpack Has Not Declined 50% After All." Station KUOW, *National Public Radio,* March 15, 2007.

34 "NASA administrator questions need to combat global warming." *ABC13.com,* May 31, 2007.

35 *Ibid.*

36 Ross Gelbspan, *The Heat Is On: The High-Stakes Battle Over Earth's Threatened Climate.* New York: Addison Wesley, 1997, pp. 33-61.

37 Sharon Begley, "The Truth about Denial." *Newsweek*, Aug. 13, 2007.

38 Robert J. Samuelson, "A Different View of Global Warming." *Newsweek*, Aug. 20, 2007.

39 Samuelson, *Ibid.*

40 U.S. General Accounting Office, "Federal reports on climate change funding should be clearer and more complete." 2005. This report notes that funding for climate-change initiatives had increased from $3.1 billion in 1993 to $5.1 billion in 2005.

41 Greenpeace, *ExxonMobil's Continued Funding of the Global Warming Denial Industry*, May 2007, p. 4.

42 Ben Pile & Stewart Blackman, "The Well-Funded 'Well-Funded Denial Machine' Denial Machine." *Climate Resistance*, Jan. 11, 2008. Greenpeace itself says it does not accept government or corporate funding. See "Questions about Greenpeace in general" at http://www.greenpeace.org/international/about/faq/questions-about-greenpeace-in

43 Wallace S. Broeker & Robert Kunzig, *Fixing Climate: What Past Climate Changes Reveal About the Current Threat—and How to Counter It.* New York: Hill & Wang, 2008, p. 58.

44 "Biography: Dr. R.K. Pachauri." U.S. Climate Change Science Program. Oct. 11, 2003.

45 Bjorn Lomborg, *The Skeptical Environmentalist: Measuring the Real State of the World.* Cambridge: Cambridge Univ. Press, 1998, p. xix.

46 Lomborg, p. xix.

47 "Misleading math about the earth: Science defends itself against *The Skeptical Environmentalist*." *Scientific American*, January, 2002.

48 Lomborg was allowed to put his full reply on the magazine's website, but with conditions.

49 Detailed information about the Lomborg case can be found in Hans Labohm, Simon Rozendaal & Dick Thoenes, *Man-Made Global Warming: Unravelling a Dogma.* Brentwood, Essex: Multi-Science Publishing, 2004, pp. 97-111.

50 "The rhetorical tone of his opponents is remarkable, whereas Lomborg answers in a cool style," says Dutch geneticist Arthur Rorsch in his assessment of the storm over the Lomborg book, "Good scientific practice and the Lomborg affair in Denmark." March 2003.

51 Arthur Rörsch, Thomas Frello, Ray Soper, & Adriaan de Lange, "An Analysis of the Nature of the Opposition Raised against the Book *The Skeptical Environmentalist* by B. Lomborg," August 11, 2003.

52 This dismissive attitude is also sometimes (although I believe less often than among alarmists) found on the skeptics' side of the argument, such as some of the pronouncements of U.S. Senator James Inhofe, former chair of the U.S. Senate Environment and Public Works Committee and a strong critic of the global warming hypothesis. As with the alarmists, Inhofe's dismissal of opponents doesn't mean Inhofe's facts are wrong.

53 Lomborg, *Skeptical Environmentalist*, p. xx.

54 Lomborg, p. 4.

55 Fuller, p. 5.

56 Earthbeat, "Skeptical Environmentalist Debates Critics." *Australian Broadcasting Corp.*, Oct. 10, 2001.

57 Stephen Schneider, "Global Warming: Neglecting the Complexities." *Scientific American*, January, 2002.

58 Lomborg, "Lomborg's comments to the 11-page critique in January 2002 *Scientific American*," Jan. 4, 2002, p. 2.

59 Lomborg, *Skeptical Environmentalist*, p. 4.

60 Roger A. Pielke, Jr., "When scientists politicize science: Making sense of controversy over *The Skeptical Environmentalist*." *Environmental Science & Policy*, Vol. 7 (2004), pp. 405–417.

61 Pielke, Jr., *Ibid.*, p. 411.

62 Pielke, Jr., *Ibid.*, p. 407.

63 Pielke, Jr., *Ibid.*, p. 413.

CHAPTER 8: CRYING WOLF: A BRIEF HISTORY OF ALARMISM

1 Gwen Schulz, *Ice Age Lost*. Garden City, NY: Anchor Press/Doubleday, 1974, p. 201.

2 Leo Hickman, "Humans are too stupid to prevent climate change." Interview with James Lovelock. *The Guardian*, March 29, 2010.

3 Lowell Ponte, *The Cooling*. Inglewood Cliffs, NJ: Prentice Hall, 1976, p. 74.

4 Kesten C. Green & J. Scott Armstrong, "Effects and outcomes of the global warming alarm: A forecasting project using the structured analogies method." Dec. 12, 2009. Available online.

5 Thomas Malthus, *An Essay on the Principle of Population*. Antony Flew, ed. Markham, ON: Penguin Classics, 1988 (1798), p. 73.

6 It's worth noting that some of this inventive spark was roused by the Little Ice Age, as European countries tried to cope with greater cold. As we've seen, global cooling is more dangerous for humanity than global warming because warming times are generally good times for agriculture. On the other hand, human beings have evolved, physically, culturally, and technologically, when conditions were harsher and colder. Our primate ancestors left the trees in response to a dryer Africa due to the onset of ice age conditions two million years ago. So perhaps, if we care about human evolution, genetically or technologically, we should welcome cooling. On the other hand, if we care about the people living now, we should prefer warming because global cooling conditions kill many more people than warming.

7 Karl Marx, *The Economic and Philosophical Manuscripts of 1844*. Dirk J. Struik, ed., New York: International Publishers, 1964, pp. 70-71.

8 Karl Marx & Friedrich Engels, *The German Ideology*. Moscow: Progress Publishers, 1968, p. 86.

9 Marx & Engels, *German Ideology*, p. 96.

10 Marx & Engels, *German Ideology*, pp. 46-47.

11 Marx & Engels, *German Ideology*, p. 45. Marx and Engels wrote: "In communist society, where nobody has one exclusive sphere of activity but each can become accomplished in any branch he wishes, society regulates the general production and thus makes it possible

for me to do one thing today and another tomorrow, to hunt in the morning, fish in the afternoon, rear cattle in the evening, criticize after dinner, just as I have a mind, without ever becoming hunter, fisherman, shepherd or critic."

12 Robert Heilbroner, "The Triumph of Capitalism." *New Yorker*, Jan. 23, 1989, p. 98.

13 Peter Foster, "From Berlin to Copenhagen." *National Post*, Nov. 6, 2009.

14 Ehrlich, *The Population Bomb*. New York: Ballantine Books, 1969 (1968), p. 108.

15 Ehrlich, *Population Bomb*. p. 11.

16 Ehrlich, *Population Bomb*, rev. edition. New York: Ballantine Books, 1978, p. xi.

17 Paul R. Ehrlich & Anne H. Ehrlich, *The Population Explosion*, Toronto: Touchstone, 1990, p. 9.

18 Ehrlich & Ehrlich, *The Population Explosion*, p. 69.

19 Ehrlich, *Population Bomb*, p. 53.

20 See, for example, "Obesity of China's kids stuns officials," *USA Today*, Jan. 9, 2007; "Morbid obesity on rise in India?" *Hindustan Times*, Sept. 1, 2006.

21 Andrew Weaver, *Keeping Our Cool: Canada in a Warming World*. Toronto, Viking Canada, 2008, p 283.

22 Ehrlich, *Population Bomb*, p. 11.

23 Ehrlich, *Population Bomb,* p. 136.

24 Ehrlich, *Population Bomb*, p. 138.

25 Eugene Linden, *The Winds of Change: Climate, Weather, and the Destruction of Civilizations*. New York: Simon & Schuster, 2006, pp. 96-97.

26 UN Department of Economic and Social Affairs, *World Population to 2030,* pp. 1,4. It's worth noting that the annual rate of population growth is estimated to fall dramatically, from 1.76 per cent from 1950-2000 to .77 per cent from 2000-2050. By 2050, the global annual population growth rate is expected to fall to only .33 per cent.

27 Again and again in their books, the Ehrlichs proclaim the need to lower the developed nations' standard of living. See, for example, Paul and Anne Ehrlich, *One with Nineveh: Politics, Consumption, and the Human Future*. Covelo, CA: Island Press, 2004, Chapter 7, "Consuming Less," pp. 206-236.

28 Paul R. & Anne H. Ehrlich, *Betrayal of Science and Reason*. Washington, DC: Island Press, 1996, p. 220.

29 Gregg Easterbrook, *A Moment on the Earth: The Coming Age of Environmental Optimism*. Toronto: Penguin Books, 1995, p. xiv. Ehrlich & Ehrlich, *Betrayal*, p. 223.

30 Ehrlich, *Population Bomb*, p. 97.

31 Ehrlich & Ehrlich, *Betrayal*, p. 223.

32 Dennis J. Flynn, *Intellectual Morons: How Ideology Makes Smart People Fall for Stupid Ideas*. New York: Crown Forum, 2004, p. 68.

33 Before his death, Simon offered a similar bet to the climate alarmists: "I'll bet a week's or a month's pay with Mr. Gore or anyone else that I've got the above matters right and he does not. And I'll go further: I'll bet that just about any broad aggregate trend pertaining to human welfare will improve rather than get worse—health, standard of living, cleanliness of our air and water, natural resource availability—you name it, and you pick

any year in the future." If Gore had taken the bet, Simon would almost certainly have won again.

34 Richard Feynman, *The Character of Natural Law*. Cambridge, MA: MIT Press, 1965, p. 156.

35 For a more complete list of Ehrlich's failures as a prophet, see Michael Fumento, "Doomsayer Paul Ehrlich strikes out again," available at http://fumento.com/bomb.html.

36 Jeremy Rifkin, *The End of Work: The Decline of the Global Labor Force and the Dawn of the Post-Market Era.* New York: Jeremy P. Tarcher/Putnam, 1995, p. xv.

37 The economic slump of late 2008 doesn't prove Rifkin right. Capitalist economies have had slumps before and recovered, and the economy will recover from this latest slump as well.

38 Jeremy Rifkin, *Entropy: Into the Greenhouse World,* rev. ed. Toronto: Bantam, 1989 (1980), p. 3.

39 Rifkin, *Entropy*, p. xi.

40 Rifkin, *Entropy*, p. 285.

41 Patrick Michaels, *Meltdown: The Predictable Distortion of Global Warming by Scientists, Politicians, and the Media.* Washington, DC: Cato Institute, 2004, p. 167.

42 Schneider is quoted in Jonathan Schell, "Our Fragile Earth." *Discover*, October, 1989, pp. 45-48. See also Stephen Schneider, "Don't Bet All Environmental Changes Will Be Beneficial," APS (American Physical Society) News, August/September 1996, p. 5.

43 Stephen Schneider, *The Genesis Strategy*, New York: Delta, 1976, p. ix.

44 Schneider, *Genesis Strategy*, p. 66.

45 David Roberts, "Al Revere: An interview with accidental movie star Al Gore." *Grist*, May 9, 2006.

46 Zoe Cormier. "Will oceans surge 59 centimetres this century—or 25 metres?" *The Globe and Mail*, Aug 25, 2007, p. F9.

47 James E. Hansen, "Can we defuse the global warming time bomb?" *Natural Science*, Aug. 1, 2003.

48 Andrew Brod, "It's not too late—yet—to counteract global warming." Greensboro *News & Record*, April 9, 2007.

49 Quoted in Richard A. Kerr, "Hansen Vs. the World on the Greenhouse Threat." *Science*, New Series, Vol. 244, No. 4908 (Jun. 2, 1989), p. 1043.

50 James Lovelock, *Gaia's Revenge,* Toronto: Penguin, 2006, pp. 58-59.

51 Thomas Homer-Dixon, *The Upside of Down: Catastrophe, Creativity, and the Renewal of Civilization.* Toronto: Vintage Canada, 2006, p. 300.

52 Quoted in Sandra McCulloch, "Lemons flourish in North Saanich yard." *Times Colonist*, May 25, 2008, p. A1.

53 Usha Lee McFarling, "Scientists alarmed by continued warming trend." *Los Angeles Times*, Dec. 12, 2002. The irony here is that, after 2001, there was no warming trend.

54 Jan Zwicky, Introduction, *Hard Choices: Climate Change in Canada.* Harold Coward & Andrew Weaver, eds. Waterloo, ON: Wilfred Laurier Press, 2004, p. 3.

55 Mike Hulme, "The IPCC, consensus, and science." Feb. 19, 2010. Available online.

56 Patrick Michaels, *Meltdown*, pp. 73-80.

57 Michaels, *Meltdown*, p. 195.

58 Carl Sagan, *The Dragons of Eden: Speculations on the Evolution of Human Intelligence*. New York: Random House, 1977, p. 183.

59 Immanuel Kant, *Critique of Pure Reason*. London: George Bell & Sons, 1887 (1787), p. 502 (Chapter II, Section 3).

60 Walter Kaufmann, *Critique of Religion and Philosophy:* Princeton, NJ: Princeton Univ. Press, 1990 (1958), p. 107.

61 Green & Armstrong, "Effects and outcomes of the global warming alarm," p. 10.

62 Seymour Garte, *Where We Stand: A Surprising Look at the Real State of Our Planet*. New York: Amacom, 2008, pp. 58-59.

63 "The Deluge of Disastermania: Armageddon seems to be just a payday away." *Time*, March 5, 1979.

CHAPTER 9:
A BRIEF HISTORY OF PROGRESS, AND WHY WE DON'T BELIEVE IN IT

1 Victor Davis Hanson, *Carnage and Culture: Landmark Battles in the Rise of Western Culture*. New York: Anchor, 2001, p. 455.

2 Robert Fulford, "Biting the (invisible) hand that feeds us." *National Post,* April 12, 2008, p. A17.

3 Richard Nixon, State of the Union Address, 1970.

4 Thomas B. Macaulay, "Southey," in *Reviews, Essays and Poems*. London: Ward, Lock, 1890, p. 143. First published in the *Edinburgh Review*, January, 1830.

5 John Fowles, "Weeds, bugs, Americans." *Wormholes: Essays and Occasional Writings*. New York: Henry Holt, 1998, p. 246.

6 Rob Gordon, "Capitalism brings global warming." Letter to the editor, Victoria *Times Colonist*, April 8, 2008.

7 For example: Alex Morales, "Rhinoceros-Poaching Nears 15-Year High on Horn Demand in Asia." *Bloomberg.com*, July 9, 2009.

8 S. Jonathan Singer, *The Splendid Feast of Reason*. Berkeley, CA: Univ. of California Press, 2001, pp. 169-170.

9 John Fowles, *Wormholes*, p. 246.

10 Robert Southey, *Sir Thomas More: or, Colloquies on the Progress and Prospects of Society*, "Colloquy III." London: John Murray, 1829. Available through the Gutenburg website and at Google Books.

11 Southey, "Colloquy IV."

12 Macaulay, "Southey," p. 143.

13 Macaulay, "Southey," p. 144.

14 Macaulay, "Southey," p. 155.

15 Macaulay, "Southey," p. 155.

16 Macaulay, "Southey," p. 154. Historian Judith Flanders, in her book *Consuming Passions*

(London: Harper Perennial, 2006, p. 25) notes that "as late as the 1690s, something as basic to us as a utensil to hold a hot drink—that is, a cup—was 'extremely rare' even in prosperous households." A poor man who died in 1648 left three possessions: an iron pot, a small skillet, and a brass skimmer. By contrast, a poor man who died in 1739, barely a hundred years later, left behind more than 30 items.

17 Macaulay, "Southey," p. 144.

18 Macaulay, "Southey," p. 156.

19 Macaulay, "Southey," p. 158.

20 Macaulay, "Southey," p. 159.

21 Macaulay, "Southey," p. 153.

22 Mark Jaccard, a leading Canadian advocate for severe carbon restrictions, quoted in Peter Foster, "Cracks in the core of sustainability," *National Post*, April 15, 2008, p. FP15.

23 Bjorn Lomborg, *The Skeptical Environmentalist: Measuring the Real State of the World.* Cambridge: Cambridge Univ. Press, 1998, p. xix.

24 Fraser Institute, "Natural Resources," *Environmental Indicators (6th edition),* April 1, 2004.

25 Indur Goklany, *The Improving State of the World: Why We're Living Longer, Healthier, More Comfortable Lives on a Cleaner Planet.* Washington, DC: Cato Institute, 2007, p. 9.

26 Goklany, p. 19.

27 Goklany, p. 54.

28 Goklany, p. 34.

29 Goklany, p. 52.

30 Judith Flanders, *The Victorian House*. London: Harper Perennial, 2003, p. 370. "The Industrial Revolution brought with it a pall that hung over all the major cities, made up of the coal smoke, dirt and dust pouring out of millions of chimneys, together with the mist that these prevented from being burned off by the sun."

31 Ted Nordhaus & Michael Shellenberger, *Break Through: From the Death of Environmentalism to the Politics of Possibility*. Boston: Houghton Mifflin, 2007, p. 6.

32 Nordhaus & Shellenberger, p. 6.

33 Nordhaus & Shellenberger, p. 27.

34 Nordhaus & Shellenberger, p. 12.

35 Lomborg, *The Skeptical Environmentalist*, p. 112.

36 Doug Saunders, "The hush-hush regreening of Europe." *Globe and Mail*, Dec. 22, 2007, p. F3. Saunders's source is the United Nations Ministerial Conference for the Protection of Forests.

37 Gregg Easterbrook, *The Progress Paradox: How Life Gets Better While People Feel Worse*. New York: Random House Trade Paperbacks, 2004, p. 109.

38 Fraser Institute, "Natural Resources," *Environmental Indicators (6th edition)*. April 1, 2004, p. 84.

39 Gregg Easterbrook, *A Moment on the Earth*. Toronto: Penguin Books, 1995, p. 14.

40 Elisabeth Rosenthal, "New jungles prompt a debate on rain forests." *New York Times*, Jan. 30, 2009.

41 Ramakrishna R. Nemani, et al., "Climate-Driven Increases in Global Terrestrial Net Primary Production from 1982 to 1999." *Science*, Vol. 330, Issue 5625 (June 6, 2003), pp. 1560-1563. See also Lawrence Solomon, "In praise of carbon dioxide." *National Post*, June 7, 2008.

42 Easterbrook, *The Progress Paradox*, p. 40.

43 Rachel Carson, *Silent Spring*. Boston: Houghton Mifflin, 2002 (1962), p. 6.

44 Easterbrook, *Progress Paradox*, pp. 41-44.

45 Fraser Institute, *Environmental Indicators (6th edition)*, 2004, p. 8.

46 Easterbrook, *Progress Paradox*, p. 63.

47 Goklany, p. 59.

48 Easterbrook, *Progress Paradox*, p. 67.

49 Kesten C. Green & J. Scott Armstrong, "Effects and outcomes of the global warming alarm: A forecasting project using the structured analogies method." Dec. 12, 2009, p. 10.

50 Julian Simon, *Hoodwinking the Nation*. New Brunswick, NJ: Transaction Publishers, 1999, pp. 2-3.

51 Thomas B. Macaulay, speech given at his inauguration as Lord Rector of Glasgow University, March 21, 1849.

52 Robert Fulford, "These days, no news can ever be good news." *National Post*, Sept. 27, 2003, p. A18.

53 Easterbrook, *Progress Paradox*, pp. 33-34.

54 Easterbrook, *Progress Paradox*, p. 107.

55 Ian Plimer, *Heaven and Earth—Global Warming: The Missing Science*. London: Quartet, 2009, p. 470.

56 David Shoenbrod, "The lawsuit that sank New Orleans." *Wall Street Journal*, Sept. 26, 2005.

57 "Hurricane Katrina." *Wikipedia*.

58 Plutonic Power Corp., "Building a Global Green Energy Powerhouse." April 22, 2009.

59 "Site C dam," *Wikipedia*. In April 2010, the B.C. government announced it was going ahead with the Site C dam, subject to environmental approvals.

60 Maggie Paquet, "Site C Dam." *Watershed Sentinel*, Jan.-Feb. 2008.

61 John Nyboer, Michelle Bennett, & Katherine Muncaster, "Energy Consumption and Supply in British Columbia: A Summary and Review." B.C. Ministry of Energy and Mines, June 2004.

62 Easterbrook, *Progress Paradox*, p 164.

63 Geoffrey Household, *Dance of the Dwarfs*. Boston: Atlantic Monthly Press, 1968, p. 165.

64 H.H. Lamb, *Climate History and the Future*. Princeton, NJ: Princeton Univ. Press, vol. 2, 1985, p. 264.

65 Easterbrook, *Progress Paradox*, p. 81. Easterbrook could have added Canada, Australia, Japan, or any of a number of the developing countries as well.

66 Easterbrook, *Progress Paradox*, p. 80.

67 Julian Simon, "Introduction," *The State of Humanity*. Oxford: Blackwell, 1995, p. 8.

68 George Orwell, *Nineteen Eight-four*. Harmondsworth, UK: Penguin, 1968 (1949), p. 239.

69 Hanson, *Carnage and Culture*, p. 455.

70 Simon, *Hoodwinking*, p. 123.

71 IPCC Working Group III, IPCC Special Report, Emissions Scenarios, Summary for Policymakers, 2000, pp. 6, 13-14.

72 Andrew Weaver, *Keeping Our Cool: Canada in a Warming World.* Toronto: Viking Canada, 2008, p. 58.

73 Jeffrey E. Foss, *Beyond Environmentalism: A Philosophy of Nature*. Hoboken, NJ: Wiley, 2009, pp. 292, 295.

74 Foss, *Ibid.,* p. 288.

75 Foss, *Ibid.,* p. 223.

76 Bryson Brown, "The 'new' atheism is precisely not a form of fundamentalism." Review, *Globe and Mail*, April 12, 2008, p. D23.

CHAPTER 10: IS CONSENSUS CLIMATE SCIENCE A 'SCIENCE'?

1 Thomas H. Huxley, "Possibilities and impossibilities," *Science and Christian Tradition.* New York: D. Appleton, 1896, p. 206

2 Karl Popper, *Conjectures and Refutations*. London: Routledge, 2007 (1963), p. 48.

3 Carl Sagan, *The Demon-Haunted World: Science as a Candle in the Darkness*. New York: Ballantine Books, 1996, p. 211.

4 Leo Hickman, "Humans are too stupid to prevent climate change." Interview with James Lovelock. *The Guardian*, March 29, 2010.

5 Paul Johnson, "The nonsense of global warming." *Forbes Magazine,* Oct. 6, 2008.

6 I have taken this list of scientific criteria from two sources: Victor J. Stenger, *God: The Failed Hypothesis: How Science Shows that God Does Not Exist* (New York: Prometheus Books, 2008, pp. 23-25), and Jay. H. Lehr, "The Scientific Process: Do We Understand It?" *Rational Readings on Environmental Concerns*, Jay H. Lehr, ed. (New York: Von Nostrand Reinhold, 1992, pp. 719-721).

7 Bertrand Russell, *The Impact of Science on Society.* New York: Routledge, 1985, p. 102.

8 Russell, *The Impact of Science on Society*, p. 101.

9 Sagan, *The Demon-Haunted World,* p. 210.

10 Bryan Magee, *The Story of Philosophy*. New York: DK Publishing, 2001 (1998), p. 220.

11 Bryan Magee, *Confessions of a Philosopher*. New York, Random House, 1997, p 51.

12 Russell, *The Impact of Science on Society*, p. 101.

13 Andrew Weaver, *Keeping Our Cool: Canada in a Warming World.* Toronto: Viking Canada, 2008, p. 22.

14 David Rind, "The doubled CO_2 climate and the sensitivity of the modeled hydrological cycle." *Journal of Geophysical Research,* 1988, Vol. 93, pp. 5385-5412.

15 This example appears in Richard F. Sanford, "Environmentalism and the assault on

reason." *Rational Readings on Environmental Concerns.* Jay H. Lehr, ed. New York: Von Nostrand Reinhold, 1992, p. 18.

16 Stephen Schneider, "Global Warming: Neglecting the Complexities." *Scientific American*, January, 2002.

17 Earthbeat, "Skeptical Environmentalist Debates Critics," *Australian Broadcasting Corp.*, Oct. 10, 2001.

18 Georges Cuvier, *Recherches sur les Ossemens*. Paris: Chez G. Dufour et d'Ocagne, Libraires, 1822, p. 292.

19 Karl Popper, "Prediction and Prophecy in the Social Sciences," *Conjectures and Refutations*. New York: Routledge & Kegan Paul, 2007 (1963), p. 454.

20 Richard Dawkins, *The Devil's Chaplain: Reflections on Hope, Lies, Science and Love*. New York: Mariner Books, 2003, p. 19.

21 Popper, *Unended Quest: An Intellectual Biography*. Glasgow: Fontana Collins, 1976 (1974), p. 38.

22 Eric D. Beinhocker, *Origin of Wealth: Evolution, Complexity, and the Radical Remaking of Economics*. Cambridge, MA: Harvard Business School Press, 2006, p. 58.

23 Dan Gardner, *Risk: Why We Fear Things We Shouldn't—and Put Ourselves in Greater Risk*. Toronto: McClelland & Stewart, 2008, pp. 347-348.

24 Mike Hulme, "The IPCC, consensus, and science." Feb. 19, 2010. Available online.

25 John R. Christy, "Briefing to the House Ways and Means Committee." Feb. 25, 2009.

26 Leo Hickman, "Humans are too stupid."

27 Kevin Trenberth to Michael Mann, Oct. 12, 2009.

28 "Flannery defends scientists in leaked emails row." *ABC News online*, Nov. 24, 2009.

29 Quoted in Patrick J. Michaels, *Sound and Fury: The Science and Politics of Global Warming*. Washington: Cato Institute, 1992, p. 83.

30 Quoted in David Rose, "Climate change emails row deepens as Russians admit they DID come from their Siberian server." *Daily Mail Online*, Dec. 13, 2009.

31 James Ihaka, "Half chance of getting weather forecast right." *New Zealand Herald*, June 6, 2007. Renwick adds, "We have a level of skill that is better than guessing for sure; it's all about probability and making maximum use of those observations."

32 D. Koutsoyiannis, A. Efstratiadis, N. Mammassis, & A. Christofides, "On the credibility of climate predictions." *Hydrological Sciences–Journal–des Sciences Hydrologiques,* 53 (2008).

33 Leo Hickman, "Humans are too stupid."

34 Stephen Schneider, *Laboratory Earth: The Planetary Gamble We Can't Afford to Lose*. New York: Basic Books, 1997, p. 67.

35 Mike Hulme, "The IPCC, consensus, and science."

37 IPCC, Climate Change 2001, Chapter 14, Advancing Our Understanding, Section 14.2.2.2.

38 Jay H. Lehr, "The Scientific Process—Do We Understand It?" In *Rational Readings on Environmental Concerns*, p. 723.

39 Michael Duffy, "Truly inconvenient truths about climate change being ignored." *Sydney*

Morning Herald, Nov. 8, 2008. For an in-depth (and disturbing) analysis of Pachauri's deliberate misrepresentation of the climate facts, see Christopher Monckton's "Open Letter to Chairman Pachauri," *Science and Public Policy*, Dec. 16, 2009.

40 Jonathan Leake, "World may not be warming, say scientists." *Times Online*, Feb. 14, 2010.

41 Karl Popper, *The Logic of Scientific Discovery*, Chap. 1, Sect. 6. New York: Routledge & Kegan Paul, 2002 (1935), p. 18.

42 Stenger, *God: The Failed Hypothesis*, p. 26.

43 For example: Guy Dauncey, "The science of climate change is settled." *Times Colonist*, June 18, 2009.

44 Schneider, *Laboratory Earth*, p. 67.

45 Scripps Institute of Oceanography, "Global Warming: The Great Experiment on Planet Earth." *Earthguide*, 2002.

46 Ian Plimer, *Heaven and Earth—Global Warming: The Missing Science.* London: Quartet, 2009, p. 456. The *Junk Science* website is offering a $500,000 to any scientist who can provide a falsifiable test proving anthropogenic warming. So far, the prize has not been collected. See http://ultimateglobalwarmingchallenge.com.

47 Stenger, *God: The Failed Hypothesis*, p. 24.

48 "Q&A: Professor Phil Jones." *BBC*, Feb. 13, 2010.

49 Mark Lynas, "Has global warming really stopped?" *New Statesman,* Jan. 14, 2008.

50 Jane Jacobs, *Dark Age Ahead*. Random House Canada, 2004, p. 70.

51 Stenger, *God: The Failed Hypothesis,* p. 24.

52 Cited emails can be read at "Climate Cuttings 33," Bishop Hill blog, Nov. 20, 2009. Available at http://bishophill.squarespace.com/blog/2009/11/20/climate-cuttings-33.html. Another very useful site is John Costella, "Climategate," available at http://assassinationscience.com/climategate.

53 Steve McIntyre, "Fresh data on Briffa's Yamal #1." *Climate Audit*, Sept. 26, 2009. McIntyre has a number of posts on the cherry-picked Yamal data.

54 David Rose, "Climate change emails row deepens as Russians admit they DID come from their email server." *Daily Mail Online,* Dec. 13, 2009.

55 Stenger, *God: The Failed Hypothesis*, p. 25.

56 For details on McIntyre's attempt to replicate Mann's work see Stephen McIntyre and Ross McKitrick, "The IPCC, the 'Hockey Stick' Curve, and the Illusion of Experience." Washington Roundtable on Science & Public Policy, George C. Marshall Institute, Nov. 18, 2003. Available at www.marshall.org/pdf/materials/188.pdf. See also Plimer, *Heaven and Earth,* pp. 87-99.

57 Andrew Orlowski, "Global Warming ate my data." *The Register*, Aug. 13, 2009. In 2009, the CRU revealed that it had destroyed the original data upon which much of the world temperature record is based, making it impossible for a third party to reconstruct the record.

58 Plimer, *Heaven and Earth*, pp. 481-482.

59 Plimer, *Heaven and Earth*, p. 453.

60 James Delingpole, "Climategate: What Gore's useful idiot Ed Begley, Jr., doesn't get about the 'peer review' process." *Telegraph.co.uk*, Nov. 26, 2009.

61 Plimer, *Heaven and Earth,* p. 98.

62 Thomas H. Huxley, "An Episcopal Trilogy," *Science and the Christian Tradition*. New York: D. Appleton, 1896, p. 135

63 Carl Sagan, "Nuclear War and Climatic Catastrophe: Some Policy Implications," *Foreign Affairs*, Winter 1983/84, pp. 257-258.

64 Weaver, *Keeping Our Cool*, p. 218.

65 Andrew Weaver, "'Environmentalists' are abandoning science." *Vancouver Sun*, March 24, 2009.

66 James E. Hansen, "Huge sea level rises are coming—unless we act now." *New Scientist*, July 25, 2007. Curiously, in this article, Hansen accused consensus climate science of being "too conservative" in their estimates of damage caused by global warming.

67 Al Gore, *An Inconvenient Truth*, p. 208-209. Gore predicted no summer Arctic ice by 2014 at his Nobel Prize acceptance speech.

68 IPCC, 2007, Summary for Policymakers, p. 13.

69 Steve Goddard & Anthony Watts, "Nansen Corrects Sea Ice Data—Sea Ice Extent Now Greater, Near Normal for Most of April/May." *Watts Up With That*, June 7, 2009. See also "Increase in Arctic sea ice confounds doomsayers—but does not spell the end of global warming, scientists say." *Daily Mail,* April 3, 2010.

70 Rex Murphy, "Armageddon theory: Vancouver." *Globe and Mail*, Jan. 10, 2009.

71 Terence Corcoran, "After Copenhagen, the end of science." *National Post*, Nov. 24, 2009.

72 Jonathan Schell, "Our Fragile Earth." *Discover*, October 1987, p. 47.

73 Friedrich Nietzsche, Part V, Section 344, *The Gay Science*. In *The Portable Nieztsche*, Walter Kaufmann, ed. Toronto: Penguin, 1976 (1954), p. 450.

74 Leo Hickman, "Humans are too stupid to prevent climate change." Interview with James Lovelock. *The Guardian*, March 29, 2010.

75 Dennis Bray & Hans von Storch, "A response to RealClimate concerning a new survey of climate scientists." *Promethus: The Science Policy Blog*. Oct. 13, 2008.

76 "Author and physicist Richard A. Muller chats with *Grist* about getting science back in the White House." *Grist.beta*, Oct. 6, 2008.

77 Examples of this trio's alarmist exaggeration can be found in detail in Chapter 8. Excerpts from the East Anglia University's Climate Research Unit, quoted in this chapter, reveal the same willingness to deceive the public and even other scientists in what is considered a good cause.

78 Walter Kaufmann, *Critique of Religion and Philosophy*. Princeton, NJ: Princeton Univ. Press, 1990 (1958), p. 104.

79 The proposed Scientific Code of Ethics is available at http://blogs.nature.com/news/thegreatbeyond/2007/09/hippocratic_oath_for_scientist.html.

80 "In U.S., Many Environmental Issues at 20-Year-Low Concern." Gallup, March 16, 2010.

81 George Monbiot, "The trouble with trusting complex science." *The Guardian*, March 8, 2010.

82 Plato, *Phaedrus*.

83 Michael Ruse, *The Evolution-Creation Struggle*. Cambridge, MA: Harvard Univ. Press, 2005, p. 35.

84 Bertrand Russell, *Mysticism and Logic and Other Essays*. Montreal: Pelican Books, 1953, p. 13.

85 Jeffrey E. Foss, *Beyond Environmentalism: A Philosophy of Nature*. Hoboken, NJ: Wiley, 2009, p. 89.

86 Norman Myers, "Specious: On Bjorn Lomborg and species diversity." *Grist,* December 12, 2001.

87 There have been many calls for Pachauri to resign.

88 Michael Duffy, "Truly inconvenient truths about climate change being ignored." *Sydney Morning Herald,* Nov. 8, 2008.

89 David Rose, "Glacier scientist: I knew data hadn't been verified." *Mail Online,* Jan. 24, 2010.

90 Foss, p. 82.

91 Northrop Frye, *The Modern Century: The Whidden Lectures 1967.* Toronto: Oxford Univ Press, 1967, p. 41.

92 Seth Borenstein, "Warming panel, under attack, seeks outside review." *Associated Press*, Feb. 28, 2010.

93 Sunanda Creagh, "UN to create science panel to review IPCC." *Reuters*, Feb. 26, 2010.

94 In 2009, Jones revealed that Climatic Research Unit materials holding decades of the raw temperature data had been destroyed. This means the original temperature record can not be reconstructed.

95 Frank. J. Tipler, "Climategate: The Skeptical Scientist's View." *Pajamas Media,* Nov. 26, 2009.

96 Mike Hulme, comment to Andrew Revkin blog Dot-Earth, *New York Times,* Nov. 27, 2009.

97 Judith Curry, "An open letter to graduate students and young scientists in fields related to climate research ... regarding hacked CRU emails." Available at *Climate Progress*, Nov. 27, 2009.

98 Institute of Physics submission to the UK House of Commons Committee on Science and Technology. February, 2010.

99 Petr Chylek, "An open letter to the climate research community." Dec. 5, 2009.

100 Richard Foot, "Canadian scientist says UN's global warming panel 'crossing the line'." CanWest News Service, Jan. 25, 2010.

101 "Climate author goes political." Book review, *Ottawa Citizen*, Oct. 12, 2008.

102 "Working Group I statement on stolen emails." Dec. 4, 2009.

103 Betwa Sharma, "Pachauri attacks the 'Climategate' affair." *Rediff Business*, Dec. 7, 2009.

104 "Q&A: Professor Phil Jones." *BBC*, Feb. 13, 2010.

105 Andrew Bolt, "Not unusual, not the hottest, not still warming." Melbourne *Herald Sun*, Feb. 15, 2010.

CHAPTER 11:
ENVIRONMENTALISM, RELIGION, AND GLOBAL WARMING

1 Rex Murphy, "Global warming more a cause than a science." *Globe and Mail*, June 8, 2007.

2 Robert C. Balling, Jr., *The Heated Debate: Greenhouse Predictions Versus Climate Reality*. San Francisco: Pacific Research Institute for Public Policy, 1992, p. 150.

3 Bertrand Russell, *The Problems of Philosophy*. Indianapolis: Hackett Publishing, 1912, p. 121.

4 The *Oxford Dictionary of Quotations* can't find an actual source for this comment by Chesterton, although it is quoted in another book and has been extensively reproduced since.

5 Gregg Easterbrook, *A Moment on the Earth: The Coming Age of Environmental Optimism*. Toronto: Penguin Books, 1996, p. 127.

6 Freeman Dyson, "The Question of Global Warming." *New York Times Review of Books*, June 12, 2008.

7 Dyson, *Ibid.*

8 Royal Society, "Climate change controversies: A simple guide." June 30, 2007.

9 Karl Popper, *Conjectures and Refutations: The Growth of Scientific Knowledge*. New York: Routledge & Kegan Paul, 2007 (1963), p. 35.

10 Trudy Thorgeirson, "Playing it safe." Letter to the editor, *Times Colonist*, Jan. 11, 2001.

11 Story repeated at Jay Nordlinger, *Impromtus. National Review Online*, Aug. 29, 2002.

12 See, for example, *But Is It True: A Citizen's Guide to Environmental Health and Safety Issues* (Cambridge, MA: Harvard Univ. Press, 1995), edited by Aaron Wildavsky. *But Is It True* began when political scientist Wildavsky challenged his students to investigate a number of environmental "dogmas," like the perils of DDT (they discovered that DDT is far more a boon than a peril to humankind). Apart from the danger of CFCs to the ozone layer, Wildavsky's students found most environmentalist fears, including human-caused global warming, scientifically dubious.

13 Don Martin, "A false mascot for climate change." *National Post*, May 15, 2008.

14 Balling, Jr., *Heated Debate*, p. 150.

15 Julian Simon, *Hoodwinking the Nation*. New Brunswick, NJ: Transaction Publishers, 1999, p. 107.

16 Quoted by Easterbrook, *A Moment on the Earth*, p. 528.

17 Simon, *Hoodwinking*, p. 113.

18 Michael Crichton, "Environmentalism as Religion." *MichaelCrichton: The Official Site*, Sept. 15, 2003.

19 Crichton, *Ibid.*

20 See, for example, Nicholas Wade, *The Faith Instinct: How Religion Evolved and Why It Endures*. New York: Penguin, 2009. For a review of this book, see "Spirit level: Why the human race has needed religion to survive." *The Economist*, Dec. 17, 2009.

21 Eleanor Wachtel, Interview with A.S. Byatt. *Writers and Company. CBC Radio 1*. I don't have the date when this interview was broadcast.

22 Baruch Spinoza, "Preface," Part 7, *Theologico-Political Treatise*, 1670.

23 Bart D. Ehrman, *God's Problem: How the Bible Fails to Answer Our Most Important Question—Why We Suffer.* New York: HarperOne, 2008. In this book, Biblical scholar Ehrman offers an intriguing discussion of the Old Testament approach to suffering. The irreconcilable contradictions in the Jewish and Christian approaches to suffering eventually persuaded him, reluctantly, to abandon religion for agnosticism.

24 See, e.g., Edgar W. Lewis, "Haiti's sin." *Louisiana Weekly*, Jan. 20, 2010. The article deals with the claim by U.S. evangelist Pat Robertson that Haiti's earthquake was God's punishment because the Haitians had made a pact with the devil in embracing voodoo.

25 Jacques Derrida, "Structure, Sign and Play in the Discourse of the Human Sciences," *Writing and Difference*. Chicago: Univ. of Chicago Press, 1978, p. 279.

26 John Dewey, *The Influence of Darwinism on Philosophy and Other Essays.* Amherst, NY: Prometheus Books, 1997 (1909), p. 1.

27 Not that the idea of natural evolution is new with Darwin. The same point—that nature is self-generating—was made by the Roman poet Lucretius (c. 95-45BC) in his poem "On the Nature of Things" (*De Rerum Natura*), in which he wrote: "Nature does all things spontaneously, by herself, without the meddling of the gods." (Book II, lines 1102-1104). Darwin supplied the scientific proof that Lucretius and other early believers in natural evolution lacked.

28 Ehrman, *God's Problem*, p. 205.

29 Ehrman, *Ibid.*, p. 259.

30 Greg Gerrard, *Ecocriticism*. London: Routledge, 2004, p. 100.

31 Andrew Weaver, "The science tells the story." *The Ring*, Jan. 23, 2003.

32 Mark Nichols & Dana Huffam, "Global warming crisis," and "The climate conundrum: Canada's Kyoto commitment." *Maclean's*, Feb. 21, 2000, p. A22.

33 Andrew Weaver, *Keeping Our Cool: Canada in a Warming World.* Toronto: Viking Canada, 2008, p. 58.

34 Christopher Monckton, "*Scientific American's* climate lies." *The SPPI blog*, Dec. 27, 2009.

35 William Glasser, *Control Theory: A New Explanation of How We Control Our Lives*. New York: Harper & Row, 1985, p. 73.

36 See, for example, Mark Schapiro, "Conning the Climate: Inside the Carbon-Trading Shell Game." *Harper's*, February 2010, pp. 31-39.

37 David Suzuki, *The Sacred Balance: Rediscovering Our Place in Nature*. Vancouver: Douglas & McIntyre, 1997, p. 191.

38 Suzuki, *Ibid.*, p. 186.

39 Quoted in Steven Pinker, *The Blank Slate: The Modern Denial of Human Nature*. Toronto: Penguin, 2002, p. 57.

40 Jared Diamond, *The Third Chimpanzee: The Evolution and Future of the Human Animal*. Toronto: Harper Perennial, 2006 (1993), pp. 312-313.

41 Peter Douglas Ward, *The Call of Distant Mammoths: Why the Ice Age Mammals Disappeared.* New York: Springer, 1988, p. 182. A Russian climatologist, Michael Budyko, has argued that the large mammals of Europe were better adapted to avoiding

human hunters than the megafauna of North and South America. However, eventually human hunting populations in Europe reached a "threshold density" that triggered mass extinctions of mammoths, mastadons, and the wooly rhinocerous. Again, climate change—the relatively sudden warming of the planet—undoubtedly also played a role, but climate was not the key factor given that large mammals survived previous interglacials.

42 Pinker, *Blank Slate,* p. 162.

43 Al Gore, *Earth in the Balance: Ecology and the Human Spirit*. Toronto: Penguin Books, 1993 (1992), p. 256.

44 Gore, *Earth in the Balance,* p. 368.

45 J.S. Mill, *On Theism.* New York: Bobbs-Merrill, 1957, p. 81.

46 Suzuki, *Sacred Balance*, p. 209.

47 Ken McQueen, "The remarkable transformation of Saint Suzuki." *Maclean's*, Oct. 25, 2007.

48 James E. Lovelock, *The Revenge of Gaia*, Toronto: Penguin Books, 2006, p. 20.

49 James E. Lovelock, *Gaia*. Toronto: Oxford Univ. Press, 1979, pp. vii, 2, 148.

50 Lovelock, *Gaia*, p. 20.

51 Lovelock, *Revenge of Gaia*, pp. 55-57, 8.

52 Lovelock, "The Earth is about to catch a morbid fever." *The Independent*, Jan. 20, 2006.

53 Lovelock, *Gaia*, p. 27.

54 Lovelock, *Revenge of Gaia*, p. 112

55 John Houghton, *Global Warming: The Complete Briefing*, 3rd edition. Cambridge: Cambridge Univ. Press, 2004, pp. 204, 208.

56 Ross Gelbspan, *The Heat Is On: The High-Stakes Battle Over Earth's Threatened Climate*. New York: Addison Wesley, 1997, p. 54.

57 Quoted in S. Fred Singer & Dennis T. Avery, *Unstoppable Global Warming: Every 1,500 Years* (Lanham, MD: Rowman & Littlefield, 2007), p. 4.

58 Roy W. Spencer, *Climate Confusion: How Global Warming Hysteria Leads to Bad Science, Pandering Politicians and Misguided Policies that Hurt the Poor.* New York: Encounter Books, 2008, p. 38.

59 Quoted in Peter Menzies, "Concerned about the cost of Kyoto." *Calgary Herald,* Dec. 14, 1998.

CHAPTER 12: GLOBAL WARMING FEARS BASED ON PSYCHOLOGY, NOT SCIENCE

1 Donald R. Prothero, *The Eocene-Oligocene Transition: Paradise Lost*. New York: Columbia Univ. Press, 1994, p. 89.

2 Carl Sagan, *The Demon-Haunted World: Science as a Candle in the Dark.* New York: Ballantine Books, 1996, p. 49.

3 William James, "The present dilemma in philosophy," in *Pragmatism and the Meaning of Truth.* Cambridge, MA: Harvard Univ. Press, 1975, p. 11.

4 Sagan, *The Demon-Haunted World:* p. 49.

5 Mark Steyn, "The 'science' of global warming." *Maclean's*, Dec. 7, 2009, p. 63.

6 Richard Courtney, "Global Warming: How It All Began." May 15, 1999.

7 David Maraniss & Ellen Nakashima, "Gore's grades belie image of studiousness." *Washington Post*, March 19, 2000.

8 Lawrence Solomon, "Gore's guru disagreed." *National Post*, May 19, 2007.

9 U.S. General Accounting Office, "Federal reports on climate change funding should be clearer and more complete." 2005. This report notes that funding for climate change initiatives had increased from $3.1 billion in 1993 to $5.1 billion in 2005.

10 John P. Costella, "Climategate Analysis." *SPPI* blog, Jan. 20, 2010.

11 Gregg Easterbrook, "Hot air and global warming: Gregg Easterbrook questions the environmentalist rhetoric that would have us believe our planet is on its last legs." *The Independent,* Aug. 7, 1992.

12 World Bank, "Adapting to Climate Change to cost US$75-100 billion a year." Sept. 30, 2009.

13 David Suzuki & Faisal Moola, "Developing countries taste effects of climate change." *Saanich News*, Oct. 21, 2009.

14 William J. Bernstein, *The Birth of Plenty: How the Prosperity of the Modern World was Created.* Toronto: McGraw Hill, 2004, pp. 15-17.

15 Quoted in Kenneth Bagnell, review of Dhambisa Moyo, *Dead Aid: Why Aid is Not Working and How There is a Better Way for Africa. National Post*, May 16, 2009.

16 See, for example, "Brown: 50 Days to Save the World." *BBC News*, Oct. 19, 2009.

17 Edward Wegman, et al., "Ad Hoc Committee Report on the 'Hockey Stick' Global Climate Reconstruction." U.S. House Committee on Energy and Commerce, 2006, p. 65.

18 James Chin, *The AIDS Pandemic: The Collision of Epidemiology with Political Correctness*. Oxon, UK: Radcliffe Publishing, 2007, p. 1.

19 Quoted in Michael Fumento, "The WHO's political pandemic." *National Post*, Oct. 22, 2009.

20 Jeffrey Simpson, "It gets harder to ignore the signs of climate change." *Globe and Mail*, Oct. 1, 2009.

21 George Monbiot, *Heat: How to Stop the Planet from Burning*. Toronto: Penguin Books, 2007 (2006), p. xix.

22 James Randerson, "Western lifestyle unsustainable, says climate expert Rajendra Pachauri." *The Observer*, Nov. 29, 2009.

23 "A hypocrite as well as a liar." *EurReferendum* blog site, Dec. 23, 2009.

24 Thomas B. Macaulay, *The History of England*, Vol. 1, Chapter 3.

25 Spencer R. Weart, *The Discovery of Global Warming*. Cambridge, MA: Harvard Univ. Press, 2003, p. 30.

26 "Climate Science Leader Rajendra K. Pachauri Confronts the Critics." Interview with *Science*, Jan. 29, 2010.

27 Edwin X. Berry, "How they are turning off the lights in America." Nov. 7, 2009.

28 Thomas Sowell, *Conflict of Visions*. New York: William Morrow, 1987, p. 14.

29 William James, "The present dilemma in philosophy," p. 11.

30 Steven Pinker, *The Blank Slate: The Modern Denial of Human Nature*. Toronto: Penguin Books, 2002, pp. 287-293.

31 Sowell, *Conflict of Visions*, p. 144.

32 Bertrand Russell, "Psychology and politics," *Skeptical Essays*. London: Unwin Books, 1928, p. 153.

33 Sowell, *Conflict of Visions*, p. 7.

34 Gregg Easterbrook, *The Progress Paradox: How Life Gets Better While People Feel Worse*. New York: Random House Trade Paperbacks, 2004, pp. 188-189.

35 Steven Pinker, *The Blank Slate: The Modern Denial of Human Nature*. Toronto: Penguin, 2002, p. 231.

36 John Dietrich, "We are doomed—again." *American Thinker,* Feb. 28, 2010.

37 Sagan, *The Demon-Haunted World*, p. 427.

38 Irvin L. Janis, *Groupthink: Psychological Studies of Policy Decisions and Fiascoes*. Boston: Houghton Mifflin, 1982, p. 2.

39 *Ibid.,* p. 3.

40 *Ibid.,* p. vii.

41 *Ibid.*, p. 9.

42 *Ibid,* p. 247.

43 *Ibid.*, pp. 174-175.

44 The Oregon Petition against the AGW "consensus", for example, has been signed by 31,000 scientists. See http://www.oism.org/pproject.

45 Janis, *Ibid.,* p. 5.

46 Andrew Potter, "The newspaper is dying—hooray for democracy." *Maclean's*, April 7, 2008, p. 17.

47 Walter Kaufmann, *Critique of Religion and Philosophy*. Princeton, NJ: Princeton Univ. Press, 1990 (1958), p. 51.

48 The text of these CRU emails and many others are available at a number of sites. Two of the best are "Climate Cuttings 33," Bishop Hill blog, http://bishophill.squarespace.com/blog/2009/11/20/climate-cuttings-33.html, and John Costella, "Climategate," available at http://assassinationscience.com/climategate.

49 Wegman, et al., p. 65.

50 *Ibid.*, p. 29.

51 *Ibid.*, p. 51.

52 Judith Curry, "On the credibility of climate research." *Climate Audit* blog, Nov. 22, 2009.

53 Andrew Revkin, "A climate scientist who engages skeptics." *Dot Earth*, Nov. 27, 2009.

54 Steve Fuller, *Kuhn vs. Popper: The Struggle for the Soul of Science*. Cambridge: Icon Books, 2006 (2003), p. 105.

55 Martin Cohen, "Beyond debate?" *Times Higher Education*, Dec. 10, 2009.

56 Stefan Theil, "The Dark Side of Green: Gaming the global-warming fight." *Newsweek*, Nov. 2, 2009.

57 For the Greenpeace figures, see *www.exxposeexxon.com/facts/ExxonSecretsAnalysis.pdf.*

58 Prothero, *The Eocene-Oligocene Transition: Paradise Lost*, p. 130.

59 Wegman, et al., p. 65.

60 Martin Durkin, "Warming swindle." *New Scientist*, May 19, 2007.

61 "The true impact of climategate and glaciergate," *The Scientific Alliance*, Jan. 22, 2010.

62 Richard P. Feynman, "Cargo cult science: The 1974 Caltech Commencement Address," in *The Pleasure of Finding Things Out: The Best Short Works of Richard P. Feynman.* Cambridge, MA: Persius Books, 1999, p. 212.]

63 For example, Vicky Pope, "Scientists must rein in misleading climate change claims: Overplaying natural variations in the weather diverts attention from the real issues." *The Guardian*, Feb. 11, 2009. In this article, Pope writes: "The scientific evidence [for AGW] is overwhelming."

64 David Hume, "Of Superstition and Enthusiasm." *Of the Standard of Taste and Other Essays.* New York: Bobbs Merrill, 1965, p. 146.

CHAPTER 13:
WHAT IS TO BE DONE—OR NOT DONE?

1 Roy W. Spencer, *Climate Confusion: How Global Warming Hysteria Leads to Bad Science, Pandering Politicians, and Misguided Policies that Hurt the Poor.* New York: Encounter Books, 2008, p. 151.

2 Jeffrey Simpson, "Ottawa blows a lot of hot air about greenhouse gases." *Globe and Mail*, Dec. 18, 1997.

3 Jack M. Hollander, *The Real Environmental Crisis: Why Poverty, Not Affluence, Is the Environment's Number One Enemy.* Berkeley: Univ. of California Press, 2003, p. 2.

4 Lydia Dotto, *Storm Warning: Gambling with the Climate of Our Planet.* Toronto: Doubleday Canada, 1999, p. 280. Dotto writes: "There is *no proof* that global warming will cause adverse, even catastrophic damages around the world." [italics the author's]

5 Dotto, *Storm Warning*, pp. 278-279.

6 Environment Canada, "The cost of Bill C288 to Canadian families and business." April 2007.

7 Environment Canada, "A climate change plan for the purposes of the Kyoto Protocol Implementation Act—2007." Sept. 6, 2007.

8 Murray Langdon, "Gas prices putting brakes on Meals on Wheels." *Globe and Mail*, June 10, 2008, pp. S1,S2.

9 CanWest News Service, "Skyrocketing oil bumps prices for fish, ferries." *Times Colonist*, June 12, 2008, p. A6.

10 Jonathan Fowlie, "Fuel prices force hike in taxi rates." *Times Colonist*, June 12, 2008, p. A6.

11 Andrew Weaver, *Keeping Our Cool: Canada in a Warming World.* Toronto: Viking Canada, 2008, p. 58.

12 Andrew Weaver, "The science tells the story." *The Ring*, Jan. 23, 2003.

13 Dotto, *Storm Warning*, p. 285.

14 Jeffrey Simpson, "Ottawa blows a lot of hot air."

15 Stephane Dion, "Toward a richer, greener Canada." *National Post*, June 6, 2008, p. A12.

16 "Canada-Kyoto timeline." CBC news backgrounder.

17 Spencer, *Climate Confusion*, p. 151.

18 Christopher Monckton, "Climate change: Proposed personal briefing." *Watts Up With That*, Jan, 3, 2010.

19 David Suzuki, *The Sacred Balance: Rediscovering Our Place in Nature*. Vancouver, Greystone Books, 1997, p. 212.

20 Dotto, *Storm Warning*, p. 280.

21 Dotto, *Ibid.*, p. 273.

22 Jeffrey E. Foss, *Beyond Environmentalism: A Philosophy of Nature*. Hoboken, NJ: John Wiley & Sons, 2009, p. 96.

23 Christopher Monckton, "Dishonest political tampering with the science on global warming." *Jakarta Post*, Dec. 5, 2007.

24 Becky Rynor, "Grizzly victim drives 25 km for help." *Times Colonist*, May 8, 2008, p. A9.

25 *BP Statistical Review of World Energy, June 2008*, pp. 10, 26, 32. The time left for a given energy source is described as the R/P (Reserves to Production) ratio.

26 Ian Plimer suggests that if we burned up all known fossil fuels, and if there were no carbon recycling by nature (which there is), the maximum level of carbon dioxide would be 2,000 ppm. Given that at these levels the additional warming of each CO_2 molecule is minimal while the effect on plant growth would be positive, 2,000 ppm would not be a disaster. However, a level this high is extremely unlikely. See *Heaven and Earth—Global Warming: The Missing Science*. London: Quartet, 2009, p. 426.

27 Plimer, *Heaven and Earth*, pp. 442-443. See also Mark Shapiro, "Conning the climate: Inside the carbon-trading shell game." *Harper's Magazine*, February, 2010.

28 Ross McKitrick, "The T3 Tax as a Policy Strategy for Global Warming." Aug. 1, 2007. A shorter version appears as "Call their tax: Why not tie carbon taxes to levels of warming?" *National Post*, June 12, 2007.

29 Colin Campbell, "What it will *really* take to stop global warming?" *Maclean's*, April 7, 2008, pp. 43-46.

30 Justin Fox, "A 6.2% drop in GDP: Is the worst yet to come?" *Time.com*, Feb. 27, 2009.

31 Chris Isidore, "The Great Recession." *CNNMoney.com*, March 25, 2009.

32 Gene Epstein, "Global warming is manageable—if we're smart: Interview with Bjorn Lomborg." *Barron's* magazine online, May 18, 2009.

33 "Climate change can be 'tackled'." *BBC News*, May 4, 2007.

34 Epstein, "Global warming is manageable."

35 Stephen Schneider, *The Genesis Strategy: Climatic and Global Survival*. New York: Delta, 1977 (1976), p. 35.

36 "Chernobyl death toll grossly underestimated." *Greenpeace*. April 18, 2006.

37 Tim Radford, "Chernobyl death toll under 50." *The Guardian*, Sept. 6, 2005.

38 U.S. Nuclear Regulatory Commission, "Fact sheet on Three Mile Island accident." March 9, 2009.

39 For example, Merrick Goodhaven, "Nuke Mark Lynas." *Bristling Badger*, Feb. 19, 2009.

40 Mark Lynas, "Nuclear power: a convert." *New Statesman*, May 30, 2005.

41 George Monbiot, "A kneejerk rejection of nuclear power is not an option." *The Guardian*, Feb. 20, 2009. See also "Nuked by friend and foe." *Monbiot.com*, Feb. 20, 2009.

42 James Lovelock, *The Revenge of Gaia*. Toronto: Penguin, 2007 (2006), p. 14.

43 Lovelock, *Revenge of Gaia*, p. 131.

44 Hollander, *The Real Environmental Crisis*, p. 163.

45 "Nuclear power in France." World Nuclear Association, July 2009.

46 In particular, *Cool It: The Skeptical Environmentalist's Guide to Global Warming (*New York, Alfred A. Knopf, 2007), Chapter 5, and *How to Spend $50 Billion to Make the World a Better Place* (New York: Cambridge Univ. Press, 2006).

47 Epstein, "Global warming is manageable."

48 Epstein, *Ibid.*

49 Lovelock, *The Revenge of Gaia*, pp. 112-115.

50 Lomborg, *Cool It*, p. 148.

51 Hollander, *Real Environmental Crisis*, p. 3.

52 Hollander, *Ibid.*, p. 11.

53 Gregg Easterbrook, *A Moment on the Earth: The Coming Age of Environmental Optimism*. Toronto: Penguin Books, 1995, p. 479.

54 Hollander, *Real Environmental Crisis*, p. 2.

55 Friedrich A. Hayek, *The Fatal Conceit: The Errors of Socialism*. Chicago: Univ. of Chicago Press, 1991 (1988), p. 9.

56 Dotto, *Storm Warning*, p. 270.

CHAPTER 14:
THE BLACK SWANS OF ALARMIST CLIMATE SCIENCE

1 Thomas H. Huxley, "Spontaneous Generation." *Lay Sermons, Addresses, and Reviews*. New York: D. Appleton, 1888, p. 356.

2 Quoted by scientist Kiminori Itoh in "Guest weblog." *Climate Science: Roger Pielke Sr. Blog*. June 17, 2008.

3 Ben Webster, "Public skepticism prompts Science Museum to rename climate exhibition." London *Times Online*, March 24, 2010.

4 Albert Einstein, *The New Quotable Einstein*. Alice Calaprice, ed. Princeton: Princeton Univ. Press, 2005, p. 291.

5 Albert Einstein, "What is the Theory of Relativity?" *Ideas and Opinions*. New York: Laurel Edition, 1954, p. 227.

6 R.A. Warrick, E.M. Barrow & T.M.L Wigley, *Climate and Sea Level Change:*

Observations, Projections and Implications. New York: Cambridge Univ. Press, 1993, p. 3.

7 "Q&A: Professor Phil Jones." BBC, Feb. 13, 2010. See also Christopher Monckton, "Climategate: Caught green-handed." *SPPI Blog*, Nov. 30, 2009, p. 24.

8 Chris Turney, *Ice, Mud and Blood: Lessons from Climates Past.* Houndmills, UK: Macmillan, 2008, p. 146.

9 "Q&A: Professor Phil Jones." BBC, Feb. 13, 2010. For the same period, the IPCC has claimed that the warming was "unequivocal." Clearly, given these two completely different stories, the science can't be *that* "settled" and "certain."

10 See, e.g., Nigel Lawson, *An Appeal to Reason: A Cool Look at Global Warming.* New York: Duckworth Overlook, 2008, p. 36. As Lawson notes, the IPCC predicts that, if nothing is done about climate change, future generations will be "only" 4.7 times as well off as we are now. See also Indur Goklany, *The Improving State of the World: Why We're Living Longer, Healthier, More Comfortable Lives on a Cleaner Planet* (Washington, DC: Cato Institute, 2007), Table 10.1, page 308. The table and accompanying text on pages 301-309 show that if we did nothing about global warming, in the warmest IPCC scenario, by 2100 the average per capita income of people in the developing world would also be the higher, at $66,000 per year in 1990 dollars, than if we took action against carbon. The average income in the developing world in 1990 was $886. Average income in the developing world in the most economically negative scenario was still $11,000, a sixth what incomes would be if we did nothing about warming.

11 Lewis Smith, "Al Gore's inconvenient judgment." London *Times Online*, April 11, 2007.

12 Walter Kaufmann, *Critique of Religion and Philosophy.* Princeton, NJ: Princeton Univ. Press, 1990 (1958), p. 107.

13 Neil Frank, "Climategate: You should be steamed." *Watts Up With That*, Jan. 2, 2010.

14 Victoria Gill, "Science damaged by climate rows says NAS chief Ciccerone." *BBC News*, Feb. 20, 2010.

15 Robert P. Murphy, "Apologist responses to Climategate misconstrue the real debate (quantitative, not qualitative)." *MasterResource* blog, Dec. 2, 2009.

16 One of James Lovelock's books is entitled *The Revenge of Gaia.*

17 Irvin L. Janis, *Groupthink: Psychological Studies of Policy Decisions and Fiascoes.* Boston: Houghton Mifflin, 1982, p. 9.

18 For example, see Scripps Institute of Oceanography, "Global Warming: The Great Experiment on Planet Earth." *Earthguide*, 2002. As we saw in Chapter 10, the Scripps Institute writes: "Thus [carbon emissions are] an experiment *whose course and outcome are uncertain. …* We are passengers on a voyage into fog-shrouded uncharted waters. Some say, there are reefs ahead. Others say, there are none. *The truth is, no one knows.*" [italics added] Lydia Dotto, in *Storm Warning*, also acknowledges that there is no *empirical* evidence supporting fears of global warming catastrophe.

19 Richard Steinberg, "Solid evidence can take heat." Letter to the editor, *Globe and Mail*, Feb. 23, 2010.

20 I would argue that this is not so, that alarmist climate science has become a religion.

21 Charles Clover, "Grandaddy of green, James Lovelock, warms to eco-sceptics." *Sunday Times,* March 14, 2010.

22 Mike Hulme, "The IPCC, consensus and science." Feb. 19, 2010. Available online.

23 Judith Curry, "An open letter to graduate students and young scientists in fields related to climate research ... regarding hacked CRU emails." Nov. 27, 2009. Available online.

24 Andrew Hough, "John Beddington: chief scientist says climate change sceptics 'should not be dismissed'." *Daily Telegraph*, Jan. 27, 2010.

25 Gallup, "Americans' Global Warming Concerns Continue to Drop." March 11, 2010.

26 Bertrand Russell, *The Impact of Science on Society*. New York: Routledge, 1985, p. 102.

27 Robert Fulford, "These days, no news can ever be good news." *National Post*, Sept. 27, 2003.

28 "Americans skeptical of science behind global warming." *Rasmussen Reports*, Dec. 3, 2009.

29 "A climate of doubt: Almost half of Britons believe global warming is NOT caused by man." *Daily Mail*. Dec. 7, 2009.

30 Lawrence Solomon, "Canadian concern over climate change plummeting." *National Post*, Nov. 2, 2009.

31 Kiminori Itoh, "Guest weblog." *Climate Science: Roger Pielke Sr. Blog*. June 17, 2008.

32 Jean-François Revel, *Neither Marx nor Jesus*. New York: Laurel, 1971, p. 15.

33 Earthbeat, "Skeptical Environmentalist Debates Critics." *Australian Broadcasting Corp.*, Oct. 10, 2001.

34 For example, see Richard B. Alley, et al. *Abrupt Climate Change: Inevitable Surprises*. Washington: National Academy Press, 2002, p. 27.

35 Quoted by Elizabeth May, "Don't shoot the messenger." *National Post*, March 11, 2010.

Concluding Unscientific Postscript: Canadian Journalism and Climate Science

1 Lawrence Solomon, "What she didn't ask." *National Post*, Nov. 21, 2009. The "she" in this case is the CBC's Anna-Maria Tremonte, who interviewed Solomon—badly, in Solomon's opinion—for *The Current*.

2 Kiminori Itoh, "Guest weblog." *Climate Science: Roger Pielke Sr. Blog*. June 17, 2008.

3 Paul MacRae, "A shaky tower of assumptions." *Times Colonist*, March 3, 2002.

4 See, for example, Dawn Walton, "The calls of the wild: Nunavut residents say polar bears are thriving." *Globe and Mail*, Jan. 8, 2009.

5 Randy Boswell, "Record-setting meltdown feared for Arctic." *National Post*, Aug. 11, 2008.

6 See "Bad reporting about Northwest Passage issue," *Classically Liberal* blog, Sept. 18, 2007, for a history of ships that have crossed the Northwest Passage before 2008.

7 "Increase in Arctic sea ice confounds doomsayers—but does not spell the end of global warming, scientists say." *Daily Mail*, April 3, 2010.

8 See, for example, Cindy Harnett, "The planet has a fever, Gore says: Environmental activist warns of a global emergency." *Times Colonist*, Sept. 29, 2007. This article reported glowingly, rather than critically, on a talk by Gore in Victoria.

9 Andrew Coyne, "The truth is out there. Somewhere. How does one distinguish between genuine authority and received wisdom?" *Maclean's*, Jan. 7, 2010.

10 On Dec. 1, 2009, the Munk Institute sponsored a debate carried nationally on the Internet, not television, between AGW believers George Monbiot and Elizabeth May and skeptics Nigel Lawson and Bjorn Lomborg. An audience vote at the end of the debate gave victory to the skeptics—that is, there was an eight per cent shift away from the alarmist position and toward the skeptical position. In almost all if not all of the debates between alarmists and skeptics where a vote is held, audiences have given victory to the skeptics.

11 The BBC climate policy reads: "The BBC has held a high-level seminar with some of the best scientific experts, and has come to the view that the weight of evidence no longer justifies equal space being given to the opponents of the consensus [on anthropogenic climate change]." It adds: "But these dissenters (or even sceptics) will still be heard, as they should, because it is not the BBC's role to close down this debate." (*From Seesaw to Wagon Wheel,* BBC Trust, June 2007, p. 40.) However, in practice, as BBC newsperson Jeremy Paxman has noted: "The BBC's coverage of the [climate] issue abandoned the pretence of impartiality long ago." ("Jeremy Paxman, the BBC, Impartiality, and Freedom of Information," *Harmless Sky* blog, Aug. 6, 2008.) A good round-up of the issue can be found at "BBC impartiality and climate change," *alexlockwood.net* blog, Aug. 14, 2008.

12 The sole exception seems to be Alberta's Wildrose Alliance party leader Danielle Smith.

13 Lawrence Solomon, *The Deniers: The World-Renowned Scientists Who Stood Up Against Global Warming Hysteria, Political Persecution, and Fraud.* Minneapolis: Richard Vigilante Books, 2008. See also the "Deniers" series in the *National Post*, available at the NP website.

14 Mick Hume, "Thou shalt not ask awkward questions." *Spiked Review of Books*, May 2008.

15 Solomon, "What she didn't ask."

16 John Dietrich, "We are doomed—again." *American Thinker,* Feb. 28, 2010.

Appendix

A DO-IT-YOURSELF CLIMATE CHANGE CHECK

It's possible to see for yourself that the climate has been cooling, not warming, over the past decade or more. Go to the National Oceanic and Atmospheric Administration (NOAA) site at http://www.ncdc.noaa.gov/oa/climate/research/cag3/na.html. There you will find a gadget that allows you to see the temperature trend in the continental United States between any two years from 1895 to 2009. Simply input whether you want the temperature comparison for any given month or annually.

If you input the years 1997-2009 and you choose "annual" in "Period", you will get the following chart:

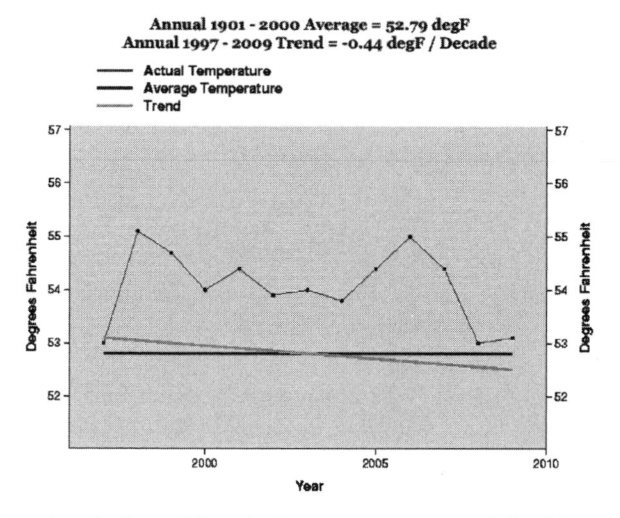

The light-colored, slanted line is the temperature trend, in this case down to the right, toward cooling. The trend in 1996 was flat-line—no warming, no cooling. The later the date after 1997, the more pronounced the cooling trend. In other words, the United States has been cooling since 1997. The U.S. has the world's best temperature records, so it's likely that since at least 1997 the rest of the Northern Hemisphere, and probably the globe as well, has also been cooling, not warming as we've been told.

ABOUT THE AUTHOR

PAUL MACRAE

Paul MacRae began his newspaper career in 1967 with the University of Toronto *Varsity* and has worked as an editor and reporter for The Canadian Press, *The Toronto Star*, *The Bangkok Post* and *The Globe and Mail*, and as an editorial writer and weekly columnist for the Victoria *Times Colonist*. In 2002 he left journalism to get an MA in English and now teaches professional writing at the University of Victoria.

He lives in Victoria, B.C., with his wife Sheila, has four children, and still mourns the death of the palm tree in his backyard during the cold winter of 2008-2009.